Public-Private Collaboration in Agricultural Research

Other related Iowa State University Press titles—for additional information on these or any Press title, contact Iowa State University Press (1-800-862-6657 or *www.isupress.edu*).

Agricultural Ethics: Research, Teaching, and Public Policy by P.B. Thompson. 0-8138-2806-6

The Development and Growth of the Soybean Industry in Brazil by P.F. Warnken. 0-8138-2196-7

Crop and Livestock Technologies: RCA III Symposium edited by B.C. English, R.L. White, and L.-H. Chuang. 0-8138-2239-4

Evaluating Natural Resource Use in Agriculture edited by T. Robertson, B.C. English, and R.R. Alexander. 0-8138-2958-5

Farm Animal Welfare: Social, Bioethical, and Research Issues by B.E. Rollin. 0-8138-2563-6

Lithuania's Accession to the European Union: Successes and Challenges for a Rural Economy in Transition edited by W.H. Meyers, N. Kazlauskiene, M.M. Guigale. 0-8138-1973-3

The Making of the 1996 Farm Act by L.P. Schertz and O.C. Doering III. 0-8138-2608-X

Natural Resource and Environmental Economics by T. Prato. 0-8138-2938-0

Rural Education and Training in the New Economy: The Myth of the Rural Skills Gap edited by R.M. Gibbs, P.L. Swaim, and R.A. Teixeira. 0-8138-2333-1

The Unheeded Cry: Animal Consciousness, Animal Pain, and Science. Expanded Edition by B.E. Rollin. Hardcover: 0-8138-2575-X; Paperback: 0-8138-2576-8

Visions of American Agriculture edited by W. Lockeretz. 0-8138-2709-4

Public-Private Collaboration in Agricultural Research

NEW INSTITUTIONAL ARRANGEMENTS AND ECONOMIC IMPLICATIONS

Edited by
KEITH O. FUGLIE AND
DAVID E. SCHIMMELPFENNIG

Iowa State University Press / Ames

Keith O. Fuglie is an economist with the International Potato Center, Bogor, Indonesia. Formerly, he was senior economist for agriculture and natural resources on the President's Council of Economic Advisers and an agricultural economist with the USDA's Economic Research Service. Fuglie, who has lived and worked in Tunisia, Thailand, Nigeria, and Kenya, received his Ph.D. in agricultural and applied economics from the University of Minnesota.

David E. Schimmelpfennig is the program leader in agricultural research and development at the USDA's Economic Research Service in Washington, DC, where he also teaches in the USDA graduate school. Schimmelpfennig has Iowa roots, lived in South Africa, and received his Ph.D. in economics from Michigan State University.

Iowa State University Press
2121 South State Avenue, Ames, Iowa 50014

Orders: 1-800-862-6657
Office: 1-515-292-0140
Fax: 1-515-292-3348
Web site: www.isupress.edu

First edition, 2000

Library of Congress Cataloging-in-Publication Data
Public-private collaboration in agricultural research: new institutional arrangements and economic implications / edited by Keith O. Fuglie and David E. Schimmelpfennig.—1st ed.

p. cm.

Includes bibliographical references.

ISBN 0-8138-2789-2

1. Agriculture—Research—United States. 2. Agriculture—Research—Economic aspects—United States. I. Fuglie, Keith O. II. Schimmelpfennig, David E.

S541.P85 1999
630'.7'2073—dc21 99-047027

The last digit is the print number: 9 8 7 6 5 4 3 2 1

CONTENTS

FOREWORD

For more than 130 years, the United States has looked to the publicly funded agricultural research institutions at the Department of Agriculture and the land-grant universities as well-springs of the scientific thought and technological innovation that have fueled development of this nation's remarkable food and agricultural sector. Estimates of the high rates of return to the public's support of this research at the state and federal levels strongly reinforce the idea that increased spending would garner an even higher level of benefits for farmers and consumers alike. However, at a time when budget austerity has been accepted as the price of achieving fiscal discipline, the prospects for increases in federal spending, anyway, are not bright. While holding steady in real terms over the past decade, public funding for agricultural research is now outstripped by outlays by the private sector, likely the only source of future dynamism and increasing resources.

Public spending on agricultural research has reached a plateau at a time of extraordinary promise and extraordinary challenge. The development of the field of molecular biology and the subsequent expansion of possibilities for agricultural science present opportunities for solving some of the most pressing problems of farm productivity, natural resource management, and the betterment of human nutrition. What lies ahead is the global challenge to feed an ever larger population adequately and with an improved quality of diet. How can one ensure that scientific possibility is translated into the reality of more and better food? At least part of the answer lies, as it has historically, in leadership from the public agricultural sector in mapping for itself a wise and productive role and in recognizing and promoting the complementary action of private sector firms.

Toward the end of better public decision making on agricultural research policy, USDA's Economic Research Service (ERS) pursues investigations of the funding, conduct, and outcome of public and private sector farm and food research. The ERS emphasizes identification of the economic causes and consequences of changes in public and private research financing and in the terms of technology transfer between firms and among nations. In 1996, ERS published a widely acclaimed report, *Agricultural Research and Development: Public and Private Investments under Alternative Markets and Institutions,* that reviewed the economic performance of past investments in agricultural research and documented the growing capacity of the private sector in developing new technology for the U.S. agricultural and food sectors. To further explore the implications of these trends for agricultural research policy, in August 1997 ERS and the Farm Foundation jointly convened the workshop Public-Private Collaboration in Agricultural Research: New Institutional Arrangements and Economic Implications. Participants in the workshop came from a range of institutions: from ERS and other USDA research agencies; from universities; from private firms; and from the nonprofit sector. The chapters immediately following document the work presented by ERS researchers and their collaborators from across the nation.

The current volume expands our knowledge of the changing structure of the agricultural research system in three important ways. First, it describes and compares public and private agricultural research investments in considerably more detail than was previously available—down to the commodity level for plant and animal breeding. Second, it assesses several new institutional innovations that have arisen to strengthen public-private collaboration in research. And third, it explores the international dimension of agricultural technology transfer and the growing global interdependence in agricultural science and technology. Together, these chapters provide a solid foundation for serious reflection on the most appropriate and effective role for public sector agricultural research. The unfolding drama documented here will no doubt prove to be interesting to practitioners in other fields of science (medicine, energy, and transportation) who face many of the same challenges as agricultural practitioners.

Susan E. Offutt
Administrator
Economic Research Service
U.S. Department of Agriculture
Washington, DC

LIST OF TABLES

LIST OF FIGURES

ACKNOWLEDGMENTS

MOST OF the papers presented in this volume were first presented at a workshop cosponsored by the Economic Research Service and the Farm Foundation and held in Rosslyn, Virginia, in August 1997. Participants from government research agencies and the private sector provided valuable input and comments to help guide the authors in their research, and many of the papers have been significantly revised since the workshop. We would like to thank all of the participants of the workshop, especially Margot Anderson, Walter Armbruster, Derek Byerlee, Peter Fuller, Richard Herret, John Pfund, Susan Offutt, Elizabeth Owens, Walter Parks, Richard Parry, and John Pfund, for serving as formal discussants at the workshop. Ellen Bielema provided excellent assistance in helping to organize the workshop.

WE WOULD also like to express our sincere gratitude to the many individuals and institutions that gave generously to help bring this volume to fruition. Especially we would like to thank Margot Anderson, William Anderson, Bruce Greenshields, Susan Offutt, John Reilly, and Robbin Shoemaker for their constant encouragement and support of the work on agricultural research policy at the Economic Research Service. We would also like to thank Emily Chalmers for her excellent technical and editorial input into the manuscript; Doug Parry, Anne Pearl, and Kathy Walker for their help with graphics; and Anne Bolen, Bonnie Harmon, Jim Ice, and Jane Zaring of Iowa State University Press for shepherding the volume through to publication.

—Keith O. Fuglie and David E. Schimmelpfennig

Introduction

1

OVERVIEW OF THE VOLUME

KEITH O. FUGLIE AND
DAVID E. SCHIMMELPFENNIG

OVER THE past several decades, significant changes have taken place in the structure of the U.S. agricultural research system. Probably the most significant development has been the growing role of the private sector in funding and performing agricultural research. Encouraged by expanding global markets, strengthened intellectual property rights, and new opportunities afforded by biotechnology, private firms have significantly increased their investment in developing new agricultural technology. In fact, private-sector expenditures for agricultural research in the U.S. now exceed those of the public sector. The growth in private-sector research capacity and changes in federal legislation have also led to new opportunities for public-private collaboration in agricultural research and development. Several new technology transfer mechanisms have been established to increase the flow of science and technology between public and private research laboratories, including jointly conducted research projects, research consortiums, and patent licensing. Another trend is the increasingly global nature of agricultural technology development and transfer. The rise of multinational agribusiness and "life science" firms and greater research capacity in other countries have served to increase the flow of new agricultural technology across national borders.

These trends have important implications for U.S. agricultural research policy and planning. They affect how research is financed, who conducts it, the type of technology developed, and who is likely to benefit from these developments. In the public sector, fiscal belt-tightening has led to greater emphasis on nongovernment sources of funding for public research institutions. Policy makers are also reexamining the agricultural research funding process for public research institutions, putting greater emphasis on competitive processes rather than block grants (formula funds) for allocating resources to agricultural scientists. In the private sector, the strengthening of intellectual property rights for biological innovations reduces technology spillovers and increases private research incentives. This allows the public sector to reallocate resources to areas of research that are neglected by the private sector.

More rapid international transmission of agricultural technology, however, weakens incentives for individual countries to pursue research. Large international spillovers and technology transfers put greater emphasis on mechanisms for international collaboration in agricultural research.

The present volume brings together recent Economic Research Service (ERS) and academic research that documents these trends and examines their policy implications. The volume is organized into four main sections, described below.

Part I examines new innovations in financing agricultural research. In chapter 2, Fuglie shows how agricultural research expenditures by the U.S. public and private sectors have evolved over the past three decades. Despite the growth in private research capacity in agriculture, there continues to be a critical role for the public sector in particular areas where private incentives remain weak, such as basic research in agricultural sciences and applied research where benefits are difficult to appropriate. The changing roles of the public and private sectors in agricultural research are not unique to the United States, however. In chapter 3, Alston, Pardey, and Smith explore the development of agricultural research policy in four industrial countries—Australia, the Netherlands, the United Kingdom, and the United States—during the twentieth century and especially since the late 1970s. In each of these countries, the role of the private sector in financing and performing agricultural research has grown, shifting the source of funds for agricultural research away from general tax revenues to the agricultural sector itself. In chapter 4, Frisvold and Vogel use a general equilibrium model of the U.S. economy to examine how the financing of agricultural research, whether by general government revenues or specific commodity taxes, affects economic efficiency and welfare. They show that the welfare effects of shifting from general taxes to commodity taxes depends crucially on assumptions about intersectoral labor supply and wage determination.

In this new research environment, effective research policy and planning requires detailed information on the allocation of public and private resources to agricultural research. With both the public and private sectors performing significant amounts of research on agricultural commodities, there is greater potential for overlap and duplication in some areas at the expense of underinvestment in other areas. In part II, the findings of new surveys of private and public research are presented and their implications for public research policy discussed. In chapter 5, Frey presents the results of a comprehensive survey of public and private plant breeding, in which scientist-years for basic and applied crop breeding are provided for each commodity. He finds that the private sector accounted for more than 60 percent of the national investment in plant breeding in 1994 and that private breeding was significantly concentrated on hybrid crops and "near market" cultivar development. His findings raise concerns about the adequacy of resources for basic breeding research and germplasm enhancement, activities required both for long-run growth in productivity and yield and for minor crop improvement. Frey then outlines research strategies for addressing these issues. In addition to these findings, Frey's work provides a valuable new model for agricultural research policy and planning in the research environment of the 1990s. In chapter 6, Falck-Zepeda and Traxler further examine Frey's data to lay out a conceptual framework for the roles of the public and private sectors in the development of new crop varieties. They also trace the evolving importance of public and private sources of germplasm for the improvement of crop varieties, using data on the parentage of new cotton varieties.

Turning to animal commodities, Fuglie, Narrod, and Neumeyer, in chapter 7, examine the roles of the public and private sectors in animal research. Private research emphasizes animal breeding, product development for animal health and nutrition, and new machinery for animal production. For breeding research, the authors present new survey evidence on private and public breeding for broilers, layers, swine, dairy, and beef. As with crops, they find that private-sector investments in applied animal breeding exceed those of the public sector for several commodities, but that the public sector is still responsible for most basic research animal breeding and genetics. They also find evidence that over time the public sector has shifted resources into basic animal sciences and into important areas of applied sciences that may be neglected by the private sector, such as food safety.

The growing capacity of the private sector in conducting agricultural research has opened new opportunities for public-private research collaboration. These opportunities were strengthened in the 1980s by federal legislation that changed patent policies and promoted technology transfer. Part III explores how these initiatives affected agricultural research. In chapter 8, Day-Rubenstein and Fuglie provide an overview of various models for public-private research collaboration and examine the research priorities of Cooperative Research and Development Agreements (CRADAs) at USDA research agencies. In chapter 9, Knudson assesses the role of research consortiums in strengthening public-private collaboration in agricultural research. In case studies of the Genetic Enhancement of Maize consortium and the Biotechnology Research and Development Corporation, she explores how these consortiums were successful in increasing private-sector support for basic agricultural research. Knudson, Lower, and Jones surveyed state agricultural experiment stations on their use of plant breeders rights and patents for crop varieties and germplasm and report these results in chapter10. Policies toward intellectual property rights in agriculture are important not only as incentives for private research but also as technology transfer and funding tools for public research.

While it has long been a major global supplier of new agricultural technology, the United States also benefits from the expanding international capacity in agricultural research. Part IV examines the significance of international "spillovers" of agricultural technology for the United States and discusses some important sources of international technology transfer. Chapters 11 and 12 present new evidence on the size and significance of international flows of agricultural knowledge into and out of the United States. Evenson, in chapter 11, distinguishes between direct and indirect technological spillovers and develops indices for their measurement. In chapter 12, Schimmelpfennig and Thirtle examine the influence of U.S. agricultural research and development on the productivity of European agriculture, and vice versa. Both chapters find evidence of significant spillovers across national boundaries.

The growing importance of international technology transfer in agriculture is due to a number of factors. In chapter 13, Pray and Fuglie examine the role of multinational firms in transferring and adapting agricultural technology for developing countries. They present new survey results on private-sector agricultural research investments in several developing countries. Ballenger and Klotz-Ingram explore the important contribution of graduate training at U.S. agricultural universities to expanding the global capacity for agricultural research in chapter 14. They find that U.S. agricultural universities have a strong international comparative advantage, evidenced by the growing share of doctorates in food and agricultural sciences that are awarded to foreign students.

Finally, in chapter 15 Reilly and Schimmelpfennig conclude the volume with a synthesis of key findings of the previous chapters. They describe the changes occurring in the U.S. agricultural research system in the context of larger social and economic forces shaping the U.S. food and agricultural economy.

SUGGESTIONS FOR FURTHER READING

Alston, J.M., and P.G. Pardey. 1996. *Making Science Pay.* Washington, DC: American Enterprise Institute.

Fuglie, K., N. Ballenger, K. Day, C. Klotz, M. Ollinger, J. Reilly, U. Vasavada and J. Yee. 1996. *Agricultural Research and Development: Public and Private Investments Under Alternative Markets and Institutions.* Agricultural Economic Report 735. Washington, DC: Economic Research Service, U.S. Department of Agriculture.

Huffman, W.E., and R.E. Evenson. 1993. *Science for Agriculture: A Long-Term Perspective.* Ames: Iowa State University Press.

Pray, C.E., and V.W. Ruttan. 1986. *Policy for Agricultural Research.* Boulder, CO: Westview Press.

Ruttan, V.W. 1982. *Agricultural Research Policy.* Minneapolis: University of Minnesota Press.

I. Innovations in Financing Agricultural Research

2

TRENDS IN AGRICULTURAL RESEARCH EXPENDITURES IN THE UNITED STATES

KEITH O. FUGLIE

DURING THE past century, investments in research have transformed agriculture in the United States from a resource-based to a science-based industry (Ruttan 1982). Since the Second World War, almost all growth in U.S. agricultural production has come from improved methods of farming—that is, from increases in output per unit of input—rather than from increases in the use of total inputs. In fact, between 1948 and 1994 the total amount of conventional inputs in use (land, labor, chemicals, machinery, seed, and livestock) actually fell by 3 percent while output increased by 137 percent (Ahearn et al. 1998). New technology resulting from research and development (R&D) has been the primary source of this growth in productivity (Huffman and Evenson 1993). Future growth in agriculture is likely to be similarly dependent on productivity improvements and investments in research.

Both the public and private sectors play important roles in developing new technology for agriculture. In recent decades private sector investments in agricultural research have risen considerably, both in absolute terms and relative to public funding (Fuglie et al. 1996). The type of research conducted by the private sector has also changed: More and more, private investors are undertaking research that once fell in the public domain. Nevertheless, there are significant gaps in the kind of agricultural research the private sector is willing to support. Thus, continued public support for agricultural research is needed to sustain productivity growth in agriculture.

The purpose of this paper is to review the recent changes in who funds and who conducts agricultural R&D and to examine some of the implications of these changes for research policy. The first section discusses the kinds of research that qualify as "public goods" and therefore require government support. The second section reviews private and public funding trends for agricultural research over the past 30 years and examines the types of research each sector undertakes. Implications of these trends for agricultural research policy are discussed in the third section.

WHY GOVERNMENTS SUPPORT RESEARCH

Historically, governments have played a major role in supporting agricultural research. The principal economic rational for such support is that the economic incentives for the private sector to invest in research are often inadequate. It is widely recognized that the private sector will underinvest in certain types of goods and services, normally classified as "public goods." Thus, a more direct role of the public sector in the provision of public goods and services is warranted. Public goods, according to the textbook definition, are goods that are nonrival (use by one person does not diminish the amount available to others) and nonexcludable (others cannot be prevented from using the good once it is first made available). The classic example of a public good is a lighthouse, which warns all passing ships of danger regardless of whether a particular ship helped finance the lighthouse or not. Investment in research is similar—it produces new knowledge, which is, once it is in the public domain, both nonrival and nonexcludable since it can be reproduced and distributed at only a fraction of the initial cost of discovery (Nelson 1959).

Research, therefore, produces benefits to society that go far beyond what a private inventor can expect to earn. Users of a new invention (or of the new knowledge embodied in the invention) can make further improvements to it, apply it to areas not thought of by the inventor, or simply copy it at a fraction of the cost it originally took to develop the invention. The benefits from research that accrue to others are called "spillovers." Spillovers include benefits to rival firms that copy the invention or use it to develop new and improved inventions. Spillovers also include benefits to consumers from lower priced or higher quality products. Spillovers can also occur across national borders when one country is able to borrow and benefit from technology developed in another country. Since a private company can be expected to invest in research only up to the point where the marginal costs of research equal the marginal *private* benefits, and not the larger marginal *social* benefits, the private sector will invest less in agricultural R&D than is optimal for society. Thus, relying only on the profit motive will not induce sufficient resources for research. More direct public-sector intervention is therefore required to assure an adequate level of support to achieve a more optimal rate of economic development and growth.

Public policies to support research can take several forms. One is to strengthen intellectual property rights (IPRs), such as patents and copyrights. Intellectual property rights are legal constructs that limit the use others can make of an invention without the permission of the inventor. In other words, IPRs serve to convert some public goods into private goods by limiting the nonexcludability characteristic of information. Intellectual property rights have long been available for mechanical and chemical inventions, but have only become available for biological inventions in the last few decades.

The establishment and enforcement of IPRs involve tradeoffs between competing goals for public policy. On the one hand, IPRs create additional incentives for private research. IPRs also encourage private companies to make public the results of their research so that it can be used by other scientists to make further technological progress. On the other hand, IPRs by their nature create limited monopoly rights to new inventions, which can raise the price of new technology and reduce its use.

A second way for governments to support science and technology is to directly fund research at public or private institutions. Even in countries with strong intellectual property rights, the private sector is unlikely to invest sufficient resources in many kinds of research. The clearest case is for direct public funding of basic research.[1] Private companies are unlikely to devote many resources to research in basic agricultural sciences because the commercial benefits are not immediately evident. While the private sector often benefits from basic research, spillovers are large and individual companies can't recoup much of their investment in it. Moreover, the outcome of basic research is often long term and quite uncertain, which further discourages the private sector from supporting it. Nonetheless, investments in basic agricultural research can offer very high returns to the economy as a whole, benefits that are widely shared among its many potential users (Huffman and Evenson 1993).

Direct public support of many areas of applied research is often warranted as well. The private sector is unlikely to invest sufficiently in research on the environment and natural resources, on food safety and nutrition, on poverty reduction, and on public policy issues (Fuglie et al. 1996). Research in these areas generates new knowledge and technology that may not result in marketable products but nevertheless contribute to important societal goals. The private sector may also underinvest in research aimed at small farmers and certain kinds of commodities. In the United States, the private sector invests considerably more research in crops produced from hybrid seed (e.g., corn), which must be repurchased by farmers every season, than in crops produced by self-pollinated seed that can be saved by farmers (e.g., wheat and soybeans). Thus, there are many areas of applied agricultural research that are likely to require direct public financing in order to sustain productivity growth in the food and agricultural sector.

Several studies have found that both public and private investments in agricultural research in the United States have resulted in an average annual rate of *social* returns of 35 percent or higher, even after accounting for potential sources of measurement error (Ruttan 1982; Huffman and Evenson 1993; Fuglie et al. 1996). The social returns include the spillover benefits of new technology to the whole economy, including benefits to agribusiness firms, farmers, and consumers. Private rates of return—what a for-profit company can expect to make from its investment in research—are typically less than half the social rates of return (Mansfield et al. 1977). The gap between private and social returns provides empirical support for the significance of spillovers and the "public good" component of agricultural research. Furthermore, the high social return to agricultural research suggests that society as a whole underinvests in agricultural research, since the social returns exceed most measures of the opportunity cost of social capital (Evenson, Waggoner, and Ruttan 1979; Fuglie et al. 1996).

EXPENDITURE TRENDS IN AGRICULTURAL RESEARCH

Between the mid-1960s and mid-1990s, total annual spending on agricultural research in the United States increased from just under $4 billion to $7 billion in constant 1995 dollars (figure 2.1). Despite the spillover problem, private investment in agricultural research grew considerably over this period, more than doubling from $2 billion in 1970 to $4.2 billion in 1996. In fact, since 1978, the private sector has been responsible for virtually all of the growth in total U.S. agricultural research expenditures.

Several factors are behind the rise in private agricultural research. First is increased demand from agricultural producers for improved farm inputs, both in the United States and abroad. Globalization and trade expansion have en-

Figure 2.1. Sources of funding for U.S. agricultural research

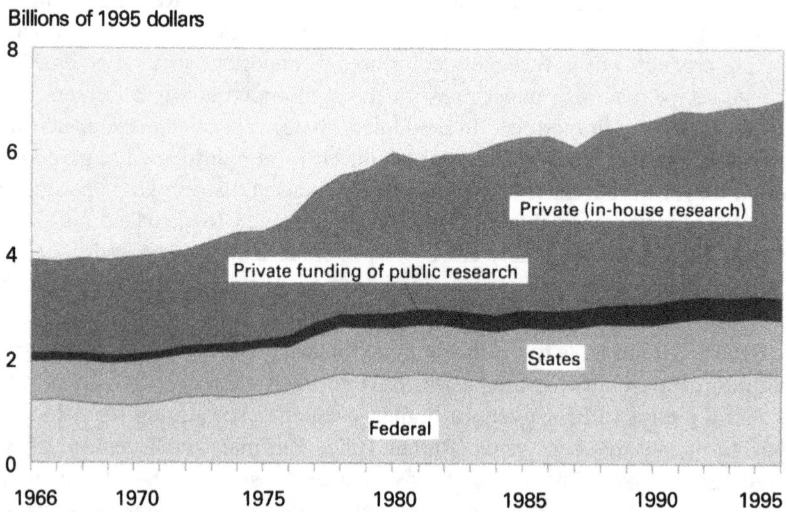

Source: Economic Research Service.

abled multinational seed, chemical, pharmaceutical, and livestock firms to greatly expand their markets for improved agricultural technology. A second factor is the new opportunities afforded by scientific advances in molecular biology and genetic engineering. The private sector is investing more than $600 million a year in developing commercial applications of biotechnology for the agricultural and food industries in the United States alone (Klotz, Fuglie, and Pray 1995). A third factor is the strengthening of intellectual property protection for new agricultural inventions, which has enabled inventors to capture a larger share of the economic benefits derived from their inventions (Fuglie, Klotz, and Gill 1996). The United States established plant breeders' rights for new crop varieties in 1970, and in 1980 the Supreme Court ruled that standard industrial (utility) patents could be awarded for new biological inventions. Furthermore, members of the World Trade Organization are required to strengthen protection of intellectual property rights under the terms of the GATT–Uruguay Round international trade agreement. In particular, the agreement requires member nations to adopt IPR protection for agricultural chemicals, veterinary pharmaceuticals, and new plant varieties. Finally, environmental and safety regulations have driven up the cost of new product development, especially for agricultural chemicals. The private sector must now fund an extensive array of tests in order to register a new pesticide, increasing its agricultural chemical R&D budget (Fuglie et al. 1996).

While private expenditures for agricultural research grew, funding of agricultural research by the federal and state governments stagnated in real terms after 1978. The federal government provided about $1.7 billion, and state governments another $1.0 billion, for agricultural research in 1995. The recently enacted Agricultural Research, Extension, and Education Act of 1998 may increase federal funding of agricultural research by up to $200 million per year during 1998 through 2002.

Funds, Flows, and Performers

In discussing the appropriate roles for public and private research, it is useful to distinguish between who funds and who performs research. Figure 2.2 maps the flow of resources between funders and "performers" of agricultural R&D for 1995. The private sector provided about 61 percent of the total funding for agricultural R&D. Federal and state governments provide another 24 and 15 percent of total funding, respectively.

The federal government maintains substantial in-house research capacity at the research agencies of the U.S. Department of Agriculture (USDA) and also transfers more than $700 million to state agricultural experiment stations (SAESs) at land-grant universities and other cooperating institutions. State governments and the private sector also provide funds to SAESs. Private funds for agricultural research at SAESs, including grants from industry and revenue that public institutions are able to raise from product sales and tech-

Figure 2.2. Agricultural research funders and performer in 1995. *Figures are in millions of dollars (percent of total research in parentheses)*

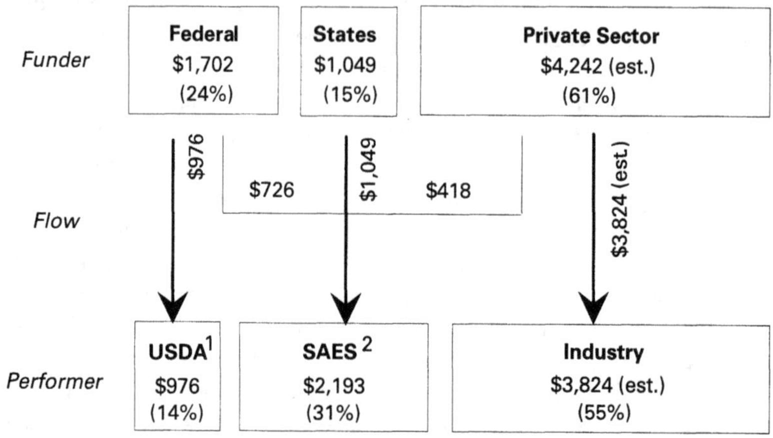

[1] Agricultural Research Service, Economic Research Service and Forest Service, excluding research contracted out.
[2] State agricultural experiment stations, 1890 schools, forestry schools, and veterinary schools.

Source: Economic Research Service.

nology licensing, accounted for about 19 percent of all SAES research resources. Of the $7 billion in total agricultural research expenditures, state research institutions account for nearly one-third. But the largest performer in agricultural research is industry, which spends about 55 percent of all food and agricultural research resources. Research agencies of the USDA accounted for the rest, about 14 percent of total expenditures.

The sources of funding and the institutions assigned responsibility for carrying out research influence the type of research undertaken. Private-sector funders are likely to demand more applied R&D that has clear market potential. State governments prefer to fund research on topics significant to a state, while federal funds are often focused on issues of national or regional importance. Motivations of scientists at performing institutions also differ. Scientists at universities and government laboratories are rewarded for academic publications and professionally recognized excellence in their fields. Scientists in industrial laboratories are more likely to be judged on the eventual commercial success of their research activities.

Who Does What

Research and development activity can be classified along a continuum from basic research, to applied research, and finally to developmental research. The public agricultural research system focuses primarily on basic, or precommercial, research. In 1992, nearly half of all research conducted at USDA and SAES labs was classified as basic research and another 45 percent as applied research. Only 7 percent was allocated to product development (table 2.1). The private sector lacks incentives to invest in basic research because such research is considerably more risky than applied research and the results are more difficult to patent. As a result, more than 40 percent of private agricultural research conducted in 1984 was classified as product development, 43 percent as applied research, and only 15 percent as basic research (table 2.1).

Basic research, though lacking immediate commercial applications, can nevertheless produce large economic benefits. Results from basic research can often open up whole new fields of applications. For example, just as Mendel's experiments in the 1860s led to the field of quantitative plant breeding, Watson and Crick's discovery of the basic structure of DNA in the 1950s helped open the door to applications of molecular genetics and modern biotechnology. Due to the potential for large spillovers, basic or precommercial research in agricultural sciences has been found to yield a high social rate of return (Huffman and Evenson 1993). However, because of the lack of clear and immediate commercial applications, incentives to fund basic research remain weak.

Another way to break down the allocation of research resources is by technology or problem areas. Public-sector research priorities by technology area have remained relatively unchanged in the last two decades (figure 2.3). Crop and animal research remain the largest components of federally and state-

Table 2.1. Percentages of public and private agricultural research expenditures devoted to basic, applied, and development research in the United States

Type of research	Public sector	Private sector
	(% of total)	
Basic research	47.3	15.0
Applied research	45.4	43.5
Development research	7.3	41.5
All research	100.0	100.0

Basic research determines cause or mechanism of why certain results occur, with no specific commercial application in mind.
Applied research develops knowledge directly relevant to technology or product development.
Development research develops new or improved product or supports market testing and introduction.

Source: Fuglie et al. (1996). Public sector data refer to expenditures for 1992, from the Current Research Information System, U.S. Department of Agriculture. Private sector data are for 1984, reported in Crosby et al. (1985).

funded research (34 percent and 23 percent, respectively, of agricultural research in 1995). A large share of public research also addresses forestry and natural resource issues. One area of growth has been food and nutrition research, which increased from about 3 percent of total public research in 1975 to 6.7 percent in 1995.

Figure 2.3. Composition of public agricultural research

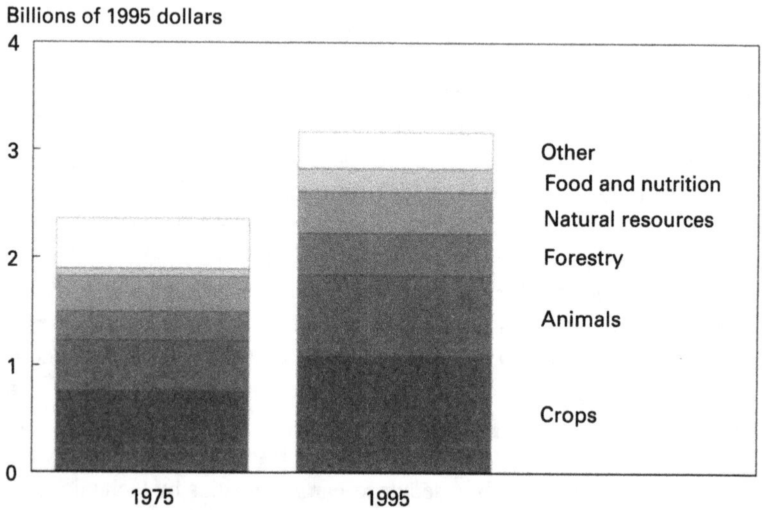

Source: Economic Research Service.

Research priorities in the private sector, on the other hand, have undergone significant changes over the last two decades. Private companies have rapidly expanded research investments into technology areas that were long the purview of the public sector, especially plant breeding and livestock improvement. Figure 2.4 shows the composition of private research in 1975 and 1995. In 1975, research on food products and processing accounted for 40 percent of all private spending on agricultural R&D, agricultural chemical research for another 23 percent, and farm machinery for 20 percent. Crop and livestock research were relatively small components of private research. In the last 20 years private investment has grown in real terms in all of these areas except for farm machinery, but growth has been most rapid in agricultural chemical research, plant breeding research, and animal health research. In plant breeding, the U.S. private sector now spends more on research than the federal and state governments combined (see chapter 5). The private sector is also playing a dominant role in developing applications of biotechnology for agriculture.

Figure 2.4. Composition of private agricultural research

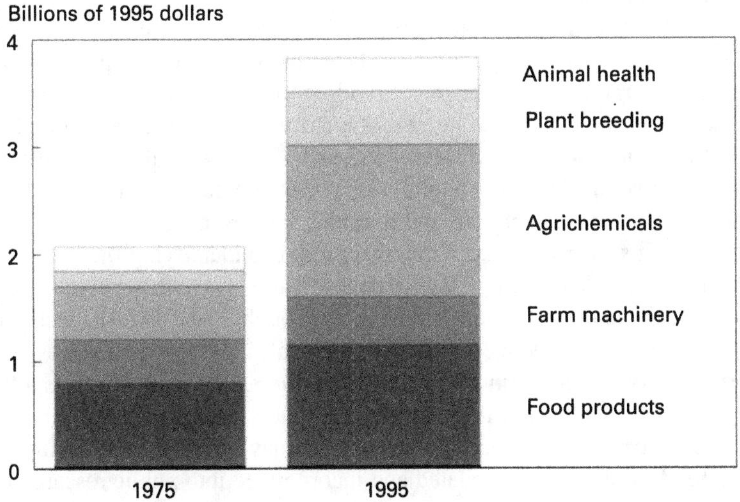

Source: Economic Research Service.

IMPLICATIONS FOR RESEARCH POLICY

The trends described above have several implications for agricultural research policy. The growth of private sector investment in agricultural research and changes in IPR laws has changed the boundary between the "public good" and "private good" components of agricultural science and technology. This affects not only priorities for public-sector research, but also the role of public-private cooperation in research and technology development.

Priorities for Public Agricultural Research

Societal demands on the public agricultural system have increased in recent years. The public is interested not only in reducing the cost of agricultural production and maintaining abundant food supplies, but also in protecting the environment, improving the quality of rural life, and enhancing the quality and safety of food products.

The lack of new public resources for agricultural research could mean that difficult choices will have to be made in addressing the growing list of public concerns. Private companies do not address many areas of research that are necessary to meet the growing list of societal demands on and expectations of the agricultural and food system. In addition to basic research, the private sector is unlikely to invest in areas of applied R&D such as natural resource conservation, environmental protection, food safety and nutrition, and the social sciences. These applied fields do not result in products that can be easily marketed, but they nonetheless provide useful information and technology to meet societal needs.

As we have seen, the private sector is developing significant research capacities in areas that have traditionally been the purview of the public sector. Continued increases in private-sector funding in these areas could allow public research to shift resources into other areas where the private sector is unlikely to provide funding. As figures 2.3 and 2.4 showed, there is more potential for overlap between public and private research than in the past, particularly in the areas of crop and livestock improvement.

Allocating scarce resources effectively requires identifying those types of R&D activities that present the strongest cases for public involvement. Stronger institutional relationships between public- and private-sector researchers can help to delineate each sector's roles and responsibilities more clearly. One of the difficulties in planning public research, however, is a lack of systematic and detailed information on private research expenditures and priorities. More formal planning mechanisms may be required to assure that this kind of information is updated and incorporated into public research priority setting.

New Mechanisms for Technology Transfer

In many technology areas, the public sector is focusing on basic or precommercial research, leaving the development of specific technological applications to the private sector. New institutional relationships are needed in order to assure that the results of science-oriented research in public labs are quickly transferred to technology-oriented research in private laboratories. Otherwise, important new scientific developments may not be incorporated into improved technologies for agriculture. Researchers in the public sector also need to be kept abreast of developments in the private labs.

Several new approaches have been developed to deal with this issue. One such approach is to establish public-private joint ventures. At the federal level, joint ventures are conducted through Cooperative Research and Development Agreements, or CRADAs (see chapter 8). A CRADA is a collaborative research project between a federal lab and a private company to develop a specific technology. Federal labs have had the legal authority to enter into these agreements since 1986, and between 1986 and early 1998 the USDA established more than 700 CRADAs. Another approach is the research consortium (see chapter 9). In the consortium model, several public and private partners join together to support precommercial research in a specific technology area. Consortium members have preferential access to develop commercial applications from the results of this research.

Land-grant universities have also increased their collaboration with agricultural industries. An important impetus was provided by the 1980 Patent

Policy Act (the Bahr-Dole Act), which allowed individuals and institutions receiving federal funding for research to own the patents on technology they developed. Most universities have established offices of technology transfer that assist in patenting and licensing inventions developed in university labs (Parker, Zilberman, and Castillo 1998). While patenting and licensing fees provide a new source of revenue for universities, they also affect incentives within the university. Faculty have a greater incentive to use research time developing inventions that private industries may be willing to market.

Another way universities are affected by the growing role of private industry in agricultural R&D is in the job market for scientists. In 1973 around 15 percent of PhDs in the agricultural sciences were employed by industry. By 1991 that figure had risen to 25 percent (National Research Council 1995). With more agricultural science graduates going into private industry rather than entering academia or government service, universities may have to adjust their curricula so that graduates have the skills required for private-sector research.

Enhancing Incentives for Private Research

Government interventions to support agricultural research have tended to rely on raising tax revenues to fund research in public institutions. An alternative approach is to consider ways to strengthen private incentives to fund and conduct agricultural research. Policies that encourage private companies to invest in more in-house private research include strengthened intellectual property protection, antitrust exemptions for research consortia, special tax incentives, and transparent and less burdensome environmental and product safety regulations. Another set of policies can encourage more private contributions to public research institutions by providing matching grants for private contributions to public research institutes—through commodity check-off funds and royalties from technology licenses, for instance.

Plant breeding provides a useful example of how enhancing private research incentives through stronger intellectual property protection affects private investment in agricultural research. Private companies can recoup their investments in hybrid seed much more easily than they can their investments in self-pollinated seed, since the progeny of hybrid seed cannot be kept for replanting. Companies therefore protect their investments in hybrid seed by restricting access to the parent inbred lines. Self-pollinated seed, however, generally breeds true to form, meaning that second-generation seed yields as well as first-generation seed. As a result, the private sector has tended to invest more in research on crops produced from hybrid seed than on crops produced from self-pollinated seed. Figure 2.5 shows the share of yield improvements from plant breeding captured by higher seed prices and the percent of seed sales companies spent on research for five major crops in the United States.[2]

Corn and sorghum are produced almost exclusively with hybrid seed, while wheat, soybean, and cotton come mainly from self-pollinated seed. For hybrid seed, private companies are able to appropriate a higher share of yield gains through higher seed prices than they are for self-pollinating seed. At the same time, companies spend two to three times as much (as a percent of seed sales) researching hybrid seed than they do researching other seed. The private sector invests more in hybrid seed research because they can capture a larger share of the economic benefits from higher yields.

Several efforts have been made to strengthen private incentives for plant breeding, most notably the 1970 Plant Variety Protection Act. This act provided modest incentives for private research on nonhybrid seed, but they were weakened by exemptions for farmer-saved seed, "brown-bagged" seed (seed sold from farmer to farmer), and research (others were free to use protected seed in breeding programs to develop new varieties) (Butler and Marion 1985). Recent amendments to the 1970 act and the Supreme Court ruling *Asgrow v. Winterboer* have narrowed these exemptions. In addition, standard industrial patents (utility patents) are now available for discoveries of new genetic sequences and other biotechnology innovations. These recent institutional innovations have strengthened the intellectual property rights of plant breeders.

Figure 2.5. Private breeders allocate a larger share of seed sales to research on hybrid seed, where they can capture a larger share of research benefits

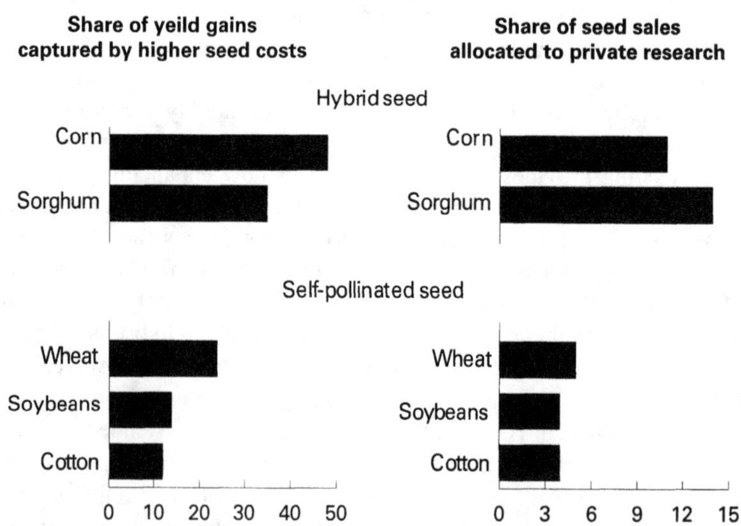

Source: Fuglie et al. (1996).

The pattern of patent protection for new agricultural biotechnology and genetic sequences is controversial. Critics have argued that in many cases broad patents have been awarded for basic tools required for genetic engineering and for genes or gene sequences in cases where the commercial value (the utility criterion) is not well established (Barton 1998). Such broad patents may restrict or slow down fundamental advances in a new fields of science such as biotechnology (Merges and Nelson 1990). Private companies with large investments in agricultural biotechnology have resisted attempts to weaken patent protection, however, preferring to access patented technology through cross-licensing or material transfer agreements, strategic alliances, mergers, and acquisitions of rival firms. Further analysis of the incentives for private breeding and the effect of patenting on the science community's access to key technology is necessary to assess the impact on the pace of agricultural science and technology development.

CONCLUSION

Government funding for agricultural research has not increased in real terms over the past three decades. Yet the need for improved agricultural technology to meet the broad development and policy goals are pressing. With adequate incentives, the private sector can play a major role in filling the gap. Public-private cooperation in research can help assure that new scientific advances are quickly incorporated into commercial technologies.

A larger role for private industry in funding and conducting agricultural research will affect the type of technology developed and the distribution of the economic benefits from research. When the cost of agricultural research is borne largely by taxpayers, the benefits of agricultural research are shared broadly by farmers and consumers. In the future, it is likely that a larger share of the benefits of research will go to private companies as an incentive and reward for their research investments. In addition, the private sector's contribution to research can be expected to be market focused, so that basic research and some essential areas of applied research will remain the responsibility of the public sector.

NOTES

1. Here, basic research is defined as research that improves our understanding of biological, mechanical, or other processes for which no commercial application may be immediately evident. Basic research can also include research to develop improved methods of conducting research, so as to improve the efficiency of research carried out further downstream.

2. Changes in crop yields and seed prices are assessed for the period between 1970 and 1990. Plant breeding is assumed to account for 50 percent of yield gains in crops. Private expenditures for plant breeding are for 1989 and are derived from Kalton, Richardson, and Frey (1989).

REFERENCES

Ahearn, M., J. Yee, E. Ball, and R. Nehring. 1998. *Agricultural Productivity in the United States.* Agriculture Information Bulletin 740. Washington, DC: Economic Research Service, U.S. Department of Agriculture.

Barton, J. 1998. "The Impact of Contemporary Patent Law on Plant Biotechnology Research." In *Intellectual Property Rights III Global Genetic Resources: Access and Property Rights.* Madison, WI: CSSA Miscellaneous Publication, pp. 85–97.

Butler, L.J., and B.W. Marion. 1985. *The Impacts of Patent Protection on the U.S. Seed Industry and Public Plant Breeding.* North Central Regional Research Publication 304, North Central Project 117, Monograph 16. Madison: Research Division, College of Agriculture and Life Sciences, University of Wisconsin.

Crosby, E.A., B.R. Eddleman, R.R. Kalton, V.W. Ruttan. 1985. *A Survey of U.S. Agricultural Research by Private Industry III.* Washington, DC: Agricultural Research Institute.

Evenson, R.E., P.E. Waggoner, and V.W. Ruttan. 1979. "Economic Benefits from Research: An Example from Agriculture," *Science* (September 14): 1101–07.

Fuglie, K., N. Ballenger, K. Day, C. Klotz, M. Ollinger, J. Reilly, U. Vasavada, and J. Yee. 1996. *Agricultural Research and Development: Public and Private Investments Under Alternative Markets and Institutions?* Agricultural Economics Report 734. Washington, DC: Economic Research Service, U.S. Department of Agriculture.

Fuglie, K., C. Klotz, and M. Gill. 1996. "Intellectual Property Rights Encourage Private Investment in Plant Breeding." *Choices* (First Quarter): 22–23.

Huffman, W.E., and R.E. Evenson. 1993. *Science for Agriculture.* Ames: Iowa State University Press.

Kalton, R.R., P. Richardson, and N. Frey. 1989. "Inputs in Private Sector Plant Breeding and Biotechnology Research Programs in the USA," *Diversity* 5(4): 22–25.

Klotz, C., K. Fuglie, and C. Pray. 1995. *Private-Sector Investment in Agricultural Research, 1960–1992.* Staff Paper 9525. Washington, DC: Economic Research Service, U.S. Department of Agriculture.

Mansfield, E., J. Rappaport, A. Romeo, S. Wagner, and G. Beardsley. 1977. "Social and Private Rates of Return from Industrial Innovation." *Quarterly Journal of Economics* 91 (May): 221–40.

Merges, R.P., and R.R. Nelson. 1990. "On the Complex Economics of Patent Scope." *Columbia Law Review* 90 (May): 839–916.

National Research Council (NRC). 1995. *Colleges of Agricultural at the Land Grant Universities: A Profile.* Washington, DC: National Academy Press.

Nelson, R.R. 1959. "The Simple Economics of Basic Scientific Research," *Journal of Political Economy* 67:351–64.

Parker, D., D. Zilberman, and F. Castillo. 1998. "Offices of Technology Transfer: Privatizing University Innovations for Agriculture," *Choices* (First Quarter): 19–25.
Ruttan, V.W. 1982. *Agricultural Research Policy.* Minneapolis: University of Minnesota Press.

3

FINANCING AGRICULTURAL R&D IN RICH COUNTRIES: WHAT'S HAPPENING AND WHY[1]

JULIAN M. ALSTON, PHILIP G. PARDEY, AND VINCENT H. SMITH[2]

PUBLIC AGRICULTURAL research institutions and the policies that surround them are at a pivotal point in their history. In many countries, in both the developed world and the developing world, public agricultural research and development (R&D) policy has changed dramatically since the early 1980s. A long period of sustained growth appears to have ended, and many perceive that we have entered a phase of general fiscal restraint, which, combined, with a more skeptical view of the social benefits from investments in science, has led to a tightening of resources available for research and calls for clearer justification for R&D funds and accountability for their use.

This article documents the nature and scope of these changes in five developed countries; Australia, the Netherlands, New Zealand, the United Kingdom, and the United States.[3] These countries are of particular interest because they have long histories of active public agricultural R&D policies, and taken together, they account for over 40 percent of total public agricultural R&D by OECD countries. In addition, they include some of the most innovative countries with respect to changes in agricultural R&D policies; lessons learned there could be useful elsewhere.

Agricultural R&D policies have often been described purely in terms of total public funding, which may be appropriate when the institutions that manage and carry out publicly funded agricultural research remain relatively stable. However, when an explicit goal of public agricultural R&D policy is to alter the structure and function of those institutions, as has been the case in all but one of the countries examined here (the exception being the United States), trends in public funding tell only a part of the story, albeit an important part. Here we consider agricultural R&D policy developments over the past 20 years in a broader sense for each of the five countries of interest, examining not only public funding trends for agricultural research, but also changes in agricultural R&D institutions.

Many aspects of the recent agricultural R&D policy changes have been remarkably similar among the countries, but there have been some important differences with respect to both the timing and the substance of policy innovations. The similarities derive from a set of common "vectors for change" that have come into play in each of the countries, including the advent of political administrations with more market-oriented, laissez-faire views of the role of government in the management of the national economy, the changing nature of scientific research, the development of a more skeptical view of the potential benefits from agricultural R&D, and the growing influence of "nontraditional" interest groups (such as agribusiness, food industry, environmental, and food safety lobbies) on the public agricultural R&D agenda.

Many of the recent agricultural science policy changes have been intended, at least ostensibly, to shift the burden for the funding and execution of some types of agricultural research from the public sector to the private sector. In most countries, aside from its own research investments, the private sector has had a longstanding active interest in public agricultural R&D. As primary users of the outputs of public R&D systems, agricultural producers, food processors, and other agribusiness firms have been involved in the management of publicly provided R&D funds. In addition, through various mechanisms, private firms have funded publicly performed agricultural R&D or have performed publicly funded agricultural research. This article pays particular attention to the degree to which each of these three roles of the private sector has become more or less important, and the extent to which changes in those roles have been linked to changes in public agricultural R&D policies.

PUBLIC INVESTMENT IN NATIONAL AGRICULTURAL RESEARCH[4]

In most countries, public-sector involvement in agricultural R&D has been important for less than 100 years. Its evolution has not been smooth.

Genesis and Historical Development of Public Agricultural Research

In many countries, formalized agricultural research institutions supported by national governments were first instituted between the middle and the end of the nineteenth century. Research developments in these countries occurred in waves, with common undercurrents in different countries. The first wave of development involved the initial establishment in the middle of the nineteenth century of various agricultural experiment stations in France, Germany, and the United Kingdom, partly in response to the emergence of the formal agricultural sciences through the work and writings of Boussingault, von Liebig, and Lawes at that time. The practice of providing public funds to support national agricultural research agencies staffed with professional scientists (introduced for the first time in Möckern, Saxony, in 1852) spread rapidly throughout Europe. By 1875 over 90 European agricultural experiment stations had been established, and the process of expansion continued throughout the rest of the century (Grantham 1984). In the Netherlands, for example, the first publicly funded agricultural experiment station was created at Wageningen in 1877, and additional public agricultural research, monitoring, and testing facilities were rapidly established between 1880 and 1900. Similar developments took place in Australia (where the Victorian Board of Agriculture took over the first experimental farm in the 1850s (see Baker, Baklien, and Watson 1990; and McLean 1982) and the United States (where the first agricultural experiment station was established at Yale in 1875 (see Kerr 1987), as well as in other countries [e.g., Japan (see Hayami and Yamada 1975; and True and Crosby 1902)].

By the turn of the century, the scientific foundations of agricultural R&D had become more fully developed. Darwin's theory of evolution, the pure-line theory of Johannson, the mutation theory of de Vries, and the rediscovery of Mendel's Laws all contributed to the rise of plant breeding. Pasteur's germ theory of disease and the development of vaccines opened up lines of research in the veterinary sciences. These developments provided the scientific basis for a second steady wave of expansion in public commitments to agricultural R&D during the first half of the twentieth century.[5]

The evolution of the science of genetics gathered pace around the middle of this century as Hersey and Chase, Watson and Crick, and others uncovered the role and structure of DNA that led directly to the modern biotechnologies based on recombinant DNA techniques, monoclonal antibodies, and new cell- and tissue-culture technologies. These changes in the practice of biological science paralleled efforts by governments to enact legislation strengthening intellectual property protection applied to living organisms (such as new plant varieties and related genetic material) and to implement a range of public science policies that stretched well beyond agriculture. Together these developments fundamentally changed the nature of the agricultural sciences, with implications for public and private roles in agricultural research and for the balance between locally provided and internationally traded R&D goods and services.[6] It was fortuitous, perhaps, that many of the discoveries on which modern agricultural science methods are based were made shortly after the end of World War II, a period when science and technology were perceived widely as a potential source of large improvements in social welfare. The result was a massive third wave of public investments in science and technology research in general, and in agricultural research in particular.

In real terms, in most developed countries public expenditures on agricultural R&D grew more rapidly between 1945 and the mid-1970s than in the rest of the World War II period.[7] Then, in the mid-1970s, rates of growth in public R&D outlays slowed quite markedly, and in the 1980s public agricultural R&D expenditures generally stagnated or declined. In the 1990s, however, public agricultural R&D expenditures recovered or began to increase again, but at much more modest rates of growth than in the 1960s and early 1970s. Changes in research focus and other elements of R&D policy were associated with these recent funding shifts, and accelerating changes in telecommunications (especially the development of the Internet) have unleashed a new wave of significant institutional and other changes in the agricultural sciences.

Trends in Public Agricultural R&D Funding in OECD Countries

Data on public agricultural R&D expenditures, measured in real, 1985 international dollars,[8] and their rates of growth over the period 1971 to 1993 are presented in table 3.1 for 22 member countries of the Organization for Economic Cooperation and Development (OECD) and for each of the five OECD countries of particular interest here (Australia, the Netherlands, New Zealand, the United Kingdom, and the United States).[9] The 22 member countries of the OECD together account for over 90 percent of all developed-country agricultural R&D; therefore, trends in agricultural R&D spending in these countries provide useful contexts for evaluating trends in agricultural R&D spending in the five countries of interest. It is also useful to compare trends in those five countries, which accounted for over 40 percent of public agricultural R&D outlays by OECD countries during this period, with trends in the other 17 OECD countries.

Table 3.1. Real public agricultural R&D spending in OECD countries, 1971–93

					Annual growth rates		
	1971	1981	1991	1993	1971–81	1981–93	1971–93
	(millions of 1985 international dollars)[b]				(percentages)		
Australia	238.7	281.9	307.8	315.4	2.1	0.3	1.2
Netherlands	134.7	202.1	216.6	229.7	4.2	0.9	1.6
New Zealand	114.3	133.8	110.2	107.3	2.2	−2.2	0.2
United Kingdom	274.5	371.0	364.4	370.8	2.6	−0.2	1.2
United States	1,235.6	1,620.4	2,023.4	2,054.3	2.4	2.3	2.1
Subtotal (5)[a]	1,997.8	2,609.1	3,022.4	3,074.5	2.5	1.4	1.8
Other OECD (17)[a]	2,300.2	3,104.3	3,919.1	4,054.9	2.9	2.1	2.6
Total OECD (22)[a]	4,298.1	5,713.4	6,941.4	7,129.4	2.7	1.8	2.2

[a]Figures in parentheses indicate the number of countries included in the respective totals.
[b]Research expenditures denominated in current local currency units are first deflated to 1985 prices using local implicit GDP deflators taken from the World Bank (1995) and then converted to international dollars (where one international dollar is set equal to one U.S. dollar) using the purchasing power parities taken from Heston et al. (1995).

Source: Pardey, Roseboom, and Craig (1997).

Between 1971 and 1993, as the data in table 3.1 indicate, spending on agricultural R&D by the 22 OECD member countries increased from about $4.3 billion to about $7.1 billion at an annual average growth rate of about 2.2 percent. However, this 22-year period can be divided into two distinct periods, 1971–1981 and 1981-1993. In the first period, essentially the decade of the 1970s, total public OECD agricultural R&D expenditures increased at an annual average growth rate of 2.7 percent; in the 1980s and early 1990s, the annual average growth rate fell by about one-third, to 1.8 percent.

The five countries of particular interest had markedly slower average annual growth rates in public agricultural R&D expenditures than did the OECD as a whole, but the general pattern of growth in the five countries was similar to that of the overall growth among the OECD countries. The differences and similarities between the five selected countries and the other OECD countries are even more apparent when growth rates in public agricultural R&D expenditures for those other 17 countries are examined. Over the period 1971–1993, total public agricultural R&D outlays in the 17 countries increased by over 75 percent. In the first subperiod, 1971–1981, the annual average growth rate in public agricultural R&D for these 17 countries was 2.9 percent; in the second subperiod, 1981–1993, it fell by more than one-fourth to 2.1 percent. However, between these two subperiods, the growth rate of public agricultural R&D expenditures fell by two-fifths in the five selected countries.

Among themselves, the five countries exhibited some important differences with respect to funding trends. In the United States, real public agricultural R&D expenditures, measured in 1985 prices, increased from about $1.2 billion in 1971 to just under $2.1 billion in 1993, an annual average growth rate of 2.3 percent (compared with 2.2 percent for all OECD countries, and 2.6 percent for the other 17 OECD countries over the same period). Moreover, the shift in the annual growth rate of public agricultural R&D expenditures in the United States, from 2.7 percent in the 1970s to 2.1 percent in the 1980s, was also closely paralleled in the other 17 OECD countries. In contrast, between 1971 and 1993, public agricultural R&D expenditures in Australia, the United Kingdom, and the Netherlands increased at much lower annual average growth rates of between 1.2 and 1.6 percent, and public agricultural R&D expenditures in New Zealand barely increased at all. These low average annual growth rates for the entire period were the result of precipitous declines in annual average growth rates in public agricultural research in Australia and the Netherlands and of absolute declines (i.e., negative growth) in New Zealand and the United Kingdom over the period 1981–1993.

Among the other 17 OECD countries, only three countries (Belgium, Canada, and Greece) also experienced reductions in real public agricultural R&D spending over the period 1981–1993. Thus, four of the five countries examined in this study experienced more dramatic changes in public agricultural R&D funding trends than most other OECD countries. During the 1980s, governments in these four countries (Australia, the Netherlands, New Zealand, and the United Kingdom) also implemented more radical changes with respect to the institutional organization and management of public agricultural research than did the U.S. government.

Data on the share of public funds provided to support research at universities are presented in table 3.2 for each of the five countries and 14 other OECD countries for which such data were available. In all of the five countries, between 1971 and 1993, the university shares of total public agricultural R&D spending increased.[10] In Australia and New Zealand, the increases were very modest (university shares rose from 10.7 percent to 11.6 percent and from 13.0 percent to 13.6 percent, respectively), but in the other three countries, the increases were substantial. In the United Kingdom, the university share increased from 2.3 percent to 14.7 percent, in the Netherlands it increased from 14.9 percent to 31.9 percent, and in the United States it increased from 67.3 percent to 74.1 percent. In the other 14 OECD countries, the average share of public agricultural R&D funds channeled to universities changed little, from 27.5 percent to 28.0 percent, over the same period. Thus there is some evidence that the five countries are somewhat atypical with respect to the reallocation of research funds toward universities.

The annual rates of growth in public agricultural R&D expenditures in both the five countries and the other 17 OECD countries were quite small in the 1970s, compared with the annual growth rates of 7 to 8 percent experienced during the 1950s and 1960s in Australia, the United Kingdom, the United States, and many other countries. During the 1950s and 1960s, many developed countries financed large-scale expansions in their national science research–education systems. Agricultural science did well, possibly because

Table 3.2. University share of public agricultural R&D spending, 1971–93

	1971	1981	1991	1993
		(percentages)		
Australia	10.7	8.2	11.5	11.6
Netherlands	14.9	22.1	31.8	31.9
New Zealand	13.0	13.0	17.8	13.6
United Kingdom	2.3	2.9	9.7	14.7
United States[a]	67.3	67.5	74.0	74.1
Subtotal (5)[b]	45.0	45.6	54.8	55.3
Other OECD (14)[b]	27.5	25.6	28.2	28.0
Total OECD (19)[b]	38.8	37.9	43.0	43.2

[a] University research in the United States includes research undertaken by the State Agricultural Experiment Stations operated in conjunction with the state land-grant universities.
[b] Figures in parentheses indicate the number of countries included in the respective totals.

Source: Pardey, Roseboom, and Craig (1997).

the agricultural sector was politically effective, and also because shortages during and immediately after World War II stimulated a demand for investment in R&D to assure security of future food and fiber supplies.

Global Perspectives on Public Agricultural Expenditure Trends

The pattern of decline in growth rates of public agricultural R&D expenditures in the 1980s was not restricted to developed economies. Nor was it entirely restricted to agricultural R&D. Some perspective is provided by considering global agricultural R&D funding patterns and funding trends for general science and technology R&D. Estimates of global public agricultural R&D expenditures, disaggregated by developing countries and developed countries, are presented in table 3.3.

Table 3.3. Real public agricultural R&D spending in developed and developing countries, 1971-91

	1971	1981	1991
	(millions of 1985 international dollars)		
Expenditures			
Developing countries (131)[a]	2,984	5,503	8,009
Sub-Saharan African (44)[a]	699	927	968
China	457	939	1,494
Asia and Pacific, excl. China (28)[a]	861	1,922	3,502
Latin America and Caribbean (38)[a]	507	981	944
West Asia and North Africa (20)[a]	459	733	1,100
Developed countries (22)[a]	4,298	5,713	6,941
Total (153)[a]	7,282	11,217	14,951
	1971–81	1981–91	1971–91
	(percentages)		
Average annual growth rates			
Developing countries	6.4	3.9	5.1
Sub-Saharan African	2.5	0.8	1.6
China	7.7	4.7	6.3
Asia and Pacific (excl. China)	8.7	6.2	7.3
Latin America and Caribbean	7.0	–0.5	2.7
West Asia and North Africa	4.3	4.1	4.8
Developed countries	2.7	1.7	2.3
Total	4.3	2.9	3.6

Regional groupings of countries are given in the appendix table in Pardey, Roseboom, and Anderson (1991); the sub-Saharan Africa group reported here includes South Africa. The data from the series published in 1991 forms the basis of the revised and updated ASTI data reported in the current table.

[a] Figures in parentheses indicate the number of countries included in the respective totals.

Sources: Pardey, Roseboom, and Craig (1997) for developed countries. The IFPRI-ISNAR Agricultural Science and Technology Indicators (ASTI) database for developing countries.

Developing-Country Trends

Between 1971 and 1991, public expenditures on agricultural R&D in developing countries increased from just under $3 billion to just over $8 billion, at an annual average growth rate of 5.1 percent. These expenditures grew most rapidly over the entire period in China (6.3 percent per year), other East Asian and Pacific Rim countries (7.3 percent per year) and West Asia and North Africa (4.3 percent per year). These are countries that, for the most part, have also enjoyed relatively rapid economic growth rates. Public agricultural R&D expenditures grew much more slowly in sub-Saharan Africa (1.6 percent per year) and Latin America and the Caribbean (2.7 percent), regions that include many countries with very low per capita incomes and low rates of economic growth. However, as in the developed countries of the OECD, public agricultural R&D funding trends in developing countries changed clearly and markedly in the 1980s. Between 1971 and 1981, public agricultural R&D expenditures in developing countries grew at an annual rate of 6.4 percent, but between 1981 and 1991 they grew at an annual rate of only 3.9 percent.

The data on public agricultural R&D expenditures by developing countries presented in table 3.3 are consistent with two conjectures. First, on average, growth rates of public agricultural R&D expenditures in developing countries between 1971 and 1991 were more than twice those experienced in developed countries. Even higher growth rates in R&D expenditure were achieved in developing countries enjoying faster economic growth, and these more rapidly developing countries appear to have been building modern national agricultural research systems in the 1970s and 1980s, just as more developed countries did in the 1950s and 1960s and prior to World War II. Second, it seems likely that at least some of the factors that led to reductions in growth rates of public agricultural R&D expenditures in the OECD countries were also at work in the developing countries.

International Research

Internationally conceived and funded agricultural research complements the work of national research agencies. This international R&D represents a relatively recent institutional innovation. The first such venture, a cooperative Mexican government–Rockefeller program initiated in 1943 to conduct wheat research, became the model for many of the subsequent programs in international agricultural research and later evolved into the International Wheat and Maize Research Center (CIMMYT). The further development of international agricultural research centers took place largely under the auspices of the Consultative Group on International Agricultural Research (CGIAR), which was established in 1971.[11] Funds for the CGIAR system are mainly obtained from the foreign aid (not agricultural R&D) budgets of developed countries, and are either given directly to CGIAR centers or through contributions to agencies such as the World Bank, the Asian Development Bank, and the European Community.[12] The chronology of funding for the CGIAR system since 1972 involves two distinct phases. In the first phase, lasting from 1971 to 1982, real spending grew by 14.3 percent per year as the CGIAR system expanded to include more centers and to cover more commodities. In the second phase, beginning in the mid-1980s, while the research mandate for the CGIAR continued to broaden to include additional commodities and a greater emphasis on environmental R&D, real spending began to stagnate, increasing by only 1.4 percent per year between 1985 and 1991 and by only 0.5 percent per year from 1991 to 1996. Thus, funding for international agricultural research, largely provided by the OECD countries, has followed a quite similar pattern to that of public agricultural R&D funding in those countries.

Public Funding for All R&D for Science and Technology

Developments in public agricultural R&D funding are intertwined with developments in public funding for R&D in science and technology more generally. Data on overall public funding for all science and technology research in OECD countries are presented in table 3.4 for the period 1981–1993. In addition, table 3.4 also includes research intensity ratios showing the value of publicly performed R&D for all science and technology relative to gross domestic product for all 22 OECD countries and each of the five countries over the same period.

Table 3.4. Real, total (agricultural and nonagricultural) public R&D spending in OECD countries, 1981–93

	1981	1991	1992	1993
		(percentages)		
Total publicly performed R&D relative to total output				
Australia	0.84	1.00[a]	0.98	0.91
Netherlands	0.95	0.99	0.99	0.99
New Zealand	0.86	0.71	0.83	0.82
United Kingdom	1.07	0.86	0.87	0.86
United States	0.78	0.84	0.84	0.83
Subtotal (5)[b]	0.82	0.85	0.85	0.84
Other OECD (17)[b]	0.74	0.84	0.86	0.89
Total OECD (22)[b]	0.78	0.85	0.86	0.87
		(billions of 1985 international dollars)		
Public funding for all R&D				
Total OECD (22)[b]	91.6	120.8	na	120.0

[a] 1990 figure.
[b] Figures in parentheses indicate the number of countries included in the respective totals.

Source: Pardey, Roseboom, and Craig (1997).

Between 1981 and 1993, in aggregate in the 22 OECD countries, public funding for all science and technology R&D increased from $91.6 billion to $120.0 billion at an annual average growth rate of 2.4 percent. Over the same period, as noted above, public funding for agricultural R&D by OECD countries increased by about 1.8 percent per year. The average science and technology research intensity ratio (measuring total public R&D relative to GDP) for all 22 OECD countries was 0.78 percent in 1981 and 0.87 percent in 1993, increasing by a much smaller proportion than the agricultural research intensities (measuring public agricultural R&D spending relative to AgGDP) for 20 OECD countries, which rose from 1.83 percent in 1981 to 2.48 percent in 1992.

According to these research intensity measures, agriculture has been treated relatively favorably in most developed countries with respect to the provision of public R&D funds over the recent past. However, this comparison is confounded by the role of overall growth in the different sectors, the denominator of the research intensity ratio. A different picture emerges when trends in agriculture's share of total publicly performed science and technology are examined. These data are presented in table 3.5 for OECD countries over the period 1981 to 1993. In the 22 OECD countries agriculture's share of the total public science and technology R&D budget declined on average from 8.9 percent in 1981 to 7.4 percent in 1993, a proportional decrease of almost 17 percent. In the five countries, the change in agriculture's share of the total public science and technology research budget was, on average, quite similar, falling from 7.6 percent to 6.6 percent, but again, these average data disguise considerable differences among individual countries. In the United Kingdom and the United States, agriculture's share of the total public R&D budget declined quite modestly between 1981 and 1993 (from 7.1 to 6.6 percent in the United Kingdom and from 6.2 to 5.6 percent in the United States). In contrast,

Table 3.5. Agriculture's share of public R&D spending in OECD countries, 1981–93

	1981	1986	1991	1993
		(percentages)		
Australia	19.9	18.0	14.1	14.8
Netherlands	14.5	12.4	12.0	12.4
New Zealand	49.1	49.5	46.7	35.5
United Kingdom	7.1	7.2	6.6	6.6
United States	6.2	5.5	5.7	5.6
Subtotal (5)[a]	7.6	6.7	6.7	6.6
Other OECD(17)[a]	10.3	9.6	8.5	8.3
Total OECD (22)[a]	8.9	8.1	7.6	7.4

[a]Figures in parentheses indicate the number of countries included in the respective totals.

in the other countries agriculture's share of the total public R&D budget declined quite dramatically (from 19.9 to 14.8 percent in Australia, from 49.1 to 35.5 percent in New Zealand, and from 14.5 to 12.4 percent in the Netherlands).

In the three countries in which agriculture had the largest shares of total R&D budgets in 1981 (Australia, the Netherlands, and New Zealand) the importance of agricultural research declined most substantially. Pressures to reallocate funds to other science R&D programs such as health were, perhaps, more severe in those cases. In fact, in all but one of the five countries (the Netherlands), between 1981 and 1993 the share of total public science R&D budgets allocated to health research increased quite substantially, at the expense of both agricultural research and defense research.

Private-Public Linkages in Science and Technology R&D Funding and Performance

Funding for agricultural R&D is increasingly intertwined with funding for research more generally. Measuring R&D investments becomes more challenging as the institutions involved in funding, managing, and performing research evolve and become more complex. Distinguishing between those who fund R&D (and, perhaps, who manage the funds) and those who perform the research provides a clearer picture of changing institutional roles within programs of national research. It also facilitates international comparisons and enables useful policy insights to be drawn from the data. The data reported in table 3.6 on total science and technology R&D expenditures by OECD countries are separated into public and private categories on the basis of both research performers and research funders. Care has been taken to maintain a consistent institutional coverage over time and to make the data comparable across countries. Although some unavoidable discrepancies remain, these data provide a reasonable basis for making broad international comparisons.[13]

Overall, public and private investment in all forms of R&D (including such fields as defense, medicine, and general science as well as agricultural science) grew quite rapidly in most OECD countries between 1981 and 1993, although more so for privately performed R&D (with an annual growth rate of 4.3 percent) than for publicly performed R&D (with a growth rate of 3.5 percent). The rates of growth of both publicly and privately performed research slowed in the late 1980s and early 1990s. In 1993 the private sector undertook $190.8 billion of research, about two-thirds of all R&D performed throughout the OECD that year.

Distinguishing between the providers of funds for all R&D and those who performed the research provides additional insights about the linkages between the public and private sectors and their respective roles in the general R&D process. It also provides evidence about the net flow of public funds to private R&D providers. Table 3.6 shows that in 1993 private agencies pro-

Table 3.6. Total R&D by performer and source of funds

	1981	1986	1991	1993	Annual growth 1981–93
	(billions of 1985 international dollars)				(percentages)
R&D performers					
Private	120.6	167.7	199.2	190.8	4.3
Public	64.4	78.1	91.3	95.8	3.5
Total	185.1	245.8	290.5	286.6	4.0
Source of funds					
Private	93.6	131.6	169.7	166.6	5.4
Public	91.6	114.3	120.8	120.0	2.4
Total	185.1	245.8	290.5	286.6	4.0
Imputed net flow of funds from public to private performers	27.0	36.2	29.5	24.2	na
Net flow of public funds to private performers as a share of private R&D			(percentages)		
	22.4	21.6	14.8	12.7	na

Derived from Gross Expenditure on R&D (GERD) data included in table 1 of OECD (1996). Data include 22 OECD countries.

vided or performed $190.8 billion of R&D and public agencies performed $95.8 billion of R&D. However, in 1993 public R&D *spending* ($120 billion) was considerably greater than public R&D *provision* ($95.8 billion), while private R&D *spending* ($166.7 billion) was correspondingly less than private R&D *provision* ($190.8 billion). The imputed net flow of public funds to private research providers shows that a considerable amount of privately performed R&D was underwritten with public dollars.[14] But the amount of public funds flowing to private R&D declined in real terms in recent years from a peak of $37.4 billion in 1987 to $24.2 billion in 1993. The publicly funded share of privately performed R&D declined steadily over a longer period, falling from 22.4 percent in 1981 to 12.7 percent in 1993.

The recent decline in public support for private research has been widespread throughout the OECD countries, and the magnitude roughly corresponds to cutbacks in defense-related R&D. For example, France, the United States, and the United Kingdom (countries where the shares of total public research dollars spent on defense research in 1985 were 67.5, 51.0, and 32.5 percent, respectively) account for the lion's share of the drop in public funds flowing to private R&D providers.[15]

Private Agricultural Research

Table 3.7 provides our current best estimates of privately provided agricultural R&D and the private shares of total agricultural R&D spending from 1981 to 1993. A common perception is that agricultural research is primarily the domain of the public sector while research in other sectors of the economy

Table 3.7. Privately performed agricultural R&D

	1981	1986	1991	1993	Annual growth 1981–93
	(millions of 1985 international dollars)				(percentages)
Privately performed agricultural R&D					
Australia	25.2	68.3	112.5	137.5[b]	15.1
Netherlands	185.2	210.7	241.2	297.9	3.8
New Zealand	9.9	12.2[a]	26.7	39.5	13.7
United Kingdom	404.2	473.6	593.5	614.3	5.0
United States	1,416.6	1,963.7	2,256.0	2,381.1[b]	4.3
Subtotal (5)[c, g]	2,041.2	2,728.4	3,229.9	3,470.2	4.7
Other (17)[g]	1,953.5	2,691.9	3,419.5	3,560.3	5.5
Total (22)[c g]	3,994.7	5,420.3	6,649.4	7,030.5	5.1
Privately performed as a share of total agricultural R&D					
	(percentages)				
Australia	8.2[d]	17.9	26.8[e]	30.3[b]	
Netherlands	47.8	51.8	52.7	56.8[f]	
New Zealand	6.9	9.1	19.5	26.9	
United Kingdom	52.1	55.8	62.0	62.4	
United States	46.6	52.1	52.7	53.7[d]	
Subtotal (5)[g]	43.9	49.3	51.7	53.0	
Other (17)[g]	38.6	43.5	46.6	46.8	
Total (22)[g]	41.1	46.2	48.9	49.6	

Data derived mainly from OECD's intramural business sector series. Calculated as the sum of R&D performed by industries classified as "agriculture, forestry, and fisheries," "food and beverages," and 10 percent of total research performed by the "chemical and pharmaceutical industries." The 10 percent share of total chemical R&D assigned to agriculture represents an approximate rule of thumb procedure that was chosen after consulting various other relevant sources. For comparability purposes, agricultural mechanization R&D was excluded from the U.S. private sector series obtained from USDA (1995). See note 4 at the end of the chapter for additional details on data coverage.

[a] 1987 figure.
[b] 1992 figure.
[c] 1981 subtotals and totals exclude Greece and New Zealand; 1985 figure excludes Greece.
[d] 1973 figure.
[e] 1990 figure.
[f] 1994 figure.
[g] Figures in parentheses indicate the number of countries included in the respective totals.

Source: Pardey, Roseboom, and Craig (1997).

are the province of the private sector. But these new data indicate that privately performed R&D is a prominent feature of agricultural R&D in rich countries. Privately performed agricultural R&D increased from $4 billion in 1981 to over $7 billion in 1993, at an annual growth rate of 5.1 percent, and now accounts for almost half of all developed-country agricultural R&D, while all science and technology research performed by the private sector grew at an annual rate of only 4.3 percent. In contrast, publicly performed agricultural R&D increased comparatively slowly at an annual growth rate of 1.8 percent.

The relative importance of private agricultural R&D in total agricultural R&D varies across the OECD countries. Among the five countries, in the United Kingdom the private sector is estimated to perform over 60 percent of all agricultural research, and in the Netherlands and the United States the private share is currently in excess of 50 percent. Australia and New Zealand still have significantly smaller private involvement in agricultural research, but privately performed R&D is becoming much more important in both countries. In New Zealand, this is partly because the government agencies that previously performed public agricultural research have been partially privatized.

The data presented in table 3.8 indicate that private and public research facilities do different types of R&D. Approximately 12 percent of private research is focused on farm-level technologies, whereas over 80 percent of public research has that orientation. Food and other postharvest research accounts for 30–90 percent of private agricultural R&D, and in countries like Australia, Japan, New Zealand, and the Netherlands, it is the dominant focus of privately performed research related to agriculture. Chemical research is of comparatively minor importance in Australia and New Zealand but accounts for more than 40 percent of private agricultural research in the United Kingdom and the United States and nearly three-quarters of private agricultural research in Germany.

Particular lines of private agricultural R&D are concentrated in particular countries. Japan, the United States, and France account for 33, 27, and 8 percent, respectively, of all food processing research carried out by the private sector in the OECD countries. Chemical research related to agriculture is even more concentrated; the United States, Japan, and Germany account for 41, 20, and 10 percent of all reported private-sector research.[16] This pattern of concentration of private agricultural research would not alter significantly if we also included counterpart research in developing countries.

Contemporary cross-country evidence on private R&D is revealing, but for the United States more detailed time-series data are available (USDA 1995). The focus of private agricultural research in the United States has changed. In 1960, agricultural machinery and postharvest food processing research ac-

Table 3.8. Focus of public and private intramural agricultural R&D, 1993

	Publicly performed		Privately performed		
	Agriculture	Food and kindred products	Agriculture	Food and kindred products	Animal health and agricultural chemicals
			(millions of 1985 international dollars)		
Expenditures					
Australia	na	na	36.7	87.4	13.4
Netherlands	224.9	14.3	60.1	173.1	64.7
New Zealand	87.7	19.7	4.8	34.2	0.5
United Kingdom	325.6	45.1	106.2	211.2	296.9
United States	na	na	314.9	817.7	1,248.5
Subtotal (5)[a]	na	na	522.6	1,323.6	1,624.1
Other OECD (16)[a]	na	na	300.4	1,736.1	1,448.5
Total OECD (21)[a]	na	na	823.0	3,059.7	3,072.6
			(percentages)		
Shares of respective subtotals					
Australia	na	na	26.7	63.6	9.8
Netherlands	94.0	6.0	20.2	58.1	21.7
New Zealand	81.7	18.3	12.1	86.5	1.4
United Kingdom	87.8	12.2	17.3	34.4	48.3
United States	na	na	13.2	34.3	52.4
Subtotal (5)[a]	na	na	15.1	38.1	46.8
Other OECD (16)[a]	na	na	8.6	49.8	41.6
Total OECD (21)[a]	na	na	11.8	44.0	44.2

See table 3.7. OECD 21 country total excludes Switzerland due to lack of data.

[a] Figures in parentheses indicate the number of countries included in the respective totals.

Source: Pardey, Roseboom, and Craig (1997).

counted for over 80 percent of total private agricultural R&D. By 1992, these areas of research collectively accounted for only 42 percent of the total, the share of total private research directed toward agricultural machinery having declined from 36 percent in 1960 to less than 12 percent. Two of the more significant growth areas in private R&D have been plant breeding and veterinary and pharmaceutical research. Spending on agricultural chemicals research grew most quickly and now accounts for more than one-third of total private agricultural R&D. These data point to a dramatic shift in the pattern of publicly and privately performed agricultural R&D in the United States over the past three decades, similar to those reported by Thirtle, Piesse, and Smith (1997) for the United Kingdom.

RECENT INSTITUTIONAL CHANGES IN PUBLIC AGRICULTURAL R&D

Shifts in the organization of public agricultural R&D institutions in Australia, the Netherlands, New Zealand, the United Kingdom, and the United States, especially on the funding side, reflect six interrelated issues: (1) trends toward using public funds to support more basic rather than more applied or "near-market" research, (2) trends toward the joint funding of near-market research through the development of industry levies and other mechanisms, (3) revamped oversight and accountability mechanisms, and other changes in management of research resources, (4) measures for introducing competition among researchers to increase research productivity and as a means of allocating research resources, (5) measures to privatize, directly or indirectly, public agricultural research institutions, and (6) trends toward the rationalization of public agricultural research facilities. These changes are linked to recent developments in agricultural policy and general R&D policy and to transformations of agriculture and the agricultural sciences themselves.

Overview of Changes in Public Agricultural R&D Institutions

Public agricultural research systems vary in terms of who funds, manages, and performs the research. Research and development is carried out by a range of government departments (some with roles extending well beyond research and well beyond agriculture), some semipublic entities, and various university faculties and departments. The five countries reviewed in detail here have quite diverse institutional setups. The public agencies undertaking agricultural R&D in the United Kingdom, the Netherlands, and New Zealand are mostly administered by national government departments, while in Australia and the United States public agricultural R&D is supported by a mix of state and national institutions. The United States has national government facilities administered by the United States Department of Agriculture, while state-level R&D is carried out in experiment stations located in land-grant colleges. The U.S. trend has involved relatively less spending in national research facilities and relatively more in state agricultural experiment stations (SAESs).[17]

Public agricultural R&D systems have experienced various degrees of restructuring in recent years. In the Netherlands, New Zealand, and the United Kingdom, the public agencies involved in carrying out research have been substantially revamped. In England and Wales, as well as in the Netherlands, the agencies that perform public R&D were consolidated and in some important instances commercialized. In New Zealand, new Crown Research Institutes (CRIs) were established in an effort to develop a public "market" for R&D services that distinguishes between buyers and sellers of such services. Comparatively little structural change has taken place in the public agencies that undertake agricultural R&D in Australia and the United States, although

Australia's funding and management structures have recently been revised substantially.

The timing of reforms in each of the five countries has been closely linked to the advent of governments with more market-oriented, laissez-faire economic philosophies about the role of government in management of the economy. In England, for example, the conservative government led by Margaret Thatcher was elected in 1979 and remained in power, with one change of leadership, until 1997. Reforms of U.K. public research institutions directed toward introducing market forces into the public research enterprise began in the early 1980s and continued into the early 1990s. In New Zealand, a reform government was elected in 1984 and confronted with a severe budget crisis; renovations of the New Zealand agricultural research system began less than one year later as part of a comprehensive package of policies. In Australia, the Hawke labor government came to power in 1981 and, beginning in 1985, progressively initiated major changes in public agricultural research management that can be seen as a part of more general microeconomic reforms. In the United States the conservative Reagan administration was elected in 1980, and cuts in the growth of public funding for research, including agriculture, followed in 1981 and 1982. And in the Netherlands a series of more market-oriented coalition governments were formed in the early 1980s.

Basic and Applied Research[18]

One important recent trend in the United States, the Netherlands and the United Kingdom has been the apparent redirection of research funds derived from general tax revenues toward basic research and away from applied research. This shift accords with the general public finance principle that public funds should be used to provide public goods. Reported patterns of change are strikingly similar across all three countries, where the share of public agricultural R&D funds allocated to universities (that typically undertake research with a more basic orientation) has increased substantially.

The evidence in relation to Australia and New Zealand is less clear. In Australia, since R&D Corporations (RDCs) and Cooperative Research Centers (CRCs) have become more important in funding agricultural research, more funds probably have been directed toward applied research. In New Zealand, the creation of the Public Good Science Fund in 1992 was intended to signal a reallocation of public R&D resources to more general pubic good and public interest research. It is not at all clear, however, that resources within the New Zealand public agricultural R&D program have actually been redirected toward more basic research (Jacobsen and Scobie 1997).

Privatizing Previously Public Roles

A related development in the organization and funding of agricultural research, common to all five countries, has been an increased emphasis on the role of the private sector and, in particular, industry and commodity groups in supporting agricultural research programs.

Levy Funds and Matching Arrangements

One of the more dramatic examples of this trend is the evolution of Australian agricultural research policy since the mid-1980s. Legislation in 1985 and 1989 created RDCs to manage research funds generated by industry levies with dollar-for-dollar matching funds to be provided by the government up to a limit of 0.5 percent of the gross value of industry production. These changes substantially increased the share of funds for publicly conducted agricultural R&D managed by the RDCs or their predecessors (Alston, Pardey, and Smith 1997). They also increased the industry representation on the RDC boards and project selection committees. A more recent initiative may turn out to be even more important. The introduction of CRCs in the early 1990s, linking research by government, universities, and the private sector, has fostered a revitalized growth in total funding for agricultural R&D.

In the Netherlands, as in Australia, the principle of joint funding for experiment station and other agricultural research programs, by the agricultural sector and the government (on a dollar-for-dollar basis), was a key element of agricultural research policy for many years. Between 1950 and 1970, the joint-funding principle became less relevant, but during the late 1970s and 1980s, confronted by successive budget crises, the Netherlands Ministry of Agriculture reemphasized its importance. In addition, after 1980 the University of Wageningen was also encouraged to seek funding from industry sources for research projects. As a result, private funding, which represented less than 25 percent of total university funds for agricultural research in the 1970s, accounted for over 40 percent of such funds by the mid-1990s.

In New Zealand, statutory bodies (typically marketing boards) have used levies to support market development and research programs since the 1920s. In 1990, the New Zealand government passed the Commodity Levies Act under which industry groups were given authority to impose mandatory levies to fund sector-specific research and other (market development) activities. As a result, new commodity group funds have been established for several agricultural products such as kiwifruit and tomatoes.

In the United Kingdom, there is no parallel history of matching fund programs. However, in conjunction with the decision to reduce funding for near-market research, the Thatcher governments of the mid-1980s passed legislation enabling the creation of commodity-specific statutory bodies, funded by industry levies, to develop markets and fund commodity-specific applied research. These bodies were to have boards of directors whose membership would be dominated by industry representatives. Statutory bodies have now been established for all major agricultural commodities, but they only account for about 6 percent of total public expenditures on agricultural research.

Developments in the United States have been less dramatic and more incremental and have not, by and large, involved any major legislative changes. Private-sector funding for public agricultural research has generally increased since 1975, largely as a result of increased private funding for State Agricultural Experiment Station (SAES) operations. However, there is no evidence of any pressure to develop levy programs to fund agricultural R&D on a widespread or substantial basis.[19]

Pricing Research

A further issue in relation to the involvement of the private sector in the publicly funded agricultural research system is the pricing of research services. In each country, the private sector supports research conducted at public research institutes. In Australia, the Industry Commission recently argued that such establishments ought to price their services to cover all long-run costs of producing the privately commissioned research unless it provides significant spillover benefits (Industry Commission 1995). Some economists have questioned this view (Watson 1996). In the United Kingdom, New Zealand, and the Netherlands, the question of full-cost pricing for the use of publicly funded agricultural research facilities has also become an important issue in the 1980s and 1990s as the share of private funding for such facilities has increased and the policies with respect to overhead charges at universities and ministry research facilities have become more formalized.

Privatization

Explicit privatization of agricultural research enterprises has been an important feature of the reorganization of agricultural research only in the United Kingdom. Indirect privatization of research, through opening up access to public funds for private and independent research institutes, has taken place in all five countries. In the United Kingdom, the direct privatization of some agricultural research institutes in the late 1980s and early 1990s was partly the result of the government's economy-wide program to privatize many publicly owned enterprises, and its intent to stop funding and managing near-market research. In New Zealand, the government opted for the comprehensive introduction of a system of quasi-private public research institutes through the creation of the system of Crown Research Institutes for all research. However, although the new CRIs must now obtain support from a general contestable public fund, they do not operate under the profitability constraints that confront private facilities and seem, implicitly, to enjoy guaranteed access to the available pool of public funds (Jacobsen and Scobie 1997). In the Netherlands, a partial form of privatization has been initiated. In the mid-1980s the government determined that agricultural research could be placed in the hands of the private sector. Since then, through a variety of mechanisms, agricultural research institutes previously funded and managed by the Ministry of Agriculture have been partially detached from its control and have become more reliant on private funding sources.

Rationalizing Government Agricultural Research

Changes in the scientific, economic, and political factors related to agricultural science have created serious problems for managers of national research organizations: How should those agencies be restructured over time? The problem is challenging for any research agency, but it is particularly acute in the public sector because of problems associated with rent seeking and policy inertia. Once public research institutes are created, the employees in those institutes establish formal or informal interest groups. The result is often tremendous inertia in publicly operated research organizations. During the late 1970s and the 1980s, circumstances arose in which the perceived costs of inefficiencies in public research systems became large, especially in the Netherlands and the United Kingdom. Thus, over this period, agricultural research in the Netherlands was continually reorganized and rationalized and the oversight process was streamlined, while in the United Kingdom the number of publicly funded research institutes fell by more than half.

Managed Competition

The introduction of managed forms of competition has been a feature of the allocation of public funds in all five countries, beginning in the mid-1980s. On balance, these changes have increased the amount of competition among researchers for those funds. In all five countries, the proportion of funds disbursed through block grants or formula funds has fallen while the proportion of funds disbursed through competitive or quasi-competitive grants processes has grown.

Although much has been said about moving toward competition, and away from formula funding or block grants, the actual movement has been uneven. New Zealand has moved fastest and furthest in this direction, but even there, there are still questions about how effective the resulting new institutions are and whether the rent-seeking costs outweigh the benefits. The United States has barely moved in this direction: in 1995, only $101 million in a public agricultural R&D budget of $3.0 billion was allocated through competitive grants (see also, National Research Council 1994). The other countries are between these extremes, in both what they have done and what they have said they might do. As well as differing in the degree of movement toward more use of competitive processes for allocating research resources, each country has adopted its own institutions for allowing competition to operate. However, to a great extent the competition has often been strictly based on ex ante claims, promised performance, rather than ex post assessments of what was actually delivered in the past. In contrast, the role for the agricultural industry to participate both as a provider of funds and in having a say as to criteria to be used to determine priorities has expanded.

A related issue has been the role of accountability processes. In several countries, the criteria for periodic reviews of block grants and the distribution of those grants among competing research programs and institutions have also tended to become more formal and more stringent, and such reviews have become more frequent. In the United Kingdom, since the late 1980s, for example, all university departments have been individually assessed with respect to their research output once every five years. Within each discipline or subdiscipline, departments are ranked on a scale of 1 to 5 (where 1 is poor and 5 is excellent), and substantial research resources for the next five years are tied to these rankings. Similarly, in the Netherlands, since the late 1980s the University of Wageningen (the country's only university center for agricultural research) has been required to submit detailed four-year research plans once every two years for each department. Block grant funding is now conditional on approval of the University's overall plan of research and each department's proposal. In Australia, while there are no similar formal mechanisms, there has been a long history of evaluating research programs, both broadly at the national level, through the Industry Commission, and more narrowly both by committees responsible for the use of levy funds and by research performers. The creation of RDCs in the 1980s enhanced this process of accountability. In New Zealand, although new research priority-setting procedures were established in the late 1980s and early 1990s, there is little evidence that the new Crown Research Institutes, which now perform the majority of publicly funded agricultural research, have become subject to more stringent evaluation processes. Finally, in the United States, block grants of federal funds for SAES research continue to be determined by formula-funding mechanisms based on socioeconomic indicators such as population and are not tied to research performance.

Broadening Research Agendas

A common thread in the evolution of agricultural policy processes in developed economies has been the increase in the influence of nonfarm interest groups on agricultural policy. These groups include providers of agricultural inputs, food processors, consumer groups, and environmental and conservation groups. In the case of European Community members, another increasingly important body has been the European Commission, both as a provider of funds and as a regulatory agency. In various ways, these groups also have all become more important in the development of agricultural research policy.

The increasing influence of the environmental and conservation lobbies on the agricultural research agenda has been particularly noticeable during the past 15 years. In the United States, the environmental lobby has greatly influenced the agricultural research agenda, leading to large publicly funded R&D programs devoted to issues such as sustainable agriculture and the maintenance of wetlands. The United Kingdom and the Netherlands have allocated substantial resources to the Ministries of Agriculture for research into environmental and conservation issues since the mid-1970s. Similar developments have also occurred in Australia.

A less widely recognized phenomenon has been the increased influence of the agribusiness lobby on agricultural R&D. Since the mid-1980s in the United Kingdom, food processors have been vocal on agricultural research oversight committees such as the Priorities Board, which in turn have recommended reallocations of public research resources to address issues of concern to that sector. A similar development has taken place in Australia, partly through the changed structure of the RDC boards and the development of CRCs with significant agribusiness involvement. In New Zealand, agribusiness representatives have also been given a larger role in determining the allocation of public agricultural research funds, and public funding for food-processing research has increased markedly in the mid-1990s.

The increasing importance of consumer groups has also been reflected in developments in agricultural research policy. For instance, funding for food-safety research has increased in both the United States and England, partly because of a series of health scares since the mid-1980s related to such things as salmonella in eggs, *E. coli* in meat, bST in milk, and mad-cow disease. Thus, just as agricultural policy in general has become the province of more than just agricultural producer groups, so too has agricultural research policy.

ASSESSMENT

The crucial questions for economists concern whether or not the recent changes in public agricultural R&D have resulted in improvements in economic efficiency and economic welfare. More specifically, the issues involve whether changes have led to an improvement in efficiency in terms of the total quantity of research being undertaken; lower-cost sources of funds; better allocation of resources among competing programs or projects, with differing emphases among commodities, natural resource management, and other areas of research focus; better allocation of research resources among institutions; better allocation between basic and applied research and extension, or between production and processing research; and more effective use of research funds and less waste in excessive administrative overhead or other transaction costs.

Reductions in growth rates for public agricultural R&D funding may be questioned in light of the broad array of empirical evidence that the rates of return associated with those investments have been relatively high. Most studies have reported annual rates of return well in excess of 30 percent.[20] However, some recent findings suggest that when lags between research and productivity effects are taken fully into account, previously very high estimates of real, long-run rates of return to public agricultural R&D investments (at least in the United States) may be reduced to about 10 percent (Alston, Craig, and Pardey 1997), still high enough to justify the investments but not the bonanza that is often claimed.

Increasing the role of the private sector in the management of public agricultural R&D has also been an important theme over the past 15 years. This has occurred either as a result of linking public and private funding through levy schemes, with the joint funds being managed by committees more heavily influenced by industry representatives (as in Australia, the Netherlands, and the United Kingdom), or through the appointment of industry representatives to committees with responsibilities for establishing research priorities. This is a double-edged sword with both potential costs and benefits. On the costs side, regulatory capture becomes a concern. When research boards that allocate public and private sector funds are dominated by industry representatives, public funds may be redirected to projects that benefit only sectoral interests or perhaps only a part of an industry (Ulrich, Furtan, and Schmitz 1986). In addition, more directed public-private ventures may crowd out industry research funds. Worse still, industry capture of research programs can result in perverse consequences (Constantine, Alston, and Smith 1994). On the other hand, especially when the focus is near-market research, increased input from the industry can reduce regulatory capture by the scientists or government bureaucrats, who also may not always have the public good in mind in making resource allocation decisions. Researchers can thus become more responsive to industry needs and concerns and carry out more effective applied research programs.

In principle, increasing competition for research funds can be an effective device for revealing information to funders about both opportunities for research and comparative advantages in its execution. Competitive processes can reduce the costs of information and of research resource misallocation (Alston and Pardey 1996), but they can also involve higher transaction costs (Huffman and Just 1994). The challenge has been to devise institutions that will minimize the total costs of research decision making, considering the invisible resource misallocation costs as well as the more obvious costs of competing for funds. Interestingly, in the United Kingdom the process of five-year research productivity assessments for university departments represents a middle ground that attempts to appropriate the benefits of some degree of competition in allocating block grant funds without incurring the costs of requiring competition on a project-by-project basis for all research resources.

The shift toward creating a competitive research supply environment by moving away from very long-term contracts (tenure) for researchers to shorter-term contracts also has its potential pluses and minuses. The major potential gains are that (1) it is easier to remove "dead wood" from the research system, (2) shorter-term contracts may provide incentives for greater research productivity among all researchers, and (3) research administrators may have greater flexibility in managing research resources. A potentially serious disadvantage is that the removal of long-term guarantees of employment may reduce incentives for competent and gifted individuals to embark on research careers that require large private investments in human capital. Perhaps insufficient attention has been given to the optimal structure of incentives for the long-run development and maintenance of productive research industries. It is certainly not entirely clear whether the short-term benefits associated with recent policy shifts in this respect outweigh any potential longer-term costs.[21]

The situation vis-à-vis rationalization of public research facilities may be less ambiguous. To the extent that rationalization has taken place in response to recent changes in scientific methods and to take advantage of new economies of size and scope, there have been clear economic efficiency gains. Where "rationalization" has simply operated as a synonym for "reductions in public R&D investments," whether the changes have been good or bad turns on whether or not the rates of return on the investments were higher than the social opportunity cost of the funds.

Broadening the research agenda may have led to some benefits. Environmental and food safety issues are often public good issues, and market failures in research often accompany the more general market failures in these areas (Antle 1995; Alston, Anderson, and Pardey 1995). Accordingly, many seem to believe that reallocations of public research resources to these issues and away from near-market research programs must enhance economic welfare (Ervin and Schmitz 1996). The answer to even this question remains unclear, however, since no formal evidence is available on the payoff to public R&D into environmental or food safety issues, achieving a payoff at all requires effective incentives to adopt results that yield social benefits, and lack of effective incentives is sometimes the central problem with such issues (see Alston and Pardey 1996).

To the extent that public resources have been diverted toward agribusiness and food processing research (as in the United Kingdom and New Zealand, and possibly in Australia), there is a further potential downside to the recent developments in agricultural research policy. It is by no means clear that projects funded in these areas more closely approximate public good projects than those they may have displaced in the area of farm productivity.[22] On a priori grounds, given that the farm sector is largely atomistic while many agribusiness and food processing markets are characterized by relatively few firms with no clear prima facie evidence of market failure in research, this shift of research resources may have reduced rather than increased the rate of return to public research investments.

Clearly then, there is considerable ambiguity about whether the changes in agricultural research policy implemented by developed countries since the early 1980s have led to net social benefits. The issues are essentially empirical, and to date, agricultural economists have not been able to provide many empirical answers. This is not to suggest that the agricultural economics profession has failed, but rather to argue that opportunities remain for agricultural economists to make important contributions to the agricultural policy debate and the formation of welfare-enhancing agricultural research policies. There are long and variable lags between changes in R&D policies and the realization of their productivity and welfare consequences. It is too soon to attempt a complete accounting of the economic consequence of these contemporary changes in agricultural R&D funding and the accompanying institutional experiments. The investigation of these issues represents an important future agenda for research policy analysts.

NOTES

1. *Australian Journal of Agricultural and Resource Economics,* March 1998. Reprinted with permission of Blackwell Publishers Ltd.

2. Authors are listed alphabetically, senior authorship is unassigned. John Mullen provided helpful comments on an earlier version, for which we are grateful. Partial support for this work was provided by the U.K. Department for International Development and the University of California Pacific Rim project.

3. The volume, edited by Alston, Pardey, and Smith (1997) documents recent changes in agricultural research institutions and investments in Australia, the Netherlands, New Zealand, the United Kingdom, and the United States in detail. The chapters on individual countries were written by Alston, Harris, Mullen, and Pardey (1997) (Australia); Alston, Christian, and Pardey (1997) (the United States); Jacobsen and Scobie (1997) (New Zealand); Thirtle, Piesse, and Smith (1997) (the United Kingdom); Roseboom and Rutten (1997) (the Netherlands). A chapter in that volume by Pardey, Roseboom, and Craig (1997) provides an overview for the OECD as a whole. The current paper draws heavily on those items.

4. There is no single, internationally accepted notion of what constitutes "agricultural R&D." Like Pardey, Roseboom, and Craig (1997) (from which much of the data reported here are drawn), this article uses a broad notion of agricultural research and follows the science and technology measurement standards laid down by OECD (1994). Research related to "agriculture" is taken to include crops, livestock, forestry, and fisheries research and related environmental R&D. It includes research targeted to the development of preproduction (e.g., input supply), on-farm, and postproduction (e.g., food processing) technologies. These data are reported primarily on a "performer" basis, but we distinguish between the public and private performance versus funding of R&D where that is possible and it makes policy sense to do so. Our series for Australia do not compare directly with the data reported by Mullen, Lee, and Wrigley (1996). Their series explicitly exclude processing R&D and are targeted to production agricultural R&D excluding forestry and fisheries research.

5. These are only a few examples of the innovations in the biological and other basic sciences that underpinned progress in the agricultural sciences and technical changes in agriculture during the earlier part of this century. See Salmon and Hanson (1964) for the details of these discoveries.

6. For example, the United States has had plant patents for asexually reproduced cultivars since the 1930s, Plant Variety Protection Certificates since the 1970s, and utility patents for living organisms since the landmark *ex parte Hibberd* decision of 1985. The agreement on Trade Related Intellectual Property Protection (TRIPS) (specifically, annex 1c of the 1995 Marrakesh Agreement establishing the World Trade Organization) mandates the extension of intellectual property protection to plants.

7. For example, between 1950 and 1970, in real terms public agricultural R&D expenditures increased at annual average growth rates of about 7 percent in Australia (Alston, Harris, Mullen, and Pardey 1997), about 7.4 percent in the United Kingdom (Thirtle, Piesse, and Smith 1997), and about 2.8 percent in the United States (Alston, Christian, and Pardey 1997). Although similar data are not available for the Netherlands and New Zealand for this period, Roseboom and Ruttan (1997) and Jacobsen and Scobie (1997) report that those countries also substantially increased public investments in agricultural R&D between 1950 and 1970.

8. All the expenditure data in this article are reported in 1985 international dollars. Expenditures were compiled in current local currency units, deflated to 1985 prices using local implicit GDP deflators obtained from the World Bank's (1995) World Tables, and then converted to international dollars using purchasing power parity rates for 1985 obtained from Heston et al. (1995).

9. The 22 member countries of the OECD included in many of the "OECD totals" reported here are Australia, Austria, Belgium, Canada, Denmark, Finland, France, Germany, Greece, Iceland, Ireland, Italy, Japan, the Netherlands, New Zealand, Norway, Portugal, Spain, Sweden, Switzerland, the United Kingdom, and the United States.

10. This parallels broader trends toward an increased share for universities in total public spending on all science research (The Economist 1997).

11. In 1996, the CGIAR spent $326 million, less than 2 percent of the public investment in agricultural R&D worldwide. For more details on institutional developments related to the CGIAR see Baum (1986), Gryseels and Anderson (1991), and Pardey et al. (1997). Gryseels and Anderson (1991) identify a further 17 multilateral agricultural research agencies that were operating outside the CGIAR system in the mid-1980s. In addition, L'Institut Français de Recherche Scientifique pour le Développement en Coopération (ORSTOM) and Centre de Coopération Internationale en Recherche Agronomique pour le Développement (CIRAD) are two French-funded institutions that undertake research of direct relevance to (sub)tropical agriculture in developing countries. In 1992 these two agencies collectively spent $368 million on R&D.

12. In 1995 the CGIAR received 31 percent of its funding from the European Community, just over 12 percent each from Japan and the United States, and 15 percent from the World Bank (Pardey et al. 1997).

13. Different agencies in different countries do not always strictly adhere to OECD standards in reporting science and technology indicators. For example,

there may be some differences in the scope of research activity included (resulting in part from different views on what constitutes agriculture, in terms of sectoral coverage and the degree to which pre- and post-farm-oriented research is included along with farm-focused research) and differences in what constitutes R&D.

14. Unfortunately the data on public funding of R&D represent "direct" government expenditures on research and appear to exclude implicit transfers to the private sector through targeted tax concessions for R&D (OECD 1994, p. 96).

15. In 1986, for example, these three countries accounted for $31.9 billion of the total of $36.2 billion of public funds flowing to private R&D providers. In contrast, in 1993 they accounted for $21 billion of a much reduced total of $24.2 billion in public funds flowing to the private sector. Correspondingly, the public funds committed to defense R&D declined from a three-country total of $42.9 billion (1985 international dollars) in 1986 to $37.8 billion in 1993.

16. These data exclude Switzerland, whose share of agricultural chemical R&D is probably substantial but is unlikely to place it in the top three performers.

17. In 1971, $2.06 was spent by the state agricultural experiment stations (SAESs) for every dollar spent on intramural R&D undertaken by the United States Department of Agriculture; in 1993, $2.86 was spent by the SAESs for every federal dollar.

18. Alston, Christian and Pardey (1997) argue that classifying research into "basic," "applied," or "developmental" activities is generally subjective and usually difficult and at times may be done in a self-serving fashion that throws doubt on the reported orientation. Moreover, these categories mean different things to different scientists, and an economic evaluation of science may not find the distinctions made by scientists especially useful.

19. As Lee et al. (1996) document, however, industry levies are used extensively to fund commodity promotion in California.

20. For instance, in a recent summary, Fuglie et al. (1996, p. 28) report: "Most studies that have estimated the aggregate social rate of return to research consistently found rates of return between 40 and 60 percent." In a forthcoming study, Alston, Marra, Pardey, and Wyatt (1997) have documented results from some 141 studies that have estimated internal rates of return to agricultural R&D, ranging from less than 10 percent to 724,323 percent per annum. Of these rates of return, 95 percent fell between 11 and 321 percent per annum, and 90 percent fell between 15 and 151 percent per annum.

21. See Schultz (1977) and Wright and Zilberman (1993).

22. Scobie (1984) provides a clear and entertaining exposition of the arguments.

REFERENCES

Alston, J.M., and P.G. Pardey. 1996. *Making Science Pay: The Economics of Agricultural R&D Policy.* Washington, DC: American Enterprise Institute Press.

Alston, J.M., J.R. Anderson and P.G. Pardey. 1995. "Perceived Productivity, Foregone Future Farm Fruitfulness, and Rural Research Resource Rationalization." In G.H.

Peters and D.D. Hedley, eds. *Agricultural Competitiveness: Market Forces and Policy Choice.* Proceedings of the Twenty-second International Conference of Agricultural Economists. Aldershot, UK: Dartmouth Publishing Co., pp. 639–51.

Alston, J.M., J.E. Christian, and P.G. Pardey. 1997. "Agricultural R&D Investments and Institutions in the United States." Chapter 4 in J.M. Alston, P.G. Pardey, and V.H. Smith, eds. *Paying for Agricultural Productivity.* Washington DC: International Food Policy Research Institute.

Alston, J.M., B.J. Craig, and P.G. Pardey. 1997. "Dynamics in the Creation and Depreciation of Knowledge, and the Returns to Public Research." EPTD Discussion Paper, Washington DC: International Food Policy Research Institute.

Alston, J.M., P.G. Pardey, and V.H. Smith, eds. 1997. *Paying for Agricultural Productivity.* Washington DC: International Food Policy Research Institute.

Alston, J.M., M.S. Harris, J.D. Mullen, and P.G. Pardey. 1997. "Agricultural R&D Policy in Australia." Chapter 5 in J.M. Alston, P.G. Pardey, and V.H. Smith, eds. *Paying for Agricultural Productivity.* Washington DC: International Food Policy Research Institute.

Alston, J.M., M.C. Marra, P.G. Pardey, and T.J. Wyatt. 1997. *Research Returns Redux: A Meta-Analysis of Agricultural R&D Evaluations.* Draft IFPRI Research Report. Washington DC: International Food Policy Research Institute.

Antle, J.M. 1995. *Choice and Efficiency in Food Safety Policy.* Washington DC: American Enterprise Institute Press.

Baker, I., A. Baklien, and A.S. Watson. 1990. *A Review of Agricultural Research in Australia.* Melbourne: Department of Agriculture, Victoria.

Baum, W.C. 1986. *Partners Against Hunger: Consultative Group on International Agricultural Research.* Washington DC: World Bank, October.

Constantine, J., J.M. Alston, and V.H. Smith. 1994. "Economic Impacts of California's One Variety Cotton Law." *Journal of Political Economy* 102 (October): 66–89.

The Economist. 1997. "A Survey of Universities: The Knowledge Factory." *The Economist*, October 4.

Ervin, D.E. and A. Schmitz. 1996. "A New Era of Environmental Management in Agriculture." *American Journal of Agricultural Economics* 78(5): 1198–1206.

Fuglie, K., N. Ballenger, K. Day, C. Klotz, M. Ollinger, J. Reilly, V. Vasavada, and J. Yee. 1996. *Agricultural Research and Development: Public and Private Investments under Alternative Markets and Institutions.* USDA-ERS-NRED Agricultural Economic Report No. 735. Washington, DC: United States Department of Agriculture, May.

Grantham, G. 1984. "The Shifting Locus of Agricultural Innovation in Nineteenth-Century Europe: The Case of the Agricultural Experiment Stations." *Research in Economic History* 3:191–214.

Gryseels, G., and J.R. Anderson. 1991. "International Agricultural Research." Chapter 9 in P.G. Pardey, J. Roseboom, and J.R. Anderson, eds. *Agricultural Research Policy: International Quantitative Perspectives.* Cambridge: Cambridge University Press.

Hayami, Y., and S. Yamada. 1975. "Agricultural Research Organization in Economic Development: A Review of the Japanese Experience." In L.G. Reynolds, ed. *Agriculture in Development Theory.* New Haven: Yale University Press.

Heston, A., R Summers, D.A. Nuxoll, and B. Aten. 1995. "The Penn World Table (Mark 5): An Expanded Set of International Comparisons, 1950–1988." Version 5.6, http://cansim.epas.utronto.ca:5680/pwt., released January.

Huffman, W.E. and R.E. Just. 1994. "Funding, Structure, and Management of Public Agricultural Research in the United States." *American Journal of Agricultural Economics* 76(4):744–59.

Industry Commission. 1995. *Research and Development, Volumes 1 to 3.* Canberra: Australian Government Publishing Service.

Jacobsen, V., and G.M. Scobie. 1997. "Agricultural R&D Policy in New Zealand." Chapter 8 in J.M. Alston, P.G. Pardey, and V.H. Smith, eds. *Paying for Agricultural Productivity.* Washington DC: International Food Policy Research Institute.

Kerr, N.A. 1987. *The Legacy: A Centennial History of the State Agricultural Experiment Station, 1887–1987.* Columbia: Missouri Agricultural Experiment Station.

Lee, H., J.M. Alston, H.F. Carman, and W. Sutton. 1996. *Mandated Marketing Programs for California Commodities.* Giannini Foundation Information Series No. 96-1. University of California, Davis, Division of Agriculture and Natural Resources, August.

McLean, I.W. 1982. "The Demand for Agricultural Research in Australia: 1870–1914." *Australian Economic Papers* 21:294–308.

Mullen, J., K. Lee, and S. Wrigley. 1996. *Agricultural Production Research Expenditure in Australia: 1953–1994.* Agricultural Economics Bulletin No. 14. Orange: NSW Agriculture, July.

National Research Council. 1994. *Investing in the National Research Initiative: An Update of the Competitive Grants Program in the U.S. Department of Agriculture.* Washington DC: National Academy Press.

Organization for Economic Co-operation and Development (OECD). 1994. *Frascati Manual 1993: Proposed Standard Practice for Surveys of Research and Experimental Development.* Paris: Organization for Economic Co-operation and Development.

______. 1996. *Basic Science and Technology Indicators.* Paris: Organization for Economic Co-operation and Development.

Pardey P.G., J. Roseboom, and J.R. Anderson. 1991. *Agricultural Research Policy: International Quantitative Perspectives.* Cambridge: Cambridge University Press.

Pardey, P.G., J.M. Alston, J.E. Christian, and S. Fan. 1997. *A Productive Partnership: The Benefits from U.S. Participation in the CGIAR.* Washington DC: International Food Policy Research Institute, unpublished draft report.

Pardey P.G., J. Roseboom, and B.J. Craig. 1997. "Agricultural R&D Investments and Impact." Chapter 3 in J.M. Alston, P.G. Pardey, and V.H. Smith, eds. *Paying for Agricultural Productivity.* Washington DC: International Food Policy Research Institute.

Roseboom, J., and H. Rutten. 1997. "Financing Agricultural R&D in the Netherlands: The Changing Role of Government." Chapter 7 in J.M. Alston, P.G. Pardey, and V.H. Smith, eds. *Paying for Agricultural Productivity.* Washington DC: International Food Policy Research Institute..

Salmon, S.C., and A.A Hanson. 1964. *The Principles and Practice of Agricultural Research.* London: Leonard Hill.

Schultz, T.W. 1977. "What Are We Doing to Research Entrepreneurship?" Agricultural Economics Paper No. 77:14, University of Chicago, March 25.

Scobie, G.M. 1984. "Agricultural Research: A Bit of Back to Basics." *The Agricultural Economist* 5(3): 10–11.

Thirtle, C., J. Piesse, and V.H. Smith. 1997. "Agricultural R&D Policy in the United Kingdom." Chapter 6 in J.M. Alston, P.G. Pardey, and V.H. Smith, eds. *Paying for Agricultural Productivity.* Washington DC: International Food Policy Research Institute.

True, A.C., and D.J. Crosby. 1902. *Agricultural Experiment Stations in Foreign Countries.* USDA Office of the Experiment Stations Bulletin No. 112, Washington DC: U.S. Government Printing Office.

Ulrich, A., W.H. Furtan, and A. Schmitz. 1986. "Public and Private Returns from Joint Research: An Example from Agriculture." *Quarterly Journal of Economics* 101(10): 101–29.

USDA. 1995. "AERI Updates: Agricultural Research." AERI Update No.5. revised, Washington DC: United States Department of Agriculture.

Watson. A. 1996. "The Price of Research: Understanding the Economics of Charging for Research." In contributed papers proceedings of conference on *Global Agricultural Science Policy for the Twenty-first Century* held in Melbourne Australia, August.

World Bank. 1995. *World Tables,* Diskette version, Washington DC: World Bank.

Wright, B.D., and D. Zilberman. 1993. "Agricultural Research Structures in a Changing World." In R.D. Weaver, ed. *U.S. Agricultural Research: Strategic Challenges and Options.* Bethesda, MD: Agricultural Research Institute, March.

4

FINANCING AGRICULTURAL RESEARCH WITH PRIOR DISTORTIONARY TAXES AND SUBSIDIES

GEORGE B. FRISVOLD AND
STEPHEN J. VOGEL

ECONOMISTS HAVE consistently estimated high rates of return to agricultural research, on the order of 30 percent or more (Alston, Norton, and Pardey et al. 1995; Fuglie et al. 1996). Despite these high returns, public agricultural research continues to face budgetary constraints. The 1980s saw a pervasive slowdown in public investment in agricultural research, affecting international research centers and national research programs in developed and developing countries alike (see chapter 3 in this volume). In the United States, real federal expenditures on agricultural research have grown little since the mid-1970s (Fuglie et al. 1996).

In response to the scarcity of research funds, economists have devoted increased attention to ways to get more out of every tax dollar raised for agricultural research. Most of this attention has focused on improving the efficiency of allocating research funds. This has generated considerable debate over how the public research system should interact with the private sector, how research projects should be selected, and which grant mechanisms are appropriate to distribute funds (Alston, Norton, and Pardey 1995; Alston and Pardey 1995; Fuglie et al. 1996; Huffman and Just 1994; Beattie and Innes 1997; U.S. Congress OTA 1995; Phillips 1997; Ulrich, Furtan, and Schmitz 1986). These debates are concerned with the costs and benefits of changing the way in which research funds are *spent.* With some notable exceptions, far less attention has been paid to the economic implications of how public research funds are *raised.*

Because taxes create distortions in the allocation of resources in an economy, the social cost of raising one dollar of revenue is greater than that dollar itself. Ballard, Shoven, and Whalley (1985) estimated that the marginal social cost per dollar raised through taxation in the United States was on the order of \$1.17 to \$1.56. Browning (1994) reported similar estimates. More recently,

Fullerton (1991) and Ballard and Fullerton (1992) reviewed the literature on the costs of taxation, reconciling different estimation procedures and results. They estimated that the marginal social cost of one dollar of revenue raised ranges from $1.07 to $1.25. Fox (1985) and Dalrymple (1990) discussed the implications of deadweight costs of taxation for estimates of rates of return to agricultural research. However, they did not address how alternative financing mechanisms might alter the costs of raising revenues for public research.

Alston and Mullen (1992) and Alston and Pardey (1995) have argued that it would be less socially costly to fund agricultural research with commodity taxes. Reliance on such taxes has increased in a number of countries, but not in the United States (see chapter 3 in this volume). In the United States, federal funding for agricultural research comes almost exclusively from general tax revenues, based largely on taxes on labor income. Alston and Pardey (1995) calculated that the deadweight loss from a tax on agricultural commodities could be well below one cent per dollar raised, much smaller than the 7 to 25 cents per dollar estimated for general revenues.[1] Thus, they argue that commodity taxes would be a more efficient way of financing agricultural research.

This argument is similar to that made by proponents of carbon taxes to reduce greenhouse gas emissions. Some economists have argued that carbon taxes would be less distortionary than general taxation and that substituting carbon or energy tax revenues for general revenues, through cuts in other taxes, would reduce the overall social cost of taxation (Pearce 1991; Repetto et al. 1992). In both cases it is claimed that commodity taxes (either on agricultural or energy commodities) raise a given level of revenue at a lower social cost than taxes on labor income.

In the case of energy taxes, however, a large body of literature tends to contradict this assertion (Bovenberg and Goulder 1996; Carraro, Galeotti, and Gallo 1996; Morgenstern 1996; Oates 1995; Parry 1995, 1998; Schob 1996; Welsch 1996). A basic finding from this literature is that a revenue-neutral tax reform, where commodity tax increases are matched, or "recycled," by reductions in other taxes in the economy does *not,* in general, improve welfare. Another key finding is that net welfare impacts are sensitive to assumptions about labor market structure and about what other taxes are cut.

What accounts for these findings and what are their implications for the efficiency of financing agricultural research through commodity taxes? First, the Alston and Pardey calculation of deadweight loss for an agricultural commodity tax is based on a standard first-order approximation that applies to small taxes when there are no preexisting taxes, subsidies or other market distortions. In the U.S. economy, however, these distortions are significant. As Oates (1995) and Parry (1998) point out, secondary welfare effects of changes in other distorted markets require second-best analysis that may yield results quite different from those implied by first-best analysis. A new commodity tax would exacerbate preexisting tax distortions and disincentives for labor supply or investment. This tax interaction effect can be large because the tax rate on labor earnings (when all types of taxes are considered) is also large, about 40 percent for the average U.S. household (Browning 1994; Parry 1998).

In sum, the carbon tax–revenue recycling literature has a number of implications for the analysis of the relative efficiency of commodity taxation versus general taxation. First, the gains or losses from shifting taxes are sensitive to preexisting distortions in multiple markets. Second, general equilibrium, second-best analyses of carbon taxes have yielded qualitatively different welfare results than analyses using simple first-best, deadweight, loss comparisons of the type employed by Alston and Pardey (1995). Third, because taxes affect labor income, results of tax shift experiments are quite sensitive to specifications of labor market behavior.

AIMS AND SCOPE OF CHAPTER

This chapter uses a computable general equilibrium (CGE) model of the U.S. economy to conduct a series of tax shift experiments to examine the welfare, distributional and other economic effects of financing public agricultural R&D through commodity taxes rather than general taxation. The U.S. CGE provides a realistic benchmark of preexisting taxes and agricultural commodity programs in the economy. Our approach follows the "tax reform" literature that focuses on the welfare effects of incremental changes from the preexisting tax structure. The preexisting tax structure serves as a benchmark for welfare comparisons rather than an optimal (Ramsey) tax structure. This approach follows Feldstein (1976), Ahmad and Stern (1984), and Schob (1996) by considering the potential for welfare-improving tax changes rather than optimal tax design.

In a fashion similar to that of the carbon tax–revenue recycling literature, we simulate the effects of a revenue-neutral tax shift, where agricultural commodity taxes are recycled through matching tax cuts in the rest of the economy. We conduct three types of sensitivity analysis. First, we consider the effects of commodity taxes placed on either agricultural producers or consumers (first purchasers) of primary agricultural commodities. Second, we examine how results are affected by changes in the structure of commodity programs. In particular, we consider the implications of the shift from target price-deficiency payment schemes to decoupled farm payments embodied in the 1996 FAIR Act. Third we consider the sensitivity of results to assumptions about labor market structure. This includes assumptions about the wage elasticity of labor supply and the existence of interindustry wage differentials. The persistence of interindustry wage differentials has been well established by numerous econometric studies dating back to the 1950s and across many countries. One implication of such differentials is that policies encouraging employment in high-wage sectors can enhance allocative efficiency (Katz and Summers 1989).

STRUCTURE OF THE U.S. CGE MODEL

This CGE model is an expanded version of the basic USDA/ERS CGE model for the United States, calibrated to a 1993 base year social accounting matrix (SAM). This version of the U.S. CGE model appeared in Hanson, Vogel, and Robinson (1995) and Hanson, Robinson, and Vogel (1996). Also see Robinson, Kilkenny, and Hanson (1990) for a detailed description of the basic model. The 51-sector aggregation includes 9 agricultural production sectors, 10 food processing sectors, 5 energy sectors, 17 other manufacturing sectors, and 10 service sectors. The disaggregation of agricultural production and processing into 19 sectors allows us to represent the essential characteristics of farm income support programs and to capture many of the forward and backward linkages among agricultural production sectors, food processing, input suppliers, and the rest of the economy. The farm sectors include dairy, poultry, livestock, cotton, food grains, feed crops, oilseed crops, sugar crops and other crops. The other crops category includes fruits, vegetables, orchard crops, honey, and greenhouse products.

Figure 4.1 presents the features and the equations of the CGE model that are key to our analysis. For each production activity, a constant elasticity of substitution (CES) production function defines domestic supply as a function

Figure 4.1. Key features of the CGE model

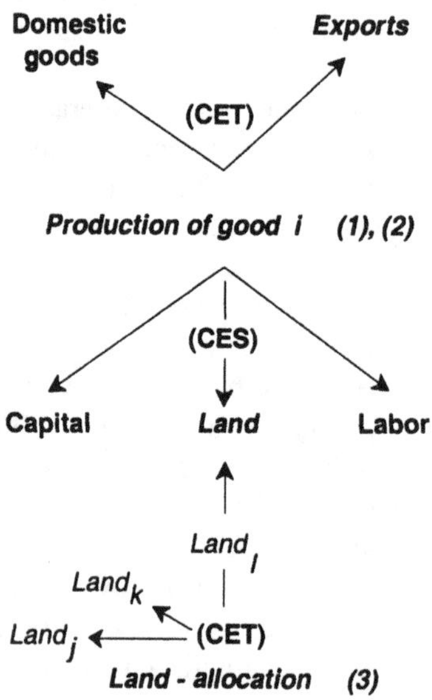

Equations:

$$(1)\begin{cases}(a)\quad P_i^X \cdot (1 + rdtr_i) = f(X_i)\\ (b)\quad P_i^{va} = P_i^X \cdot (1 - itax_i - rdtr_i) - \sum_j \left(a_{ji} \cdot P_j^q\right) + SPIE_i\end{cases}$$

$$(2)\quad SPIE_i = \theta_i \cdot \left(TP_i - P_i^X\right)$$

$$(3)\quad Land_i = g\left(\sigma_L, P_i^L\right)$$

$$(4)\quad W_h = \frac{\sum_i P_i^q \cdot (C_{ih} - CSUB_{ih})}{\prod_i \left(P_i^q\right)^{u_{ih}}}$$

of the primary inputs: labor and capital in nonagriculture, and land, labor, and capital in agriculture. Intermediate inputs are used in fixed proportions to output, except for the livestock sector, which can substitute feed crops and processed feed. Sectoral input demands are derived from first-order conditions for profit maximization.

The ease of shifting acreage among crops influences the producers' supply response. We adopt a medium-run supply elasticity of 0.7 reported in Sumner (1995). We calibrate the supply response in our model by specifying a constant elasticity of transformation (CET) land-allocation function, which allows for imperfect mobility between different crops (figure 4.1). Equation 3 in figure 4.1 represents the supply of land ($Land_l$) for each crop as a function of the elasticity of substitution (σ_L) and sectoral price of land (P_l^L).

Aggregate domestic demand has five components: consumption, intermediate demand, government, fixed investment, and inventory accumulation. Households are disaggregated into 11 income classes and each household class's expenditure behavior is modeled by a linear expenditure system. Intermediate demand is calculated from sectoral output, using fixed input-output coefficients, except for the livestock feed mix. For the government, real aggregate spending on goods and services is fixed, and its sectoral composition is given by fixed shares. Aggregate investment and inventory demand are fixed in real terms.

Both imports and exports are imperfect substitutes with respect to domestically produced goods. Following the Armington assumption, the model treats domestic absorption as a CES composite of imports and domestically produced goods. Similarly, domestic supply in each sector is transformed according to a CET function into domestic and exportable goods (figure 4.1). This treatment of imports and exports partially insulates the domestic price system from changes in world prices of sectoral substitutes. We make the small-country assumption that the United States cannot affect world prices of its imports. We assume a high import substitution elasticity for wheat. On the export side, we assume downward-sloping world demand functions for four U.S. agricultural exports: cotton, food grains, feed crops, and oilseed crops. All other exports follow the small-country assumption.

Treatment of Commodity Programs

We consider two cases that reflect the evolution of U.S. farm policy. In the first case, farm income is supported through a target price–deficiency payment scheme. For producers of program crops (food grains, feed crops, and cotton), the market-clearing, producer-incentive price equals the market price plus an endogenous deficiency payment rate. Equation 2 in figure 4.1 generates the program payment rate ($SPIE_i$) as a function of the difference between the fixed target price (TP_i) and the market-clearing producer price (P_i^X). In the second case, lump-sum transfers to households replace price supports for program crops. This reflects changes embodied in the 1996 FAIR Act where commodity program payments are (for the most part) no longer tied to current planting and production decisions.

The deficiency payment and FAIR Act cases represented here should be viewed as stylized polar examples. The stylized deficiency payment scheme assumes that 100 percent of program output is eligible to receive support prices. It does not, therefore, account for the substantial decoupling of program payments from production decisions that was introduced in the 1985 and 1990 farm bills (through, for example, introduction of nonpayment acres).[2] Likewise, under the FAIR Act, loan rates for rice and cotton remain high enough so that they may still act as price supports in some years.

Multisector agricultural models and CGE models frequently represent commodity programs as ad valorem subsidies. Although not strictly accurate, this may be a reasonable approximation in analyses that consider effects of removing or altering the levels of farm income support. This assumption is less appropriate if one is considering how other changes, such as changes in taxes, interact with commodity programs. This is precisely the issue we wish to address. Our modeling of the deficiency payment program induces a producer response not captured by a fixed ad valorem wedge (Kilkenny and Robinson 1990). The same is true for the specification of FAIR Act type income transfers. Thus, we are able to more accurately capture interactions between commodity taxes and farm programs.

The Labor Market

Labor market closure plays a significant role in the welfare outcomes of tax shift experiments. Analytical and numerical models used in the public finance literature focus on labor-leisure trade-offs and the wage elasticity of labor supply (Bovenberg and Goulder 1996; Parry 1995). These models assume full employment with competitively determined wage rates. Welsch (1996) uses a Phillips curve approach to wage formation, with wage rates rising as actual employment approaches full employment. Carraro, Galeotti, and Gallo (1996) model labor markets in a CGE model as segmented into primary and secondary markets. In the primary market wages and (un)employment levels are modeled as a result of a Nash bargaining game between employers and trade unions in the primary labor market. Wages in the secondary labor markets are set based on empirically observed differentials between primary and secondary market wage rates. In all these papers, the welfare results of tax shifts are sensitive to the specification of labor market parameters.

In this chapter, we adopt three forms of labor market closure. In the first case, aggregate labor supply is assumed to be fixed, although labor can move freely between sectors. Labor markets clear at the aggregate level. All labor receives the same equilibrium wage regardless of the sector where it is employed. The second case relaxes the assumption of a fixed aggregate supply of labor. The third case relaxes the assumption that labor receives the same wage across all sectors of the economy.

In this third case, the model is calibrated to reflect the empirical fact of persistent interindustry wage differentials. Several econometric studies dating back several decades and across countries have consistently found persistent differences in wage rates by industry that are not explained by worker characteristics (see for example, Blackburn and Neumark 1992; Dickens and Katz 1987; Gera and Grenier 1994; Katz and Summers 1989). A general finding of these econometric studies is, "differences in unobserved ability explain relatively little of inter-industry or inter-occupation wage differentials" (Blackburn and Neumark 1992). Simple competitive models do not explain this persistence. Instead, sources of these wage differences appear to be capital intensity, union intensity, firm size, search-turnover costs, and monopoly power (Thaler 1989; Montgomery 1991). Katz and Summers (1989) summarize these results and draw a policy conclusion that will be shown numerically in this chapter:

> The available evidence suggests that pervasive industry wage differentials of up to 20 percent remain even after controlling for differences in observed measures of workers' skill and the effects of unions. Theoretical analysis indicates that given noncompetitive wage differentials of these magnitudes *policies directed at encouraging employment in high-wage sectors could significantly enhance allocative efficiency* (emphasis added).

The critical labor market equations are presented in table 4.1. Equations 1–3 are first-order conditions for the model's heterogeneous, nested labor market. In equation 1, the equilibrium sectoral wage for labor of skill type *s*, defined as the average wage W_s times its sectoral proportionality constant ($W_s^* \cdot wldist_{is}$), equals the composite wage times the marginal product of labor by skill type *s* (Lc_{is}^*). Equation 2 aggregates the labor demanded by skill type into composite labor (Li^*) using a CES aggregation function. Equation 3 states that the equilibrium quantity of composite labor is supplied at a level such that its similarly defined sector wage ($W_L^* \cdot wfdist_{iL}$) equals its value marginal product.

Both $wldist_{is}$ and $wfdist_{iL}$ are sector wage proportionality constants that deviate from 1 if the sectoral market-clearing wage does not equal the economy-wide average wage for labor. This means that high-wage sectors (where $wfdist_{iL}$ and $wldist_{is} > 1$) use labor more productively than average and will attract free labor at the expense of low-wage sectors (where $wfdist_{iL}$ and $wldist_{is} < 1$). They are calibrated from U.S. data on employment and earnings by sector and skill level used in our 1993 SAM. These coefficients capture the empirical

Table 4.1. Labor market equations of the U.S. CGE model

First order conditions for nested labor market:

(1) $$W_s^* \cdot wldist_{is} = W_L^* \cdot wfdist_{iL} \cdot MP\left(Lc_{is}^*\right)$$

(2) $$L_i^* = a_i \cdot \left(\sum_S \alpha_{is} \cdot Lc_{is}^{*\,-\rho} \right)^{-(1/\rho_s)}$$

(3) $$W_L^* \cdot wfdist_{iL} = P_i^{va} \cdot wfdist_{iL} \cdot MP\left(L_i^*\right)$$

Labor supply equations:

(4) $$\overline{Lc_s} = \sum_i Lc_{is}^*$$

(5) $$\overline{L}_{TOTAL} = \sum_S \overline{Lc_s}$$

(6) $$Lc_s^* = lconst \cdot \left(W_s^*\right)^{\rho}$$

fact of interindustry wage differentials, but do not provide a causal mechanism for explaining them—such as efficiency wages (Katz 1986), search costs, or union bargaining.

In the first labor market closure in equation 4, aggregate labor supply of each skill level $(\overline{Lc_s})$ is assumed to be fixed. Aggregate labor supply $(\overline{L}_{TOTAL})$ is simply the sum of aggregate high-skilled and low-skilled labor in equation 5. In our first case, wage differentials are assumed not to exist; we set the $wldist_{is}$ and $wfdist_{iL}$ coefficients equal to 1. This specification depicts the homogeneous labor market.

In the second labor market closure, we relax the assumption that the aggregate labor supply is fixed. For each skill level, equation 6 now replaces equations 4 and 5. In equation 6, equilibrium labor supplied by skill level (Lc_s^*) is a simple function of the average wage by skill level (W_s^*), that is, aggregate labor supply curves are now upward sloping. The wage elasticity of labor supply ρ is set equal to 0.3 for low- and high-skilled labor. Base levels of labor supply are the same under all labor market closures. In this case, the $wldist_{is}$ and $wfdist_{iL}$ coefficients also equal 1.

In the third labor market closure, aggregate labor supplies are again fixed (equations 4 and 5). In this case however, interindustry wage differentials are accounted for by allowing the $wldist_{is}$ and $wfdist_{iL}$ coefficients to diverge from 1.

Model Closure and Welfare Estimation

We specify a form of neoclassical closure of the model which allows us to calculate the economy-wide equivalent variation of a policy change. We fix aggregate real investment, real government expenditure, and the real balance of trade. Changes in government expenditures from taxing agricultural commodities are returned to households in the form of lower taxes in proportion to taxable income. Enterprise savings are assumed to adjust to finance the exogenous level of real investment.

Under this closure, changes in real household consumption capture all aggregate welfare effects. Using the appropriate cost-of-living index consistent with consumer utility functions, these changes are an exact equivalent variation measure of domestic welfare changes. Equation 4 in figure 4.1 defines the welfare measure for a single household as nominal consumption out of supernumerary income,

$$\sum_i P_i^q (C_{ih} - CSUB_{ih})$$

(P_i^q—market price of good i; C_{ih}—consumption of good i by household class h; $CSUB_{ih}$—subsistence minimum of good i by household class h), deflated by a Cobb-Douglas utility-weighted price index,

$$\prod_i (P_i^q)^{\mu_{ih}}$$

(P_i^q—market price of good i; μ_{ih}—utility weight on good i for household class h). We use these household welfare measures to compute the equivalent variation as the change in household welfare, and we sum them across households to obtain the equivalent variation measure of the change in aggregate welfare.

Following standard practice in the cost-benefit literature, Gardner (1983) and Just, Hueth, and Schmitz (1982) compute welfare changes by summing net changes of consumer and producer surplus across as many markets as are relevant to the shock. In our general equilibrium approach, this is unnecessary. Given our model closure, the change in aggregate real consumption will exactly equal the sum of changes in consumer and producer surplus, accounting for all market interactions and income effects.[3] As Shoven and Whalley (1992) note, in the CGE framework, the equivalent variation welfare measure focuses on comparisons between two states of equilibrium for the entire economy, not on a partial-equilibrium approximation of what may happen from a policy shock. Although we assume downward-sloping world demand functions for U.S. agricultural export crops, we do not derive welfare measures of the foreign welfare impacts of changes in these export prices.

Finally, we fix aggregate GDP at factor cost as the numeraire. This is consistent with the practice of the Congressional Budget Office and the Department of Treasury to provide a no-inflation benchmark for comparing the impact of changes in government programs on nominal revenues and expenditures.

EXPERIMENT DESIGN

The first step in shifting the tax burden from households to producers is to specify how commodity taxes are imposed to finance public agricultural research. We assume that only commodity-specific research is funded through commodity taxation. Other research areas, such as natural resource management, rural development, and food safety are still funded by general revenues. Data on commodity-specific research comes from the Current Research Information System (CRIS) published in the *Inventory of Agricultural Research* (USDA 1993) is reported in table 4.2. In 1993, public sector expenditures on commodity-specific research were more than $1.56 billion.

In the model experiments, commodities are taxed at differential rates so that revenues raised per commodity match research expenditures. The taxes are thus structured like an industry checkoff system where each commodity group finances its own commodity-specific research. Livestock, feed crops, and other crops account for two-thirds of total R&D expenditures. Except for cotton and other crops, imputed sectoral tax rates on production remain in a narrow range between 0.5 percent and 0.8 percent (table 4.2).

Table 4.2. Public research expenditures and imputed tax rate by sector

Sector	Public research expenditures ($ millions)	Public research expenditures (percent)	Imputed tax rate (percent)
Dairy	136.6	8.7	0.566
Poultry	89.4	5.7	0.566
Livestock	393.9	25.2	0.518
Cotton	66.6	4.3	1.172
Food grains	92.0	5.9	0.745
Feed crops	214.7	13.7	0.489
Oilseeds	115.8	7.4	0.795
Sugar	21.2	1.4	0.871
Other crops	433.6	27.7	1.077
Total	1,536.8	100.0	

Source: For research expenditures, *Inventory of Agricultural Research.*

We considered two alternatives for placement of the taxes. In equation 1a of figure 4.1, under a consumer tax (or more accurately, a first-purchaser tax), the research tax rate for each commodity, $rdtr_i$, is placed directly on the producer price (P_i^X). In contrast in equation 1b in figure 4.1, a producer tax is modeled as an ad valorem wedge on the cost-price of production. In equation 1b, the unit cost-price (P_i^{va}) equals the producer price (P_i^X) net of the marginal indirect business tax ($itax_i$) and this same marginal research tax ($rdtr_i$) minus a weighted average of prices of intermediate goods

$$\sum_j (a_{ji} \cdot P_i^q)$$

where a_{ji} is the intermediate use coefficient of good j by industry i and P_i^q is the market price, plus a marginal farm program subsidy ($SPIE_i$). The producer tax would be most similar to a commodity group checkoff program. In a simple partial-equilbrium model with no preexisting distortions, equilibrium output, producer price, consumer price, and economic welfare would be independent of whether the tax is levied on the demand side or the supply side (Varian 1984, pp. 90–91). In a general equilibrium setting with multiple preexisting market distortions, this will not necessarily be the case. In particular, the placement of the tax has important implications for interactions with commodity programs. A demand-side tax increases deficiency payments while a supply-side tax reduces them.

The experiments assume that revenues from the commodity taxes are recycled as lump-sum transfers to households. This specification may understate net welfare gains of the tax shift. Other revenue recycling studies have demonstrated that gains are higher (or losses lower) if commodity tax revenues are recycled by lowering marginal tax rates rather than through lump-sum rebates. This is because reducing distortionary taxes has consequences for efficient resource allocation that are superior to those of lump-sum payments (see Bovenberg and Goulder 1996; Parry 1998; and Schob 1996).

The tax shift experiments are designed to be revenue neutral ex ante. Because of the general equilibrium effects of the tax shift on the overall economy, actual revenues collected ex post may differ such that the tax shift can have budgetary consequences. Secondary effects would likely not be considered if alternative financing mechanisms were being considered for budget scoring purposes.[4] We, thus, maintain the ex ante revenue-neutral tax shift.

Twelve experiments in all are considered, looking at different combinations of taxes, commodity program regimes and labor market specifications. There are two types of taxes (consumer and producer), two commodity program regimes (deficiency payment and FAIR Act) and three labor market specifications (inelastic aggregate labor supply without interindustry wage differentials, an upward-sloping aggregate labor supply curve without interindustry wage differentials, and inelastic aggregate labor supply with interindustry wage differentials).

NUMERICAL RESULTS

Farm Sector Impacts

Selected farm sector impacts of the 12 tax shift experiments are reported in table 4.3. In all cases, farm income, production, and exports fall. Farm sector impacts are insensitive to assumptions about labor market structure. They are more sensitive to placement of the tax (whether the tax is a consumer tax or a producer tax) and to the commodity program regime. Although farm sector income falls, it falls by less than the amount of revenues raised by the commodity tax. The cost to the farm sector, in lost income, is 40 to 60 percent of the over $1.5 billion in revenue raised for commodity-specific research (see table 4.3).

Under deficiency payments, the placement of the tax has a relatively large impact on farm income and farm program outlays. The producer tax reduces farm output and income more than the consumer tax. The producer tax reduces the wedge between the target price and market price, thereby reducing deficiency payments; the consumer tax increases this wedge, thereby increasing deficiency payments. The percent of the tax absorbed by agriculture is also greater under a producer tax.

Table 4.3. Farm sector impacts of tax shift ($ in millions except where otherwise stated)

	Deficiency payments		FAIR Act	
Change in:	Consumer tax	Producer tax	Consumer tax	Producer tax
Fixed aggregate labor, no wage differentials				
Farm production	–638	–850	–761	–787
Farm exports	–217	–361	–288	–300
Farm sector income	–631	–943	–709	–735
Deficiency payments	157	–231	—	—
Percent of tax absorbed by agriculture	42	60	45	47
Upward-sloping labor supply curves, no wage differentials				
Farm production	–638	–850	–760	–786
Farm exports	–218	–362	–289	–300
Farm sector income	–630	–941	–707	–733
Deficiency payments	156	–232	—	—
Percent of tax absorbed by agriculture	40	60	45	47
Fixed aggregate labor, interindustry wage differentials				
Farm production	–638	–861	–792	–819
Farm exports	–226	–395	–330	–344
Farm sector income	–634	–977	–753	–781
Deficiency payments	158	–205	—	—
Percent of tax absorbed by agriculture	41	62	48	50

Under the FAIR Act, the impacts of the two taxes converge. Also note that the percent of the tax absorbed by agriculture under a producer tax falls from around 60 percent under deficiency payments to around 50 percent under the FAIR Act. This suggests that the evolution of farm policy to decoupled payments reduces the costs of commodity checkoffs to program crops.

Macroeconomic Impacts

The tax shift of $1.56 billion dollars is a small percent of total tax revenues and induces very slight macroeconomic effects. The fall in farm exports of $226 million to $395 million (see table 4.3) causes the nominal exchange rate to rise by 0.05 percent.[5] This terms of trade effect increases the domestic prices of imports and exports, partially contributing to a rise in the consumer price index of 0.02 percent. Seasonal fluctuations in economic activity could easily overwhelm the effects of these tax shift simulations.

Labor Market Impacts

The tax shift causes jobs to move out of agriculture and food processing and distribution, while new jobs are created elsewhere in the economy. In table 4.4, we report the average change in number of jobs for each type of labor market structure, since there is very little variation across tax and farm policy regimes.[6]

In a homogeneous labor market, all jobs receive the same wage, regardless of occupational skill level or level of sector productivity. Under this assumption, firms respond only to relative changes in output prices. With a fixed aggregated supply, losses of 4,000 jobs in agriculture and 1,000 jobs each in the food processing and trade and transportation sectors are offset by gains of 6,000 jobs in manufacturing. With upward-sloping labor supply curves, a similar gain of 6,000 jobs in manufacturing and 2,000 jobs in services offset the losses in agriculture, food processing, and trade. The additional employment gain in services results in a net increase in aggregate employment of 2,000 jobs.

Table 4.4. Sectoral employment impacts of tax shift (thousands of jobs)

	No wage differentials		*Interindustry wage differentials*
Average change in:	Fixed aggregate labor supply	Upward-sloping labor supply curves	Fixed aggregate labor supply
Agriculture	–4	–4	–13
Food processing, distribution	–1	–1	–1
Manufacturing	6	6	8
Trade/transportation	–1	–1	2
Services	0	2	4
Net employment impacts	0	2	0

In a labor market with interindustry wage differentials, firms must compete unequally for factors *and* simultaneously respond to changes in relative prices in output markets. Consequently, since the more productive sectors will attract labor by offering higher sectoral wages, we observe a larger change in the composition of labor force. In this regime, there are losses of about 12,000 jobs in agriculture and 1,000 jobs in food processing. Given the full-employment assumption, manufacturing absorbs 7,000 new jobs, trade and transportation absorb 2,000, and services absorb 4,000.

In all experiments, average wages rise, but by less than 1 percent of 1 percent. In an economy with 123 million jobs, these employment changes are small both in percentage terms and relative to normal seasonal fluctuations in employment.

The impacts of the tax shift on aggregate employment and labor income is shown in table 4.5. In all scenarios, the tax shift increases labor income. Labor income increases the most in scenarios that account for interindustry wage differentials, between $570 and $600 million. This is over $200 million more than in scenarios that do not account for wage differentials. In the presence of wage differentials, the tax shift induces labor to be allocated away from agricultural sectors that have lower than average wages and labor productivity and toward other sectors with higher wages and labor productivity. Wages in agricultural sectors range from 29 to 34 percent of the average wage. We emphasize that these wage differentials are not simply an artifact of the model but are based on empirical evidence.

Compared to the existence of wage differentials, placement of the tax and commodity program regime have relatively less impact on labor income. The wage elasticity of labor supply has virtually no impact on labor income. In the case where the wage elasticity of labor supply is 0.3, rising average wages in-

Table 4.5. Aggregate labor market impacts of tax shift

	Deficiency payments		FAIR Act	
Change in:	Consumer tax	Producer tax	Consumer tax	Producer tax
Fixed aggregate labor, no wage differentials				
Labor income ($ million)	370	287	303	314
Upward-sloping labor supply curves, no wage differentials				
Labor income	366	287	302	314
Labor supply (thousands of jobs)	2.4	1.6	1.8	1.9
Fixed aggregate labor, interindustry wage differentials				
Labor income	577	598	571	592

crease aggregate employment by 1,600 to 2,400 jobs. In the case where the wage elasticity of supply is zero (and without wage differentials), no new jobs are created. However, in this latter case, average wages rise slightly more. So, with upward-sloping labor supply curves, there are slightly more jobs; with vertical labor supply curves, there are slightly higher wages; the overall effects on labor income are roughly equal.

Budgetary and Welfare Effects

The importance of labor market structure in determining the overall welfare effects of the tax shift is highlighted in table 4.6. In the scenarios with wage differentials, welfare gains range from $190 to $274 million. Under the FAIR Act regime, funding research through commodity taxes rather than through general revenues reduces the social cost of financing research by about 15 cents per dollar of revenue raised.[7] The effect of the tax shift is to enhance allocative efficiency in the economy by reallocating employment to higher-wage sectors.

Under the scenarios that do not account for wage differentials, the tax shift causes welfare to fall, however. In the case with upward-sloping labor supply curves (elasticity of labor supply with respect to the wage rate, $\rho = 0.3$), the tax shift has a positive effect on aggregate employment (see table 4.4). However, other negative tax interaction effects cause welfare to decline overall. In the case with no wage differentials and fixed aggregate labor, there is no positive aggregate employment effect. Consequently, welfare losses are even

Table 4.6. Welfare and budgetary impacts of tax shift (in $ million)

	Deficiency payments		FAIR Act	
Change in:	Consumer tax	Producer tax	Consumer tax	Producer tax
Fixed aggregate labor, no wage differentials				
Domestic welfare	–105	–114	–106	–111
Deficiency payments	157	–231	—	—
Deviation from revenue neutrality	59	335	89	125
Upward-sloping labor supply curves, no wage differentials				
Domestic welfare	–13	–48	–33	–34
Deficiency payments	156	–232	—	—
Deviation from revenue neutrality	—	402	143	176
Fixed aggregate labor, interindustry wage differentials				
Domestic welfare	190	274	234	241
Deficiency payments	158	–205	—	—
Deviation from revenue neutrality	–24	372	135	172

greater. These results are similar to those obtained by Welsch (1996) in a CGE study of an energy-labor tax shift in the European Community. A key parameter in his study was the elasticity of the wage respect to employment—equivalent to $1/\rho$ in this study. He found that the benefits of the tax shift were greater as values of $1/\rho$ decreased (ρ increased).

In most scenarios, the ex post effect of the tax shift is to increase total revenues raised. In the model, these extra revenues are returned to households. Table 4.6 illustrates how commodity program structure and tax placement affect deviations from revenue neutrality. Under a deficiency payment regime a producer tax reduces deficiency payments, and ex post, the tax shift raises between \$335 and \$402 million in revenues. Under a consumer tax, deficiency payments rise. The tax shift comes close to hitting the ex ante target (deviating from –\$24 to +\$59 million). Under the FAIR Act, farm payments are independent of the tax shift. Both the budgetary and welfare impacts of the taxes lie within the ranges under the deficiency payment regime.

CONCLUSIONS

This chapter uses a computable general equilibrium (CGE) model of the U.S. economy to conduct a series of tax shift experiments to examine the welfare, distributional, and other economic effects of financing public agricultural R&D through commodity taxes rather than general taxation. We summarize some of our major findings.

First, the chapter illustrates the importance of noncompetitive, interindustry wage differentials in determining the welfare effects of shifting the method of financing public agricultural research. Katz and Summers (1989) have demonstrated theoretically that given the existence of pervasive wage differentials, policies that encourage "employment in high-wage sectors could significantly enhance allocative efficiency." The CGE model used in this chapter confirms this result numerically. Financing agricultural research through commodity taxes rather than general revenues induces a reallocation of labor from lower-wage to higher-wage sectors, raising welfare by over \$200 million.

The numerical simulation results suggest that funding research through commodity taxes, rather than general revenues, reduces the social cost of financing research by about 15 cents per dollar of revenue raised. Sensitivity analysis reveals that the existence of wage differentials is a critical factor driving these welfare gains. Under alternative specifications that assume no such differentials, the welfare effects of the tax shift are negative.

Financing commodity-specific research through commodity taxes would reduce farm income by about half the amount of revenues raised by the taxes. One way to interpret this result is that if agricultural research were funded by a 1 percent commodity checkoff, half the cost would be borne by agriculture and the other half by the rest of the economy. The ad valorem tax rates necessary to finance commodity-specific research range from 0.5 to 1.2 percent for different commodity groups. The numerical results also suggest that the movement toward decoupled farm payments embodied in the FAIR Act reduce the share of the cost of commodity checkoffs borne by agriculture.

Finally, we should emphasize that we have examined the gross social costs associated with how research funds are *raised.* The analysis in this chapter does not consider the long-term impacts of agricultural R&D; nor have we explored how the method of financing research might influence what type of research is done and ultimately the net benefits of that research. This latter issue, dealt with in more detail by Ulrich, Furtan, and Schmitz (1986) and by Alston, Pardey, and Smith (chapter 3 of this volume), remains an important area of future research.

NOTES

1. Their deadweight loss calculations are based on the reasonable assumption of a 1 percent tax on agricultural commodities. A 1 percent tax would be sufficient to raise revenues equivalent to current federal expenditures on commodity-based research.

2. This specification also has implications for countries that maintain farm income support programs that are not decoupled.

3. In a general equilibrium setting, "producer surplus" translates into a change in factor income earned by households, which is thereby spent on goods and services. So it is not necessary nor possible to compute "producer surplus" as a separate conceptual category apart from consumer surplus.

4. When legislation is passed or being considered, the Office of Management and Budget and the Congressional Budget Office are required by law to make official calculations of the resultant effects on government outlays and revenues. This process is called "budget scoring." These calculations serve to inform the Executive Branch and Congressional Budget committees about the budgetary impacts of legislation and to indicate if appropriation authority is being exceeded.

5. The real value of exports and imports measured in world prices is held constant in the model, i.e., the real trade balance is fixed. As a trade spillover responding to a set of domestic shocks, the nominal exchange rate adjusts to maintain the real balance of trade by allowing the domestic (nominal) prices of these goods to adjust.

6. Except for the mean job loss in agriculture in a heterogeneous labor market, all standard deviations for the other mean job loss statistics are very much less than unity.

7. Under the FAIR Act regime, the estimated welfare gains of the tax shift are \$234 to \$241 million (table 4.6), about 15 percent of the revenue target of \$1,537 million (table 4.2) to fund commodity-based research.

REFERENCES

Ahmad, E., and N. Stern. 1984. "The Theory of Tax Reform and Indian Indirect Taxes," *Journal of Public Economics* 25:259–98.

Alston, J., and J. Mullen. 1992. "Economic Effects of Research into Traded Goods: The Case of Australian Wool," *Journal of Agricultural Economics* 43:268–78.

Alston, J., and P. Pardey. 1995. *Making Science Pay: Economics of Financing, Organizing and Managing Public-Sector Agricultural R&D.* Washington, DC: American Enterprise Institute Press.

Alston, J., G. Norton, and P. Pardey. 1995. *Science under Scarcity: Principles and Practice for Agricultural Research Evaluation and Priority Setting.* Ithaca, NY: Cornell University Press.

Ballard, C.L., and D. Fullerton. 1992. "Distortionary Taxes and the Provision of Public Goods,"*Journal of Economic Perspectives* 6:117–31.

Ballard, C.L., J.B. Shoven, and J. Whalley. 1985. "General Equilibrium Computations of Marginal Welfare Costs of Taxes in the United States," *American Economic Review* 75:128–38.

Beattie, B.R., and R. Innes. 1997. "Federal Funding of Agriculture Research, Education, and Extension in the Land Grant Universities: An Economic Perspective," *Choices* 12:8–12.

Blackburn, M., and D. Neumark. 1992. "Unobserved Ability, Efficiency Wages, and Interindustry Wage Differentials," *Quarterly Journal of Economics* 107:1421–36.

Bovenberg, A., and L. Goulder. 1996. "Optimal Environmental Taxation in the Presence of Other Taxes: General Equilibrium Analysis," *American Economic Review* 86:985–1000.

Browning, E. 1994. "The Non-Tax Wedge," *Journal of Public Economics* 53:419–33.

Carraro, C., M. Galeotti, and M. Gallo. 1996. "Environmental Taxation and Unemployment: Some Evidence on the 'Double Dividend Hypothesis' in Europe," *Journal of Public Economics* 62:141–81.

Dalrymple, D. 1990. "The Excess Burden of Taxation and Public Agricultural Research." In R. Echeverria, ed. *Methods for Diagnosing Research System Constraints and Assessing the Impact of Agricultural Research, Volume 2, Assessing the Impact of Agricultural Research.* The Hague, Netherlands: International Service for National Agricultural Research.

Dickens, W.T., and L.F. Katz. 1987. "Inter-Industry Wage Differences and Theories of Wage Determination," NBER Working Paper No. 2271. Chicago: National Bureau of Economic Research.

Feldstein, M.S. 1976. "On the Theory of Tax Reform," *Journal of Public Economics* 7:77–104.

Fox, G. 1985. "Is the United States Really Underinvesting Agricultural Research?" *American Journal of Agricultural Economics* 67:806–12.

Fuglie, K.O., N. Ballenger, K. Day, C. Klotz, M. Ollinger, J. Reilly, U. Vasavada, and J. Yee. 1996. *Agricultural Research and Development: Public and Private Investments under Alternative Markets and Institutions,* Agricultural Economic Report Number 735. Washington, DC: Economic Research Service, U.S. Department of Agriculture.

Fullerton, D. 1991. "Reconciling Recent Estimates of the Marginal Welfare Cost of

Taxation," *American Economic Review* 81:302–08.

Gardner, B.L. 1983. "Efficient Redistribution through Commodity Markets," *American Journal of Agricultural Economics* 65:225–34.

Gera, S., and G. Grenier. 1994. "Interindustry Wage Differentials and Efficiency Wages: Some Canadian Evidence," *Canadian Journal of Economics* 27:81–100.

Hanson, K., S. Robinson, and S. Vogel. 1996. "Economywide Effects of the Export Enhancement Program under Alternative Policy Regimes." Food and Rural Economics Division, Economic Research Service, U.S. Department of Agriculture. Unpublished mimeo, December.

Hanson, K., S. Vogel, and S. Robinson. 1995. *Sectoral and Economywide Impacts of Eliminating the Export Enhancement Program.* ERS Staff Paper No. AGES-9529. Economic Research Service, U.S. Department of Agriculture.

Huffman, W.E., and R.E. Just. 1994. "Funding, Structure, and Management of Public Agricultural Research in the United States," *American Journal of Agricultural Economics* 76:744–759.

Just, R.E., D. Hueth, and A. Schmitz. 1982. *Applied Welfare Economics and Public Policy.* Englewood Cliffs, NJ: Prentice-Hall.

Katz, L. 1986. "Efficiency Wage Theories: A Partial Evaluation." In S. Fischer, ed. *NBER Macroeconomics Annual 1986.* Cambridge, MA: MIT Press.

Katz, L., and L. H. Summers. 1989. "Industry Rents: Evidence and Implications." In *Brookings Papers: Microeconomics 1989,* pp. 209–75.

Kilkenny, M., and S. Robinson. 1990. "Computable General Equilibrium Analysis of Agricultural Liberalization: Factor Mobility and Macro Closure," *Journal of Policy Modeling* 12:527–56.

Montgomery, J. 1991. "Equilibrium Wage Dispersion and Interindustry Wage Differentials," *Quarterly Journal of Economics* 106:163–79.

Morgenstern, R.D. 1996. "Environmental Taxes: Is There a Double Dividend?" *Environment* 38:16–20, 32–34.

Oates, W.E.. 1995. "Green Taxes: Can We Protect the Environment and Improve the Tax System at the Same Time?" *Southern Economic Journal* 61:915–22.

Parry, I.W.H. 1995. "Pollution Taxes and Revenue Recycling," *Journal of Environmental Economics and Management* 29:S64–S77.

——— 1998. "Carbon Abatement in a Second-Best World." In G. Frisvold and B. Kuhn, eds. *Global Environmental Change and Agriculture: Assessing the Impacts.* Aldershot, UK: Edward Elgar Publishing.

Pearce, D.W. 1991. "The Role of Carbon Taxes in Adjusting to Global Warming," *Economic Journal* 101:938–48.

Phillips, M.J. 1997. " Land Grant Colleges of the Future: Recommendations of the National Research Council," *Choices* 12:4–7.

Repetto, R, R. Dower, R. Jenkins, and J. Geohagen. 1992. *Green Fees: How a Tax Shift Can Work for the Environment and the Economy.* Washington, DC: World Resources Institute.

Robinson, S., M. Kilkenny, and K. Hanson. 1990. *The USDA/ERS Computable General Equilibrium (CGE) Model of the United States.* Staff Report No. AGES 9049, Economic Research Service, U.S. Department of Agriculture.

Schob, R. 1996. "Evaluating Tax Reforms in the Presence of Externalities," *Oxford Economic Papers* 48:537–55.

Shoven, J.B., and J. Whalley. 1992. *Applying General Equilibrium.* New York: Cambridge University Press.

Sumner, D. A. 1995. "Export Subsidies, Deficiency Payments and Set-Asides: The EEP in Context." Draft Conference Paper, May 1995. In *Agricultural Trade Policy: Letting Markets Work.* Washington: AEI Press.

Thaler, R.H. 1989. "Anomalies: Interindustry Wage Differentials," *Journal of Economic Perspectives* 3:181–93.

Ulrich, A., W.H. Furtan, and A. Schmitz. 1986. "Public and Private Returns from Joint Research: An Example from Agriculture?" *Quarterly Journal of Economics* 101:101–29.

U.S. Congress Office of Technology Assessment (OTA). 1995. *Challenges for U.S. Agricultural Research Policy,* OTA-ENV-639. Washington, DC: U.S. Government Printing Office.

U.S. Department of Agriculture (USDA). 1993. *Inventory of Agricultural Research, Fiscal Year 1993.* Washington, DC: Cooperative State Research, Education and Extension Service, USDA.

Varian, H.R. 1984. *Microeconomic Analysis,* second edition. New York: W.W. Norton & Co.

Welsch, H. 1996. "Recycling of Carbon/Energy Taxes and the Labor Market: A General Equilibrium Analysis for the European Community," *Environmental and Resource Economics* 8:141–55.

II. Public and Private Investments in Plant and Animal Research

5

A NATIONAL STRATEGY FOR PLANT BREEDING IN THE UNITED STATES

KENNETH J. FREY

PLANT BREEDING in the United States has produced crop cultivars that, when coupled with appropriate management systems, have given spectacular advances in agricultural productivity. Corn yields, for example, have quadrupled from 30 to 120 bushels per acre during the past 60 years, and more than half this gain is attributable to improved hybrids. Yield plateaus have not yet occurred, but the rate of productivity gains from new cultivars has slowed. While first-generation "Green Revolution" wheat and rice cultivars gave double the yields of traditional cultivars, second- to fourth-generation cultivars seldom provide more than a 10 percent yield increase per cultivar generation.

Future gains in genetic productivity potential, which must be in harmony with a sustainable environment, will require more research input per unit of productivity gained. The cost to Pioneer Hi-Bred International, Inc., of developing a new corn hybrid, for example, increased from $4 million in 1965 to $48 million in 1990 (in current dollars) (Oscar Smith, personal communication, 1996). Given the drop in yield increases and the rising costs of developing new cultivars, plant breeding faces a significant long-term challenge across all crops: the development of cultivars that will ensure sustainable agricultural production, meet the demands of industry, and underpin the ever-increasing need for food. Meeting these demands will require nothing short of a national strategy for plant breeding that includes both the public and private sectors, employs cutting-edge science and technology, and makes use of genetic resources from around the world.

The National Plant Breeding Study was undertaken to work toward such a strategy (Frey 1996, 1997, 1998). The study had three goals:

- To develop a comprehensive database of the human and financial inputs currently devoted to plant breeding;
- From this database and other information, to assess the adequacy of various plant breeding activities; and
- To develop national plans to assure that minor crop breeding and genepool

enrichment for all crops meet the needs of society for food, fiber, and industrial and export demands two to four decades in the future.

The study met its first goal with a comprehensive survey of the human and financial resources devoted to plant breeding in 1994 in the United States (Frey 1996). Following publication of the survey results, researchers turned to the goals of developing national plans for a genepool enrichment program and of promoting breeding programs for minor crops. The first step in developing these plans was a workshop (held in October 1996) that brought together public- and private-sector research administrators, bench scientists, and professionals in plant breeding, biotechnology, agronomy, economics, sociology, and law. The work of the participants in this workshop has resulted in (1) a *National Plan for Promoting Breeding Programs for Minor Crops in the United States* (Frey 1997) and (2) a *National Plan for Genepool Enrichment of U.S. Crops* (Frey 1998).

The proposed genepool enrichment program, which is discussed here, is designed to assure a continuous flow of cultivars for U.S. crops that will enhance environmental quality, provide an affordable, high-quality supply of diverse crop products, and offer ready sources of raw materials for domestic and export industries. The proposed minor crop breeding program aims to sustain and expand breeding projects on these crops in order to meet the need for the increased research and development (R&D) that will keep these crops competitive.

This paper seeks to demonstrate the need for these programs and offers data on U.S. public and private plant breeding efforts to support its arguments. The first section discusses changes in plant breeding in the United States in the last three decades. Section two presents supporting data from the survey of the human and financial resources devoted to plant breeding. The third section focuses on the need for crop genepool enrichment, and the final section deals with minor crops. A brief summary presents some goals for the future.

THE CHANGING SCIENCE AND STRUCTURE OF PLANT BREEDING IN THE UNITED STATES

The context in which plant breeding takes place in the United States has changed dramatically in recent years, for a number of reasons. The changes suggest that plant breeding efforts must be increased in order to preserve the economic and agricultural well-being of the country. To this end, plant breeders will need strong public- and private-sector support and cooperation to capitalize on the complementary research and development abilities of both sectors. The discussion that follows highlights the major changes that have taken place in the plant breeding industry—changes that will require a corresponding response from policy makers and public and private breeders if the accompanying challenges are to be met.

First, *the structure and availability of funding for public-sector plant breeding activities have changed.* Collecting and conserving plant germplasm are national priorities for the Agricultural Research Service (ARS) of the U.S. Department of Agriculture (USDA). However, these activities are grossly underfunded, resulting in a loss of accessions in U.S. crop germplasm collections, incomplete conserved crop germplasm collections, and too little germplasm evaluation. Funding for genepool enrichment and plant breeding is nearly nonexistent at the national level, where, for efficiency's sake, germplasm conservation and genepool enrichment must be supported. Funding for state agricultural experiment stations (SAESs) has not kept pace with the inflation of agricultural R&D costs over the past 25 years, so state-level funds for plant breeding are also lacking. The private sector cannot be expected to make up these deficiencies. Private-sector breeding concentrates on crops for which breeders can produce hybrid cultivars that are easily commercialized, while the public sector concentrates on crops for which the predominant cultivar types are pure line and clonal.

Second, *there has been increased emphasis on intellectual property right (IPR) protection for plant genes and genotypes.* Strengthened IPRs for plant genes and genotypes bring more private-sector investment into plant breeding—for example, in the transformation of maize and cotton with the *Bacillus thuringiensis* (*Bt*) gene for resistance to Lepidopterous insects. But IPR protection has both positive and negative repercussions for plant breeding. One positive repercussion that has already been mentioned is private-sector input into plant breeding. However, IPR protection can also limit public-sector access to material such as protected genes and sequences. This dilemma needs to be resolved.

Third, *biotechnologies for use in plant breeding are changing the face of cultivar development.* Transformation and plant regeneration technologies for transferring a gene from almost any organism to a crop plant have opened up a new avenue for plant breeders. As these technologies are applied, they make previously inaccessible genes available for cultivar development. These genes will empower plant breeders to make even more effective use of genetic diversity; that is, the genes and knowledge from biotechnologies and enriched genepools will be synergistic to cultivar development, a concept that is rarely discussed.

Fourth, *molecular biology research is providing new knowledge about crop genome organization that could have an immense impact on plant breeding.* For example, many genes are highly conserved across species of certain plant families, such as grass (Moore et al. 1995). Once cloned from one species in a plant family, such genes could (with some molecular fine-tuning) be used to improve other crops in that family. Further, a gene from a very exotic source (such as the *Bt* gene) that is adapted to function in one species of a family could easily be adapted to function in other species of the same family. And the old dilemma of determining which germplasm bank accessions carry genes useful for improving quantitative traits could be solved with molecular markers (Tanksley and McCouch 1997).[1]

Fifth, *concerns about the environmental impact of agrochemical inputs are growing.* The demands on plant breeding to develop crop cultivars that produce high-quality products without agrochemicals are huge. The track record of plant breeding in producing cultivars with pest resistance and tolerance to aluminum toxicity, iron chlorosis, and drought shows the potential for alleviating environmental stresses via breeding. However, a dramatic increase in germplasm evaluation and prospecting will be required if plant breeding is to meet the goal of reducing agrochemical use without compromising production demands.

Sixth, *there has been an explosion in information-sharing technologies.* Research findings and information can be shared easily and quickly with electronic data transfer and correspondence. Knowledge and information about genetics and germplasm can thus be more widely disseminated, but only if evaluation data for germplasm are collected and transmitted. There is considerable concern that private-sector plant breeding, driven by profit motives, may discourage or eliminate the sharing of both information and germplasm.

Seventh, *the roles of public institutions and private industry in plant breeding have changed.* In fact, these roles continue to change daily in plant breeding R&D. The private sector devotes 80 percent of its human inputs (scientist-years, or SYs) for plant breeding to cultivar development. But SAESs and the ARS/USDA devote only 41 percent and 12 percent, respectively, of their plant breeding SYs to this activity (Frey 1996). Genepool enrichment has historically been a public-sector activity, but as the *Bt* cases demonstrate, by using transformation technology, the private sector is now involved in specialized genepool enrichment. The private sector's goals, however, tend to be short term and motivated by profit, so that its contribution to genepool enrichment is likely to involve only individual genes that qualify for IPR protection. General genepool enrichment for multigenic traits is a high-risk, long-term investment that must be undertaken by the public sector. New vehicles for public-private cooperation must be developed to ensure that this research is efficiently and successfully carried out.

Finally, *attitudes and policies about access to genetic resources have changed on the global as well as the national level.* The ARS/USDA, which administers the National Plant Germplasm System, has large and small germplasm collections for some major and minor U.S. crops. But the United States is a country with a germplasm deficit for nearly all domestically grown crops. Political instability and changes in the policies of developing countries affecting the accessibility of germplasm may make it difficult to expand or initiate U.S. germplasm collections. Thus, the United States must make more efficient use of the crop germplasm collections it possesses. Otherwise, agriculture, industry, and consumers will remain burdened with a genetic deficit, despite large national collections.

HUMAN AND FINANCIAL RESOURCES DEVOTED TO PLANT BREEDING IN THE UNITED STATES

The 1996 survey highlights the trends in U.S. plant breeding that support the need for a national strategy to improve the crop genepools and increase crop diversity. The data show clear divisions between public- and private-sector activities in plant breeding. Public-sector breeding has been on a slow decline, while private breeding has increased. Most breeding efforts, and especially those in the private sector, concentrate on cultivar development, leaving relatively few resources for important long-term research on genepool enhancement and breeding methods. Private breeding efforts also concentrate on the most economically important crops, especially those grown from hybrid seed (see chapter 6 for further analysis of the data and an assessment of the roles of the public and private sectors in seed technology generation in the United States).

How the Survey Was Set Up

The human and financial resources devoted to plant breeding in the United States in 1994 were evaluated with a questionnaire that was distributed to both the public and private sectors (Frey 1996). For the public sector, questionnaires were sent to the 50 SAESs, two universities not associated with SAESs, six experiment stations in U.S. territories, and 17 1890s colleges. The questionnaire was also sent to national program staff of the ARS/USDA and to 28 Plant Materials Centers (PMCs/USDA) that select species and accessions for land conservation. Replies were obtained from all SAESs, the two universities, five of the territorial experiment stations, six 1890s colleges, 26 PMCs, and the ARS.

For the private sector, questionnaires were sent to 690 individuals and companies. Of these, 106 provided no replies. Of the 106, 69 could not be contacted by either mail or phone, so it was assumed that they were no longer in the plant breeding business. Two refused to supply data, 22 had been bought by other companies on the list of questionnaire recipients, and 13 did not supply data in spite of verbal promises to do so. Of the 584 individuals and companies that responded to the questionnaire, 329 reported SY data, and 255 reported having no in-house plant breeding R&D.

Results of the Survey

The total number of SYs devoted to plant breeding by the ARS/USDA, SAESs, and the private sector in the United States in 1994 was 2,205 (table 5.1). The USDA's Plant Material Centers provided another 36 SYs (table 5.2), for a grand total of 2,241. Some 67 percent of these were in private industry; the other 33 percent were divided among SAESs (23 percent), the ARS/USDA (8 percent), and PMCs/USDA (2 percent). This discussion focuses only on private industry, SAESs, and the ARS/USDA.

Table 5.1. Scientist-years devoted to plant breeding arranged by employer and crop categories

	Number of SYs employed by			
Crop category	SAESs	ARS/USDA	Private industry	*Total*
Cereal	155	34	703	892
Fiber	20	13	103	136
Forage	38	33	51	122
Fruit-vegetable	38	8	167	213
Grain-legume	67	14	126	207
Lawn and turf	15	—	41	56
Leafy, bulbous, and stem vegetables	16	2	77	95
Medicinal, spice, and special crops	6	4	5	15
Mushroom	1	—	2	3
Oilseed	24	6	74	104
Ornamental	18	5	64	87
Root and tuber	45	12	24	81
Stimulant	13	2	5	20
Sugar	4	15	25	44
Temperate fruit and nut	50	23	32	105
Tropical fruit and nut	10	6	—	16
Miscellaneous	9	—	—	9
Totals	529	177	1,499	2,205

Table 5.2. Number of plant breeding scientist-years employed in the Plant Materials Centers/USDA

Crop group	Number of species	Number of SYs
Grasses	54	24.5
Legumes	21	3.2
Shrubs	20	3.6
Trees	25	3.1
Miscellaneous	19	1.8
Totals	139	36.2

The survey highlighted several important trends in plant breeding and research. First, it indicated that the percentage of SYs the private sector devotes to plant breeding has increased, while public-sector SYs have decreased slightly—but not as much as had been predicted by Brooks and Vest (1985), Collins and Phillips (1991), and James (1990). Second, it showed that while cultivar development received the most attention, germplasm enhancement accounted for one-fifth of all SYs. And third, it demonstrated that private-sector financing supports more than two-thirds of all SYs in plant breeding, with the public sector accounting for the other third.

Employment and Financial Resources

Several previous surveys had predicted that public-sector employment for plant breeding would decrease moderately to precipitously in the last half of the twentieth century (Brooks and Vest 1985; Collins and Phillips 1991; James 1990). (They made no predictions about employment trends in the private sector.) This survey bore these predictions out, but to only a small degree. Respondents to the questionnaire were asked to estimate the percentage change that had occurred in their plant breeding R&D activities over the five-year period 1990–94. These data, when converted to SYs, showed that SAESs experienced a net loss of 2.5 plant breeding SYs, or 0.5 percent, per year. Of the 329 private companies that reported on plant breeding R&D, 137 (42 percent) reported growth during 1990–94, and 20 (6 percent) reported decreases. The net change in the private sector during 1990–94 was thus an increase of 32 plant breeding SYs per year, or a gain of 2.4 percent per year.

Survey respondents allocated the SYs they reported by activity—either plant breeding research, germplasm enhancement, or cultivar development. As expected, the private sector placed a much greater emphasis on cultivar development than SAESs and the ARS/USDA (table 5.3). Of plant breeding SYs in the private sector, 80 percent were devoted to this activity. SAESs devoted 41 percent of plant breeding SYs to this activity, but the ARS/USDA only 12 percent.

Table 5.3. Number of plant breeding scientist-years devoted to plant breeding research (PBR), germplasm enhancement (GE), and cultivar development (CD) arranged by employer (percentages in parentheses)

Category	SAESs		ARS/USDA		Private industry		Activity totals
CD	217	(41%)	22	(12%)	1,191	(80%)	1,430
GE	153	(29%)	85	(48%)	165	(11%)	403
PBR	159	(30%)	70	(40%)	143	(9%)	372
Totals	529		177		1,499		2,205

Germplasm enhancement is particularly important to cultivar development because it generates the genepools that will be used in one to three decades to produce superior cultivars. More than 400 SYs, or 19 percent of the total plant breeding SYs, were devoted to this activity. Eleven percent of the SYs in private industry were devoted to germplasm enhancement.

Total annual estimated expenditures for plant breeding in the United States in 1994 were $551 million, with 61 percent attributable to the private sector and 39 percent to the public sector. Overall, the 329 companies that engaged in plant breeding R&D spent an estimated $338 million. The cost per plant breeding SY in the private sector varied according to the number of SYs a company used (table 5.4). The cost per SY was $290,000 for companies with 25 or more SYs and $148,000 for companies with fewer than 2.9. The public sector spent an estimated $213 million, with SAESs accounting for nearly three-quarters of the total (table 5.5).

Table 5.4. Cost per scientist-year, number of companies, total scientist-years, and dollar input into plant breeding R&D by private industry arranged according to company SY size

Company class in SYs	No. of companies	Cost per SY	Total no. of SYs in class	Dollar input for class
25.0–up	10	$290,000	466.20	$135,198,000
15.0–24.9	5	$228,000	90.60	$ 20,657,000
10.0–14.9	20	$240,500	235.75	$ 56,697,875
5.0–9.9	43	$186,000	286.70	$ 53,326,200
3.0–4.9	43	$217,000	153.10	$ 33,222,700
0–2.9	208	$148,000	265.95	$ 39,360,600
Totals	329		1,498.30	$338,462,375[a]

[a]If the estimated 36 SYs for companies that did not reply (see Introduction) were included, $7.8 million would be added to the private sector expenditure.

Table 5.5. Cost per scientist-year, total scientist-years, and dollar input into plant breeding R&D by the public sector arranged according to source of public-sector employment

Employment source	Cost per SY	Total no. of SYs	Dollar input for source
ARS/USDA	$300,000	177.00	$ 53,100,000
Plant Materials Center	$125,000	36.20	$ 4,525,000
SAES	$293,500	529.90	$155,525,650
Totals		743.10	$213,150,650

How Scientist-Years Are Distributed across Crops

In general, private industry undertakes plant breeding only on profitable crops—that is, crops from which commercially useful cultivars can be developed. Private firms devote themselves heavily to hybrid cultivar development, an area from which the public sector has removed itself almost completely. Of the 219 crops on which breeding is being done in the United States, the private sector is engaged in R&D on 144 of them. In fact, private industry reported plant breeding on 42 species for which there is no public-sector breeding. These 42 species were spread over nine crop categories, but a large proportion of them are ornamentals (Frey 1996).

Of the 17 crop categories reported in table 5.6, cereal crops account for 40 percent of U.S. plant breeding SYs. Among employers, cereal crops account for 47 percent of plant breeding SYs in the private sector, 29 percent in SAESs, and 19 percent in the ARS/USDA. Other categories with more than 200 SYs are fruit-vegetable and grain-legume crops. Four other crop categories—fiber, forage, oilseed, and temperate fruit and nut crops—have between 100 and 200 plant breeding SYs.

Table 5.6. Crops for which 25 or more breeding scientist-years are employed in the United States

	Number of SYs			
Crop	SAES	ARS/USDA	Private industry	*Total*
Alfalfa	15	12	41	68
Barley	16	2	14	32
Canola	6	1	28	35
Cotton	19	12	103	134
Field corn	27	8	510	545
Onion	4	1	21	26
Pepper	5	1	38	44
Potato	31	10	9	50
Rice	14	6	22	42
Sorghum	12	3	41	56
Soybean	45	10	101	156
Sugar beet	1	11	24	36
Sunflower	1	3	31	35
Sweet corn	5	1	27	33
Tomato	21	4	60	85
Wheat	65	12	54	131

Crops with the most SYs in plant breeding. Overall, 545 SYs are devoted to field corn, and 94 percent of these are in the private sector (table 5.6). Of the 35 field corn SYs in the public sector, only 3 are devoted to cultivar development. In contrast, 418, or 82 percent, of field corn SYs in the private sector are devoted to this activity. Other crops with more than 100 SYs are soybean, cotton, and wheat. The distribution of SYs for soybean and cotton differ from the distribution of SYs for wheat, however. For soybean, 65 percent of SYs are in private industry; for cotton, 77 percent; for wheat, only 41 percent. Commercial acreages of all these crops are dominated by pure-line cultivars, so the contrast in public and private employment in soybean, cotton, and wheat is a marked variance.

Crops with between 50 and 100 plant breeding SYs are potato, sorghum, alfalfa, and tomato. These crops also present a contrast between public and private SY inputs. For sorghum, alfalfa, and tomato, private companies account for 60 to 73 percent of the plant breeding SYs, but for potato, only 18 percent. These four crops also differ in terms of the kinds of cultivars used on commercial acreages in the United States. The commercial acreage of sorghum is sown entirely to hybrid cultivars, that of alfalfa to synthetic cultivars, that of tomato to around half hybrid and half pure-line cultivars, and that of potato to clonal cultivars.

Table 5.7 shows crops to which private industry allocates 20 or more plant breeding SYs. Some 91 companies devote 510 SYs to field corn breeding. Private industry also devotes more than 100 plant breeding SYs to cotton and to soybeans. Overall, the private sector accounts for 20 or more breeding SYs on just 16 different crops, for a total of 1,142 SYs, or 76 percent of the total plant breeding SYs (table 5.7). The number of SYs private industry devotes to hy-

Table 5.7. Crops for which 20 or more breeding scientist-years are employed in private industry in the United States

Crop	Cultivar type	No. of SYs	No. of companies
Field corn	H	510	91
Cotton	PL	103	35
Soybean	PL	101	38
Tomato	½H / ½PL	60	24
Wheat	PL	54	27
Alfalfa	S	41	12
Sorghum	H	41	19
Pepper	½H / ½PL	38	27
Sunflower	H	31	14
Canola	PL	28	4
Sweet corn	H	27	12
Sugar beet	H	24	7
Rice	PL	22	8
Onion	½H / ½PL	21	13
Muskmelon	H	21	15
Lettuce	PL	20	18

H = hybrid; PL = pure line; S = synthetic.

brid cultivar development in these 16 crops shows that hybrid breeding is a significant factor in the success of private-sector plant breeding.

Plant breeding by private companies began with hybrid corn, primarily because the seed of hybrid cultivars must be purchased each year, allowing private companies to maintain monopolies on their hybrids by not releasing parent stocks. Of the 16 crops listed in table 5.7, hybrid cultivars are sown on the entire commercial acreage of field corn, sorghum, sunflower, sweet corn, sugar beet, and muskmelon, which collectively account for 654 private industry SYs. The commercial acreages of tomato, onion, and pepper are sown about half to hybrid and half to pure-line cultivars. These crops account for 119 breeding SYs in private companies. Commercial acreages of cotton, soybeans, wheat, canola, rice, and lettuce are sown with pure-line cultivars, accounting for 328 SYs of private-sector plant breeding.

Crops each with 15 or more plant breeding SYs in public-sector employment are presented in table 5.8. Pure line is the cultivar type for 7 of 11 crops, which together account for a total of 233 SYs . Few of the 35 public-sector SYs for field corn are devoted to developing hybrids. In fact, 32 SYs are devoted to plant breeding research and germplasm enhancement for corn. Potatoes are unique in public-sector agricultural R&D, in that clonal cultivars are sown on commercial acreages. Of the 50 breeding SYs given to the potato crop, 41 are in the public sector, and 18 of these are devoted to cultivar development.

Table 5.8. Crops for which 15 or more breeding scientist-years are employed in SAESs and ARS/USDA

Crop	Cultivar type	SAESs	ARS/USDA	Total	No. of projects
Wheat	PL	65	12	77	42
Soybean	PL	45	10	55	32
Potato	C	31	10	41	28
Field corn	H	27	8	35	32
Cotton	PL	19	12	31	20
Alfalfa	S	15	12	27	25
Tomato	½H / ½PL	21	4	25	28
Rice	PL	14	6	20	15
Barley	PL	16	2	18	24
Peanut	PL	14	3	17	11
Bean	PL	13	2	15	14

H = hybrid; PL = pure line; S = synthetic; C = clone.

Scientist-years for minor crops. Some 250–300 different crops are grown and used in the United States for food and other purposes. Historically, the public sector conducted the research necessary to develop superior cultivars and appropriate production practices. Just a few decades ago, research on many of today's minor crops was conducted only by the ARS/USDA, because their production was so diffuse that research could be justified only on a regional or national basis. For example, when soybean was a minor crop (from 1920 until 1940), soybean breeding research was conducted almost exclusively by the ARS/USDA.

It is possible that the private sector may become more active in breeding some minor crops (Allen Stevens, private communication, 1996). Until recently, private-sector breeding of horticultural crops had been fragmented among many companies, each with a small share of the sales of seed or propagation materials. This fragmentation meant that each plant breeding SY in these companies was involved in cultivar development on several to many crops. The current trend toward company amalgamation means that the market will be split among a few companies, with each breeder responsible for only one or two major crops. In this scenario, breeders may devote 10–15 percent of R&D efforts to hobby or minor crops.

STRENGTHENING GENEPOOL ENRICHMENT FOR U.S. CROPS

This section discusses the proposed national program for enriching the U.S. crop genepools.[2] Crop genepool enrichment in the U.S. has generally been sporadic and has lacked systematic planning at the national level. But genepool enrichment is essential, because the resulting plant populations are the raw materials from which cultivars will be developed in the next 30 years. In the United States, food security, food safety and nutrition, sustainable agricultural production, and competitiveness in world trade depend on the continual evolution of improved cultivars. U.S. cultivars are also important to maintaining the food supply for much of the world. Plant breeding is the key to assuring that these dependencies are met with genetic solutions, and the U.S. plant breeding community needs new and useful gene combinations in order to continue its work.

The Current Status of Crop Genepool Enrichment

Genepool enrichment has been quite extensive for tomato, potato, wheat, sorghum, maize, and cotton. Introgressed genes have been released in new cultivars for these species. Genepool enrichment has been less extensive for oats and soybeans. The United States' most systematic evaluation and genepool enrichment program is the Latin American Maize Project/Germplasm Enhancement of Maize (LAMP/GEM) project (see chapter 9). Led by the ARS/USDA, this international program is funded in part by the private plant breeding sector and encompasses the transfer of genes for both qualitative and quantitative traits.[3] The donor maize lines are all *Zea mays* accessions, so that genepool enrichment is carried out in the traditional manner.

The development of techniques for transformation in plants has created a flurry of activity as researchers rush to transfer genes into plants from unrelated sources. The most highly publicized case of transforming plants involves the *Bt* gene, which provides resistance to insects.[4] The ability to transform crop plants with genes from almost any living organism revolutionizes genepool enrichment possibilities and opens up untold opportunities for tailoring crop products to meet the demands of domestic and foreign markets. It provides the potential for producing exotic and unusual compounds and products in crop plants that are better adapted and more productive than the source organism of the gene(s). Importantly, it has generated strong private-sector involvement in genepool enrichment, although this involvement is so far narrowly focused. Transformation technology also permits the use of genes from very remote sources to customize plant products. Technologies that are now emerging will provide knowledge about the utility of specific areas of the genome. This knowledge, along with other technologies, may enable researchers to combine specific genetic traits efficiently into new cultivars.

The ethical, legal, economic, and social implications of working with plant genetic resources need to be addressed, however. Issues in this category that need to be considered include the structure of private industry, domestic and exporting opportunities, changes in consumers' food preferences, and genetic resources. For instance, private industry is constantly restructuring itself to obtain access to biotechnologies. These technologies are an addendum to the traditional plant breeding programs. The private sector must marry biotechnology products to useful, broad-based, and adapted genepools in order to utilize all possible resources to meet huge future demands for agricultural products. The vertical integration of biotechnology, plant breeding, and seed companies will result in a few very large firms controlling most of the germplasm that is enriched through biotechnology. How to combine the plant breeding activities of the private and public sectors is therefore an issue of paramount concern.

Emphasis also needs to be placed on developing cultivars with special traits to meet new food preferences and to support sustainable production practices. Breeding crop cultivars for high productivity in sustainable crop production systems will require genetic resources that have not yet been tapped. Until recently, genetic resources were considered natural resources, like air and water, that were held in the public domain for the good (and use) of all. This attitude led to the free movement of genetic resources globally and gave plant breeders access to state-of-the-art breeding materials. However, the Convention on Biodiversity gives countries sovereign rights to genetic resources within their borders. Under the convention, there will certainly be a more formal system of exchanging germplasm. As the system develops, the United States will be in a better bartering position if it has well-evaluated germplasm collections to exchange.

Plan of Action for a National Genepool Enrichment Program

A national strategy for genepool enrichment requires a well-defined program to mobilize the necessary human, physical, and financial resources. The program would identify, access, discover, transfer, and make available genes essential to plant breeding efforts. The steps involved in implementing such a program follow.

Organize a group to assess needs and promote and coordinate research activities. This group—the Crop Genepool Enrichment Coalition (CGEC)—would be organized and coordinated by the Cooperative States Research, Education, and Extension Service of the USDA (CSREES/USDA), with help from the Experiment Station Committee on Organization and Policy (ESCOP) of the SAESs. Because SAESs would carry out most of the work, the CSREES/USDA, which provides SAESs with partial funding and acts as a liaison among them, would be the natural choice to act as coordinator for the CGEC. The coalition itself would be composed of persons from the CSREES/USDA, ARS/USDA, SAESs, and private industries who have expertise in organizing, funding, and sustaining genepool enrichment programs for U.S. crops.

Prioritize crops according to need. The CGEC would use databases developed by crop germplasm committees[5] to prioritize crops according to their requirements for genepool enrichment.

Fund genepool enrichment for U.S. crops. The national plan would be supported by a federally funded initiative of $50 million annually. These funds would be used as follows:

- $5 million to the National Plant Germplasm System for making needed plant genetic resources available and for evaluating and distributing them.
- $25 million for a competitive grants program for five-year projects, to be administered by the National Research Initiative. This program would focus on the use of basic science in developing genepool enrichment practices and programs and would emphasize multidisciplinary research, international collaboration, and training.
- $20 million for grants to be administered by the CSREES/USDA. These grants, which would be awarded competitively, would be based on the priorities set by the CGEC. These awards would also emphasize multidisciplinary research, international collaboration, and training. This portion of the program would mandate that collectively, over all projects, federal support is to be matched by equal support from SAESs and from the private sector, with a goal of $60 million annually.

Build a consortium to support the CGEC's agenda. The CGEC would involve public institutions and private companies, nongovernmental organizations

(NGOs), farmers, and end-user groups in educational programs and consortia to build broad-based public support for the program. The CGEC would create a "crop genepool enrichment committee" made up of representatives from these groups to help develop educational programs, inform governmental agencies, and create public awareness programs. The coalition would provide leadership, develop educational materials, and organizational expertise.

Develop models for distributing enriched genepools. In principle, crop genepools enriched in projects supported entirely or largely by public funds should be distributed freely to collaborative breeders in the United States. However, some genepool enrichment is carried out by private industry and individual companies, and these cases are subject to strict IPR protection. A major task for the CGEC would be to provide "consensus models" for the release of cultivars developed under the program.

Develop human resources. The CGEC would develop short courses to train plant scientists in crop genepool enrichment. In particular, the coalition would develop workshops on new technologies in plant breeding.

Evaluate the impact and performance of the program. At five-year intervals, a task force made up of scientists and administrators who are not members of the CGEC would evaluate the program. This task force would determine whether the CGEC is meeting its goals, assess its administrative effectiveness, evaluate the general progress of the national program, and judge the success of the educational, human resource, and funding programs.

Projects for individual crops would be chosen on the basis of competitive proposals and would run for an initial period of five years. One year before the initial grant period ended, the CGEC would assign a committee to evaluate the progress of each project. This evaluation would be used to determine whether the project should be continued, modified, or moved to a new location.

PROMOTING BREEDING PROGRAMS FOR MINOR CROPS[6]

Minor crops differ from major crops in one or more of five important ways. First, they are cultivated on limited acreage, either in parcels or concentrated in restricted areas. Second, they are produced as strains of major crops for niche markets. Third, they provide relatively small gross revenues. Fourth, both the public and private sectors invest very little (or nothing at all) in research for these crops. Thus, lastly, plant breeding activities for minor crops are diffuse or nonexistent.

Minor crops are important, however, if current and future demands for a stable, safe, and diverse food supply and a sustainable environment are to be met. Specifically, sustained R&D efforts on minor crops can provide

- Enhanced environmental quality (the result of crop rotation and fewer agrochemicals);
- New sources of vitamins, minerals, and micronutrients essential to human health;
- A stable and diverse food supply that meets the changing ethnic and demographic patterns of the U.S. population;
- A more affordable supply of high-quality foods;
- New sources of income for farmers, especially those working on small holdings;
- Improved ability to meet niche-market demands for special strains of major crops;
- An expanded and dependable supply of agricultural products for industrial stocks;
- Enhanced economic opportunities from new crops; and
- An improved supply of aesthetically important nonfood and ornamental crops.

Resources for a National Minor Crop Breeding Program

The United States is fortunate in the real and potential resources it has available for breeding R&D on minor crops. It has genetic resources in the form of germplasm collections (maintained by the National Plant Germplasm System of ARS/USDA) for many minor crops. These collections tend to be small, and some working collections are held by public breeders. Genetic advances in major crops made via the application of biotechnologies may be adaptable to minor crops. In the short term, genome mapping may be more important than gene transformation for minor crops, but the potential for both is there.

Minor crop R&D needs more funding, however. If provided with additional funding, some breeders currently researching major crops could add compatible minor crops to their programs. Where these funds should come from is a point of contention, however. The private sector is investing as much as can be justified in minor crop breeding, given the limited profitability of this kind of R&D. The real inadequacy may therefore be in the public sector. Increased funding would need to come from a number of sources:

A redirection of existing human and financial resources among crops. As private industry takes over more responsibility for the breeding of major crops, SAES and USDA human and financial resources devoted to breeding these crops can be redirected to breeding minor crops.

Checkoff programs. These self-imposed taxes are generally applied at the producer level. Checkoff funds, which are usually crop specific, are allocated by a board that considers R&D proposals from research institutions and individual researchers. Checkoff funds are often used to support R&D in the public sector (see chapter 3).

Industry grants. Grants of money provided by commercial companies are quite effective in leveraging other sources of money for specific R&D areas.

Cooperative research and development agreements (CRADAs). CRADAs are joint projects between federal agencies and private companies, usually created to commercialize products from public-sector research (see chapter 8).

Regional and national joint projects between the public and private sectors. Such projects draw on the strengths of both sectors. When designed to develop crop cultivars for the public good rather than for profit, these projects could be particularly useful.

Regional and national projects among SAESs. Cooperative regional and national plant breeding projects are especially appropriate for minor crops. Individual SAESs can justify using only small amounts of funding for minor crop projects, but a critical mass of funding may be available for cooperative projects.

Base allocations of Hatch formula funds from CSREES to SAESs. As SAESs breed fewer and fewer major crops (for example, SAESs allocate only 3.0 SYs to developing field corn hybrids), money formerly used for major crop R&D can be reallocated to minor crop breeding projects.

Active commodity support groups. Many major crops have individual commodity support groups. Similar groups can be organized for families of minor crops (such as crucifers) rather than for individual crops (such as cabbage or cauliflower).

Global programs. The International Agricultural Research Centers (IARCs) of the Consultative Group on International Agricultural Research (CGIAR) have developed a model for global plant breeding that includes R&D for major crops. Genetically variable plant populations are created and selected for major traits at the IARCs, and samples of the resulting populations are distributed to national plant breeding programs for cultivar selection. This model has many advantages, including a structure for using biotechnology, and may be applicable to regional or national SAES projects in the United States.

Multinational companies already conduct plant breeding on a global basis. This global activity means that these companies can gain a competitive advantage by conducting R&D in low-cost areas of the world. IPR protection and biotechnology have given these companies an additional edge. The globalization of plant breeding by multinational companies will intensify as regulations among countries become homogeneous, trade barriers decrease, and IPR enforcement becomes more uniform among countries under the North American Free Trade Agreement and the World Trade Organization.

Goals of a National Minor Crop Breeding Plan

Modern plant breeding needs to provide superior cultivars of major and minor crops to meet society's demands not only for food security but also for medicinal and industrial uses and environmental remediation or enhancement. This plan focuses only on crops identified as minor in agriculture today and would have four major goals.

First, *it would establish an organization to support the breeding of minor crops.* This organization would monitor, organize, and promote activities for the sustained breeding of minor crops; facilitate the use of biotechnology in minor crop breeding programs; and improve information transfer for minor crop research through an electronic database.

Second, *it would improve the economic viability of minor crops.* It would exploit niche markets, add value to minor crops, reduce costs for consumers with improved cultivars, and contribute new plant products to international trade.

Third, *it would promote awareness of the significance of minor crops.* The program would work to gain industry support, develop commodity support groups, ensure that minor crops are included in the Research Title of the Farm Bill, and enhance society's understanding of the role of minor plant breeding in the agricultural research mission.

Finally, *it would locate and cultivate funding sources.* Generating funding would require linking funding sources with breeders of minor crops, fostering commodity organizations, and promoting public-private sector collaboration; conducting genepool enrichment programs for minor crops; and promoting international cooperation on minor crop breeding.

Plan of Action for Promoting Breeding Programs for Minor Crops

Promoting breeding programs for minor crops would require a coordinating group and the resources to identify minor crops that need breeding programs, foster cooperative projects, generate funding, and forge alliances. This action plan sets out the steps involved in the design and development of the program.

A coalition to promote and coordinate research and development activities would be organized and administered by the Experiment Station Committee on Organization and Policy (ESCOP). The group would be called the National Coalition for the Improvement and Use of Minor Crops (MCIC). It would be composed of persons from SAESs, the ARS/USDA, private industry, and the agricultural production and user communities. Its mission would be to organize a plan to maintain and expand breeding projects on minor crops in order to sustain environmental quality, assure an affordable, high-quality supply of diverse crop products for society, and provide ready sources of raw materials for industry.

To identify those minor crops that most need breeding programs, the MCIC would compile a list of minor field, forage, and horticultural crops essential to both the agricultural sector and society. The coalition would then divide these crops into three groups: minor crops with adequate current breeding activity in the United States, minor crops without adequate U.S. breeding activity but with enough foreign breeding activities to meet U.S. needs (perhaps through transnational corporations), and minor crops with inadequate breeding activity anywhere in the world. Minor crops in the first two categories would be reconsidered periodically, but most of the coalition's efforts would be devoted to the third category.

Fostering cooperation on breeding projects for minor crops would require working not only with institutions that have autonomous administrations but also with both public and private organizations. One useful model for multistate or national projects among SAESs is the IARC model. Under this model an IARC provides coordination, and the work is divided between the IARC and the national program. Adapting this model to the United States, the MCIC would be responsible for bringing together SAESs and private companies interested in a minor crop breeding project. One SAES would provide leadership and scientific staff to carry out genepool development and primary selection for major traits. The plant populations derived at this lead institution would be distributed to the other SAESs and participating private companies. Each minor crop proposed for a regional or national breeding program would require a separate consortium of institutions.

This model has several advantages. First, each minor crop breeding program would be led by a scientist. Second, an SAES participating in several projects might be the lead institution for only one of them but would receive populations of other minor crops from other institutions. And private companies that could not afford to construct their own breeding programs for minor crops might be interested in cultivar development for the crops if enhanced populations were available.

Fostering cooperation between the public and private sectors would require a slightly different approach. In general, there has not been an entirely free flow of germplasm between the public and private plant breeding sectors, and the flow that does exist has been inhibited by IPR protection and restrictions on confidential business information. But minor crop breeding represents a specific situation. Since projects to breed these crops aim to develop improved cultivars for both the agricultural sector and consumers, private companies may be willing to contribute resources, knowledge, and infrastructure even if the profit is marginal. A slightly modified IARC model should therefore be adequate for joint public-private projects. A public institution (an SAES or the ARS/USDA) would create genepools based on input from all parties and conduct selection for highly heritable traits for a minor crop. The resulting plant populations would be distributed to one cooperating company (if the crop is grown regionally) or to several companies (if the crop is grown nationally). Working models of this arrangement for horticultural crops are currently in operation at Cornell and Texas A&M universities.

The model for cooperation in minor crop breeding programs is based on the assumption that the cooperative agreements would require little, if any, new research monies beyond those from redirected funds and income from the cultivars developed under the projects. However, the MCIC would need to address the funding issue aggressively, especially in the public sector. Several alternatives exist:

Self-imposed checkoffs or taxes. Under the "Australian model," researchers, growers, and processors/users form commodity associations for several minor crops with similar breeding, cultural, and use characteristics (for instance, root and tuber crops). In the United States, such groups could be a strong source of funding for minor crop breeding projects.

Federal and state funds. The MCIC, which would work closely with ESCOP, SAESs, the ARS/USDA, and private companies, would have the stature to compete with other interest groups and commodity associations for new and reallocated federal and SAES funds.

Alliances with nutrition, health, conservation, and environmental groups. Much agriculture in the United States today is concentrated on several major crops grown with mechanization, large inputs of agrochemicals, and monoculture. While this type of production has succeeded in providing inexpensive and abundant food, the system's sustainability has come into question. A number of groups are now promoting alternative production systems in which minor crops could play an important role. Similarly, society has become more interested in the problem of pesticide contamination and in the nutritional value of food. Groups have been organized to promote pesticide- and additive-free foods and to lobby for truthful food labeling. Minor crops can also play an important role in furthering the agendas of these groups.[7]

The MCIC could promote alliances with activist groups working on nutrition, health, conservation, and environmental issues. Such alliances could help educate society about the role of minor crops in sustainable agriculture, in providing nutritious and healthy food, and in protecting the environment. At the same time, society would learn about the problems of minor crop R&D in the United States.

Biotechnology and IPR protection for minor crops. Because individual minor crops generally are not high revenue earners, breeders working on them often do not have access to the biotechnology applied to major crops. By concentrating on minor crops as a group, the MCIC could serve as an advocate and a clearinghouse for biotechnology for minor crop breeding. The coalition would also serve as a clearinghouse for information about biotechnology, IPRs, and models for interaction to breeders of minor crops.

Miscellaneous activities. The MCIC leadership would set up a committee to monitor miscellaneous funding opportunities for minor crop breeding programs. The committee would also initiate symposia and colloquia about minor crop breeding at meetings of scientific societies and promote educational programs about the role of minor crops in agriculture and research.

SUMMARY

Plant breeding will be called upon to play a major role in providing the immense increases in crop production that will be needed to meet the demands of society and industry over the next half century. And more than ever before, plant breeding will be asked to provide genetic solutions to stress problems in crop production so that agriculture will be more sustainable. Fortunately, biotechnology is providing new tools for plant breeding to accomplish its goals with greater speed and accuracy. And vast new financial and human resources are being invested in plant breeding, primarily by the private sector. So the tools and resources for plant breeding to meet the future demands upon it are at hand, but the national planning and cooperation to make the most efficient use of these tools and resources is deficient. The National Plant Breeding Study has constructed and published national plans for promoting breeding programs for minor crops in the United States and for genepool enrichment of U.S. crops, two neglected areas in the total plant breeding arena. Without adequate attention to improving minor crops, the diversity in U.S. crop production will wither, and without immediate attention to crop genepool enrichment, there will be a deficiency of raw materials needed for crop cultivar development one to three decades in the future. Bringing these national plans to fruition is critical to plant breeding's future success.

NOTES

1. Tanksley and McCouch (1997) propose that quantitative trait loci (QTL), which often are not expressed in a plant phenotype, can be "prospected for" in an accession once they are associated with a molecular marker, without concern about the adaptation of the accession. Desirable QTLs could then be cloned and used in genepools of the same and related species.

2. Crop genepool enrichment is the transfer of useful genes from unadapted lines of the same species; related species and genera; or any plant, animal, or microbe species by sexual or asexual technology (including transformation) into plant populations that are useful for developing new crop cultivars.

3. More than 20 genes for disease and insect resistance have been transferred to cultivars of tomato from wild *Lycopersicon* species. And recently, genes for total fruit solids and salinity tolerance, which are quantitative traits, have been transferred to tomatoes.

4. Cultivars of potato, maize, and cotton with *Bt* genes are being grown on several million acres in the United States. The primary target insects are the European corn borer (maize), several boll worm species (cotton), and the Colorado potato beetle (potatoes).

5. Crop germplasm committees consist of leaders of plant breeding programs at USDA and SAES agricultural research institutes. They meet on a regular basis to share information on their respective breeding programs.

6. For more on minor crops, see Frey (1997).

7. It might be well to place different emphases on horticultural minor crops and on field and forage minor crops in seeking to form alliances. For example, oat and pearl millet, which are minor field crops, and rutabaga and artichoke, which are minor horticultural crops, would have very different clienteles and supporters.

REFERENCES

Brooks, H.J., and G. Vest. 1985. "Public Programs on Genetics and Breeding of Horticultural Crops in the United States." *HortScience* 20:826–30.

Collins, W.W., and R.L. Phillips. 1991. *Plant Breeding Training in Public Institutions in the United States.* Report 591 (Rev.). Washington, DC: National Plant Genetic Resources Board, Office of Under Secretary for Science and Education, U.S. Department of Agriculture.

Frey, K. J. 1996. *National Plant Breeding Study I. Human and Financial Resources Devoted to Plant Breeding Research and Development in the United States in 1994.* Special Report 98. Ames: Iowa Agriculture and Home Economics Experiment Station, Iowa State University.

———. 1997. *National Plant Breeding Study II. National Plan for Promoting Breeding Programs for Minor Crops in the United States.* Special Report 100. Ames: Agriculture and Home Economics Experiment Station, Iowa State University.

———.1998. *National Plant Breeding Study III. National Plan for Genepool Enrichment of U.S. Crops.* Special Report 101. Ames: Iowa Agriculture and Home Economics Experiment Station, Iowa State University.

James, N.I. 1990. "A Survey of Public Plant Breeding Programs in the United States, 1989." *Diversity* 6:32–33.

Moore, G., K.M. Devos, Z. Wang, and M.D. Gale. 1995. "Grasses, Line Up and Form a Circle." *Current Biology* 5 (7): 737–39.

Tanksley, S.D., and S.R. McCouch. 1997. "Seed Banks and Molecular Maps: Unlocking Genetic Potential from the Wild." *Science* 277:1063–66.

6

THE ROLE OF FEDERAL, STATE, AND PRIVATE INSTITUTIONS IN SEED TECHNOLOGY GENERATION

JOSE FALCK-ZEPEDA AND GREG TRAXLER

TWO MAIN policy instruments are used to influence the direction and level of agricultural research in the United States: direct public investment, and laws that protect intellectual property rights (IPRs). These policy options tend to be fairly blunt instruments intended to have three primary effects: to increase public and private investment in agricultural research; to affect the distribution of research activity by geographic area, commodity, and discipline; and to influence the level of competition in plant seed markets. Only the first effect has been considered in the agricultural economics literature (see Huffman and Evenson 1993; Alston and Pardey 1996). Literature has shown that the effects of direct investment are not transparent, and the response of any research institution to a change in public policy is poorly understood. Furthermore, no formal mechanism exists for coordinating research activities among federal, state, and private agricultural research institutions.

Similarly, it is known that strengthened IPRs for plants have affected private sector investment in agricultural research and development (R&D): Private investment increased fourfold between 1980 and 1992 (Fuglie et al. 1996). But much less is known about the effect of stronger IPRs on the distribution of those investments, despite numerous concerns (Knudson and Pray 1991). One such concern is that investments will flow to those research activities with the most commercial potential, upsetting the balance between pretechnology research and cultivar development research. Public as well as private institutions may be lured away from research that has public goods characteristics (Duvick 1984; Stallman 1990). Strengthened IPRs for plants could also make plant breeders less willing to share germplasm, threatening the expansion of the informal breeder-to-breeder germplasm exchange that began in the 1970s (Duvick 1984). Finally, since strengthened IPRs are a means of creating monopoly power over the sales of plant varieties, they may alter the market structure of seed industries.

Our paper focuses on the effects of federal policies on the distribution of resources across agricultural research activities, using as a case study the flow of germplasm in the cottonseed market. This work is in a preliminary stage. The objective of our long-term research project is to address directly the cause-and-effect relationship between policy instruments and research outcomes. Here we have a more modest objective: to present some baseline information on the distribution of research capacity and to explore avenues for future analytical research. To that end, we first present information about crop improvement research institutions in the United States and attempt to identify their roles. We also discuss the nature of public-private sector interaction in the development of improved crop varieties. We then present a case study of germplasm exchange among research institutions in the cotton sector. The concluding section offers some suggestions for future analytical research.

THE CROP IMPROVEMENT RESEARCH SYSTEM IN THE UNITED STATES

The U.S. crop improvement research effort is shared by three sets of institutions: the various arms of the U.S. Department of Agriculture (USDA), state agricultural experiment stations (SAESs), and private companies. Within the private sector, there are three classes of firms: leader firms, small firms, and life science or biotechnology firms.[1]

How Institutions Specialize

Huffman and Evenson (1993) present a simple structural model describing key research activities, the roles of the various institutions within the agricultural research system, and the linkages among the organizations. In this model the USDA, SAESs, and private research institutions collaborate across the spectrum of research—from basic to applied—to generate improved technology. Two facts stand out in this model. First, research institutions tend to specialize in order not to duplicate the research of other institutions. Specifically, the public system's mandate is to conduct upstream research from which it is more difficult to appropriate benefits, while the private sector focuses on applied technology. Second, this tendency toward specialization makes linking basic to applied research more difficult and demands that institutions be in close communication with each other in order to fully utilize the results of research.

Building on this model, we looked for evidence of research specialization among plant breeding institutions by examining a survey of plant breeders conducted by Frey in 1994. First, we examined total investments in crop improvement research. We found that the private sector dominates these investments, employing 68 percent of the 2,205 plant breeders in the United States in 1994, followed by SAESs (24 percent) and the USDA (8 percent). These figures show that the federal government is a relatively small provider of crop improvement research. It does, however, play a large role in financing agricultural research by providing formula funds for SAESs (National Research Council 1995).

Next, we divided plant breeding activities into three classes: basic plant breeding research, prebreeding or germplasm enhancement, and cultivar development (Frey 1996).[2] For all crops, nearly two-thirds of all plant breeders focused their efforts on cultivar development research. The remaining scientists were fairly evenly divided, with 16 percent dedicated to basic research and 18 percent to germplasm enhancement research. The fact that proportions of breeders in each of the three types of research are stable across crops suggests that the public and private sectors are able to coordinate their investments.

Based on our impressions about each of the institutions' concerns for client groups and the need to appropriate research benefits, we generated some expectations for the role of each participant (figure 6.1). The relatively small scientific staff of the USDA is expected to be deployed primarily in basic research and in an important activity that we call genetic resource management (GRM), which includes maintaining gene banks.[3] Basic research and GRM are appropriate federal research activities because the benefits of both are nei-

Figure 6.1. Expected focus of seed technology improvement at research institutions

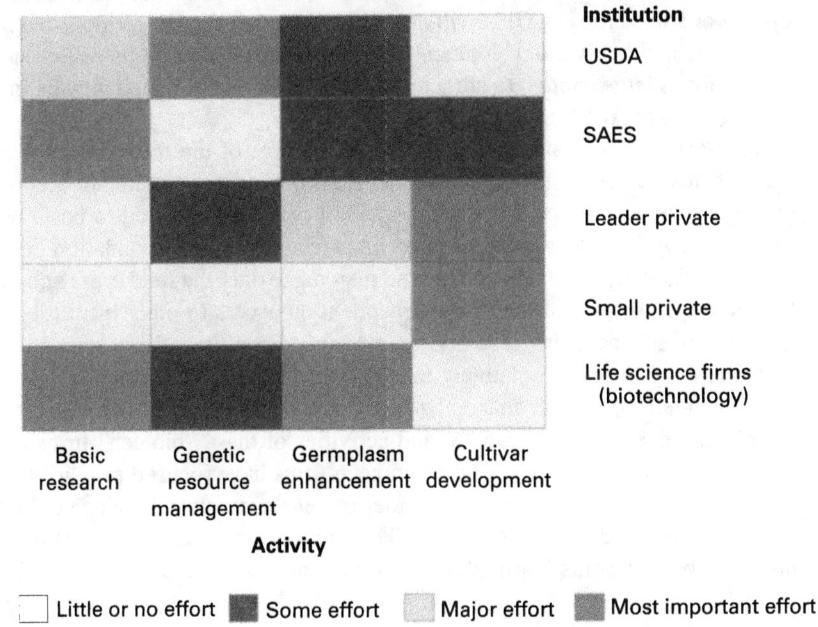

Table 6.1. Research focus of plant breeding scientists by sector of employment (percent of total plant breeding scientists in sector)

	USDA	SAES	Private	All sectors
		(percent)		
Hybrid cereals[a]				
Development	0	18	82	76
Enhancement	57	45	7	11
Basic	43	37	11	13
Total	100	100	100	100
Pure-line cereals[b]				
Development	11	44	82	38
Enhancement	38	24	13	12
Basic	51	32	5	50
Total	100	100	100	100
All crops				
Development	12	41	79	65
Enhancement	48	29	11	18
Basic	40	30	10	17
Total	100	100	100	100

[a] Dent corn, sweet corn, millet, sorghum.
[b] Wheat, oats, barley, rice, wild rice, popcorn, triticale, rye.

Source: Frey (1996).

ther geographically concentrated nor easily appropriated. State agricultural experiment stations are expected to provide primarily basic research, a focus that is consistent with the dual teaching and research mission of state universities. For some crops, SAESs will also be expected to supply germplasm enhancement and cultivar development research to adapt local crops that do not have markets large enough to attract private investment but which may be important to that SAES's interest group.

The seed corn industry provides a clear example of the different roles of leader firms and small firms. Leader firms such as Pioneer maintain a comprehensive research capacity, with significant resources devoted to basic research. Small firms obtain their inbred lines from Holden's Foundation Seed Company (now a part of Monsanto) and may make only the final cross and do field testing. These small firms are dependent on access to other institutions' germplasm enhancement research to remain competitive. Life science or biotechnology firms are relatively new players in the seed industry but possess the intellectual and financial resources to alter crop improvement research significantly. The structure and activities of these "biotech" firms are evolving (Begemann 1997). To date, biotech firms have focused primarily on basic research and on building inventories of gene constructs, leaving cultivar development to existing seed firms. Through acquisitions and investment, however, biotech firms (particularly Monsanto and Dupont) are becoming increasingly important players in financing and conducting plant breeding research.[4]

Finally, we identified the actual research concentrations of the USDA, SAESs, and private institutions, using data from Frey (1996). The USDA concentrates its efforts on genetic enhancement and basic research, employing nearly equal numbers of scientists for basic research and for germplasm enhancement (table 6.1).Only 12 percent of USDA scientists work in cultivar development, and those are concentrated in pure-line cereals. SAES scientists are relatively evenly divided across research areas, but the distribution varies among crops. Over all crops, 41 percent of SAES plant breeders work in cultivar development, 29 percent in germplasm enhancement, and 30 percent in basic research. Not surprisingly, 79 percent of private sector scientists are in cultivar development while only about 11 percent are in germplasm enhancement and 10 percent in basic research. Interestingly, there seem to be only minor differences in the allocation of private sector scientists in hybrid and pure-line crops.

The previous discussion looks at the research priorities for each sector. Examining the relative importance of each type of institution in providing each type of research provides another perspective on the composition of the total research effort (table 6.2). The private sector dominates cultivar development research in all crops. Even for pure-line cereals, the SAESs provide less than a third of all plant breeders devoted to cultivar development. As expected, the

Table 6.2. Relative importance of each sector in cultivar development, germplasm enhancement, and basic research (percentages of total plant breeding scientists in research focus)

	USDA	SAES	Private	*Total*
		(percent)		
Hybrid cereals[a]				
Development	2	15	83	100
Enhancement	21	38	41	100
Basic	19	43	38	100
Pure-line cereals[b]				
Development	2	31	67	100
Enhancement	17	52	31	100
Basic	17	52	31	100
All crops				
Development	2	15	83	100
Enhancement	21	38	41	100
Basic	19	43	38	100
All research	8	24	68	100

[a]Dent corn, sweet corn, millet, sorghum.
[b]Wheat, oats, barley, rice, wild rice, popcorn, triticale, rye.

Source: Frey (1996).

SAESs provide a larger overall share of the germplasm enhancement and basic science research than they do for cultivar development. But the private sector also makes a significant contribution to germplasm enhancement research in all crops, particularly in hybrid cereals. Similarly, SAESs provide a significant amount of basic research in pure-line cereals, but the private sector dominates basic research for hybrid cereals. The small USDA system's relative importance is greatest in germplasm enhancement, but as we have seen, the USDA remains the smallest provider of all types of research across nearly all crops.

Implications for Public Policy

The data confirm that the USDA, SAESs, and private sectors assume different roles in the crop improvement research effort in the United States. The most striking conclusion we draw from the data concerns the relative importance of the private sector across all research areas and crops. To find that 80 percent of scientists working in cultivar development are in the private sector is perhaps not surprising. It is surprising, however, to discover that the private sector is the largest provider of germplasm enhancement research and that it nearly equals the SAESs in the provision of basic plant breeding research. Two factors underscore the importance of this conclusion: First, the data do not include figures for basic research on plant molecular biology in any firms; and second, biotech firms such as Dupont and Monsanto were not surveyed. The true amount of private sector basic research activity is therefore underestimated.

Given the importance of the private sector in agricultural research, the impact of policies on private-sector incentives must be a major concern. Policy instruments such as IPR laws could restrict the flow of information and germplasm among researchers and institutions, ultimately affecting the long-term sustainability of yield increases. They could also reduce access of smaller firms to needed upstream research, affecting the firms' ability to compete and possibly forcing them to exit the industry (Butler and Marion 1985). The long-term effects of these laws on the exchange of germplasm are currently unclear, but many crop breeders report that they have already begun to limit the free exchange of material. As a result, legally enforceable material transfer agreements have become more common. These agreements allow public and private institutions to generate income from a range of germplasm products, including prebreeding and unfinished materials, and may ultimately increase the availability and exchange of germplasm, particularly among private firms. It remains to be seen whether the contributions of the USDA's Agricultural Research Service (ARS) and SAES germplasm to commercial varieties will diminish over time.

Investigating the actual flow of germplasm among institutions helps to determine the effects of current IPR laws (and the potential effects of new legislative proposals). The next section looks at the flow of cotton germplasm in an effort to determine how heightened IPR protection is affecting agricultural R&D in the United States.

THE CASE OF COTTON

After corn and soybeans, more plant breeders are employed in cotton than in any other crop (Frey 1996). Within cotton research, the private sector dominates all areas of research (table 6.3). The USDA effort is focused on germplasm enhancement research, while SAES scientists are evenly allocated across research foci. Private firms and public institutions exchange germplasm, and many smaller firms depend on larger firms and public institutions. The cottonseed sector thus provides a useful example of the flow of germplasm among institutions and firms. We include in our analysis data on the areas planted, both by institution and by region.

Table 6.3. Number of plant breeding scientists working on cotton by sector of employment, 1994

Crop improvement activities	USDA	SAES	Private	*Total*
Basic plant breeding research	4.5	6.0	9.4	19.9
Genetic enhancement research	7.2	6.6	18.7	32.5
Cultivar development research	0.0	6.5	73.4	79.9
Total	11.7	19.1	101.5	132.3

Source: Frey (1996).

To analyze the flow of germplasm, we use the *coefficient of parentage* (CP) estimated by Bowman, Calhoun, and May (1995) and Calhoun, Bowman, and May (1994) for 256 varieties released in the United States from 1970 through 1990. The CP is a measurement of the degree of relatedness between different institutions' or firms' varieties or cultivars. The CP is defined as the probability that two cultivars have identical alleles in a random locus in a chromosome. Roughly, the CP indicates on a zero to one scale the similarity of the genetic composition of the pair of varieties being compared. Two varieties with a CP of zero have no common parents, and varieties with a CP of one represent a variety compared to itself. This strict definition can be relaxed so that the CP estimates the degree of exchange a variety has on other varieties in a pedigree. This alternate definition allows us to trace the exchange of varieties released by public and private institutions.[5]

Data and Methodology

To examine the extent of spillovers, we tabulate the CP using a parentage table estimated by Bowman, Calhoun, and May (1995). This study estimates the CPs of 260 cotton cultivars released in the United States between 1970 and 1990. It is based on earlier work that traced the pedigrees of 290 upland and Pima cotton varieties released in the United States. For the CP study, the researchers discarded 30 varieties whose pedigrees could not be established.[6] To facilitate manipulation of the data set in a spreadsheet, we separated out another 11 minor varieties that are not widely sown in the United States, originate outside the country, or both. We calculate the contribution made by the USDA, SAESs, and private firms to varieties produced by other institutions using the average CP.

Areas Planted to Cotton Varieties

Private firms have assumed a dominant role in cotton in the United States. Private firms produce varieties that are sown in 86 percent of the total area devoted to this crop, while public institutions have lost ground (figure 6.2). Leader firms' varieties have increased from 28 percent to 72 percent of the area planted to cotton, while small firms' share of the total area planted has decreased from 33 percent to 16 percent. Varieties produced by the public sector increased until 1982,[7] decreasing steadily afterwards. The areas planted to Delta & Pineland (D&PL), Paymaster, and Stoneville varieties have increased significantly in the last 30 years (table 6.4).[8]

The Acala and Tamcot programs are the only public-sector programs producing varieties that are currently planted in sizable areas in the United States, but even these varieties are losing market share (table 6.4). Since they are tailored to local needs, they may be widely used in some states but represent

Figure 6.2. Area planted to cotton varieties by supplier

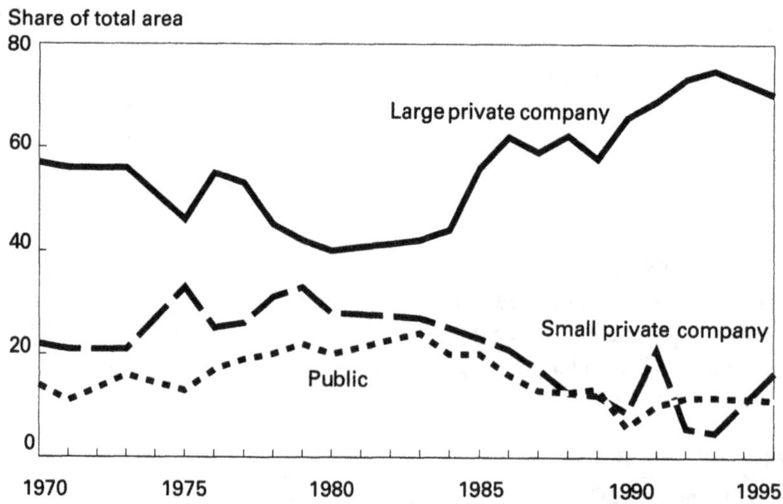

Table 6.4. Average percent area planted to cotton by variety in the United States, 1966–96

Period	Acala	Coker	D&PL[a]	Lankart	Paymaster[b]	Stoneville[b]	Tamcot	Other
1966–70	17	6	22	11	6	20	0	19
1971–75	13	7	24	11	5	21	0	18
1976–80	11	0	17	12	8	19	6	22
1981–85	13	0	16	9	9	16	7	29
1986–90	7	0	30	6	13	11	6	26
1991–96	3	0	44	<1	23	7	1	20

[a]Delta & Pineland.
[b]Paymaster and Stoneville varieties are also owned by Delta & Pineland.

Source: USDA-AMS, Cotton Varieties Planted, various issues.

only a small fraction of the total area planted to cotton nationwide. The Acala varieties are produced almost entirely by the California Planting Cotton Seed Distributors (CPCSD) program in Shafter, California, and the New Mexico Agricultural Experiment Station (NMAES). The CPCSD, a private board, regulates cotton planting in the San Joaquin Valley and was established after the dissolution of the Shafter program. The CPCSD's Acala varieties are sown on 83 percent of the area planted to cotton in California, but this areas represents just 6 percent of the total area nationwide (table 6.5). The Acala varieties produced by the NMAES are sown exclusively in New Mexico, where they

Table 6.5. Percent of total area planted to Tamcot and Acala varieties for specified areas, 1996

Brand	Variety	Originator	Oklahoma	Texas	New Mexico	California	All US
Tamcot	CAB-CS	TAES[a]	10	1	—	—	<1
	HQ-95	TAES	3	<1	—	—	<1
	SP21	TAES	—	<1	—	—	<1
	SP-BNX	TAES	<1	<1	—	—	<1
Acala	1517–75	NMAES[b]	—	—	6	—	<1
	1517–77BR	NMAES	—	—	9	—	<1
	1517–88	NMAES	—	—	47	—	<1
	1517–91	NMAES	—	—	10	—	<1
	1517–92	NMAES	—	—	—	—	—
	1517–SR2	NMAES	—	—	—	—	—
CPCSD	Acala Maxxa	CPCSD[c]	—	—	—	80	6
	Acala Prema	CPCSD	—	—	—	<1	<1
	Acala SR-2	CPCSD	—	—	—	3	<1

[a]Texas Agricultural Experiment Station.
[b]New Mexico Agricultural Experiment Station.
[c]California Planting Cotton Seed Distributors.

Source: USDA-AMS, Cotton Varieties Planted (1996).

accounted for 72 percent of the area planted to cotton in the state for 1996, but only 0.2 percent of the total area planted to this crop in the United States (table 6.5).

Tamcot varieties are produced by the Texas Agricultural Experiment Station at Texas A&M University. They are sown on only 1 percent of the total area planted to cotton in the United States and on only 2 percent of the total crop area in Texas. But they account for 13 percent of the area devoted to the crop in Oklahoma (table 6.5).

Cotton varieties are agroclimatic dependent, although germplasm has been shared among agroclimatic regions (May, Bowman, and Calhoun 1995). Table 6.6 disaggregates the 1996 crop by growth areas. Delta & Pineland dominates in all regions except the West, where it ranks second. The dominance of Delta & Pineland in different agroclimatic regions raises two questions: First, can Delta & Pineland maintain programs for the different agroclimatic regions and still increase yields? Second, will instruments such as patents that reinforce a firm's dominance in a market stimulate or undermine the current trend of yield increases? It remains to be seen if pricing above marginal cost will be justified by the gains in lint yield produced through the introduction of better varieties.

How Cotton Germplasm Is Shared

The average CPs of cotton varieties between institutions show the importance of germplasm exchange to institutional breeding programs (table 6.7). The diagonal represents the average "own" CP, or the average exchange from all

Table 6.6. Estimated percent area of upland cotton planted to leading varieties by region, 1996

South-East	D&PL[a] 75	Sure-Grow 11	Hyperformer 6	Stoneville[b] 5	D&PL[a] and Germain's <1	Paymaster[b] <1
South-Central	D&PL[a] 59	Stoneville[b] 23	Sure-Grow 15	Paymaster[b] 1	D&PL and Sure-Grow <1	Hyperformer <1
South-West	Paymaster[b] 57	D&PL[a] 18	All-Tex 8	Stoneville[b] 3	Tamcot 3	AFD 2
West	CPCSD 65	D&PL[a] 19	Germain's 4	Sure-Grow 2	Acala 2	Stoneville[b] 2
United States	D&PL 43	Paymaster 22	Stoneville 9	Sure-Grow 8	CPCSD 7	All-Tex 3

[a]Delta & Pineland.
[b]Paymaster and Stoneville varieties are also owned by Delta & Pineland.

Source: USDA-AMS Cotton Varieties Planted (1996).

other varieties released by the same institution (a symmetric diagonal matrix). The USDA has the highest degree of related varieties, with an own CP of 0.256. This result is not surprising, because most of the USDA/ARS varieties are from the western Acala and Pima and the eastern Pee Dee programs. The Acala and Pima programs tend to use the same types of varieties in their breeding programs, owing both to the high-quality characteristics imbedded in the Acala cottons and to institutional arrangements that restrict the use of other types of cottons in the western region.

Table 6.7. Average coefficient of parentage of cotton varieties between plant breeding institutions

	SAES	USDA-CPCSD	Joint USDA-SAES	Small private	Leader private
SAES	**0.098**	0.056	0.086	0.057	0.078
USDA/CPCSD	0.056	**0.256**	0.080	0.051	0.035
Joint USDA-SAES	0.086	0.080	**0.121**	0.055	0.068
Small private	0.057	0.051	0.055	**0.057**	0.054
Leader private	0.078	0.035	0.068	0.054	**0.100**

Source: Calculations based on Bowman, Calhoun, and May (1995).
Bold numbers indicate an institution's own average coefficient of parentage of cotton variety.

Figures 6.3 to 6.7 examine how the sources of germplasm used to develop new cotton varieties have changed over time. The figures show the trend in the coefficient of parentage for germplasm from each institution. For most institutions, there is a tendency to rely most heavily on sources of germplasm from within their own institution.

First consider sources of cotton germplasm used by public breeders. United States Department of Agriculture breeders have used germplasm from other USDA varieties the most, whereas they have used small private and SAES varieties at almost the same rates (figure 6.3). Use of germplasm from various institutions has remained fairly constant over time.

For SAES varieties, the highest relation is with SAES varieties (figure 6.4). The SAES varieties had similar exchanges with joint USDA/ARS and leader firm varieties. The figure for joint USDA/SAES varieties is somewhat suspect, however, because of the small number of releases from joint programs. The average CP between SAES and other SAES varieties has decreased over time, while that between SAES and USDA varieties has increased.

For joint release varieties, breeders' use of other joint release varieties has increased, while their use of USDA/CPCSD varieties use has decreased (figure 6.5). This finding indicates that breeders are using more SAES varieties. There has also been a slight increase in use of leader firm varieties, while the use of small firm varieties has remained fairly constant.

Now lets examine the sources of cotton germplasm used by the private sector. Small private firms are the only institution that relied most heavily on germplasm from other sources (figure 6.6). They appear to have used USDA/CPCSD varieties more than other institutions' germplasm. However, the USDA curve presents an anomaly that should be viewed with caution, because the USDA released only one variety between 1981 and 1985. More generally, the CPs of varieties produced by small private firms are about equally distributed among all institutions, indicating that small firms take advantage of spillovers from all sources—larger private firms, the USDA, and state breeding programs.

For leader firm varieties, breeders use germplasm from the leader firm more than any other type and USDA/CPCSD varieties the least (figure 6.7). Because USDA/CPCSD breeders likewise use few leader firm varieties, it is clear that there is very little germplasm exchange between the USDA and private firms. Leader firms have used some varieties from the Acala program, because Acala and Pima cotton introduce desirable characteristics into high-yielding varieties used in other agroclimatic regions. CPs between leader firms and all institutions' varieties have remained constant over time. This fact suggests that the diversity of leader firm varieties has not declined, as some researchers had suggested it would because of these firms' dominant market position.

Figure 6.3. Sources of cotton germplasm for USDA breeding programs

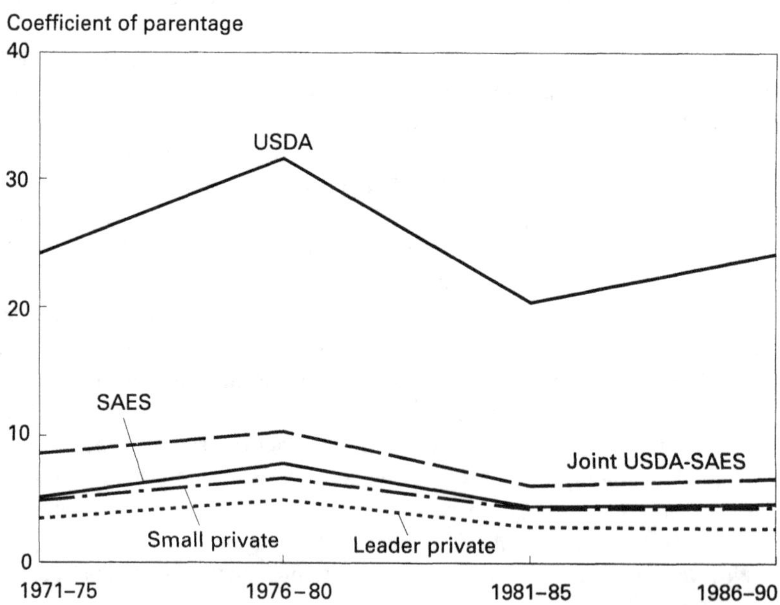

Figure 6.4. Sources of cotton germplasm for SAES breeding programs

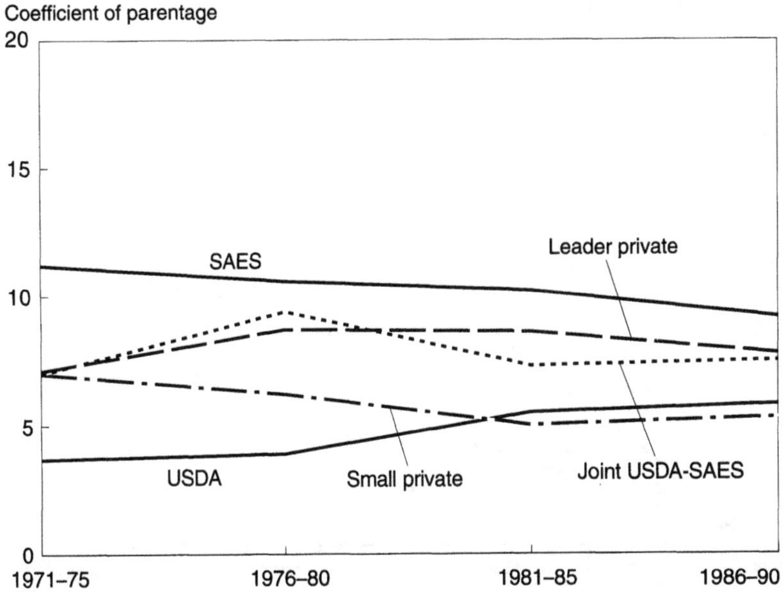

Figure 6.5. Sources of cotton germplasm for joint USDA-SAES breeding programs

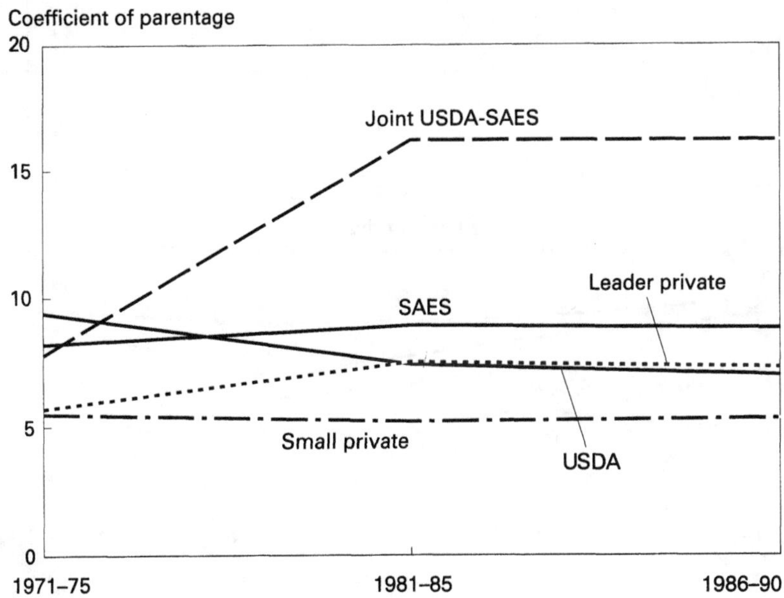

Figure 6.6. Sources of cotton germplasm for small private breeders

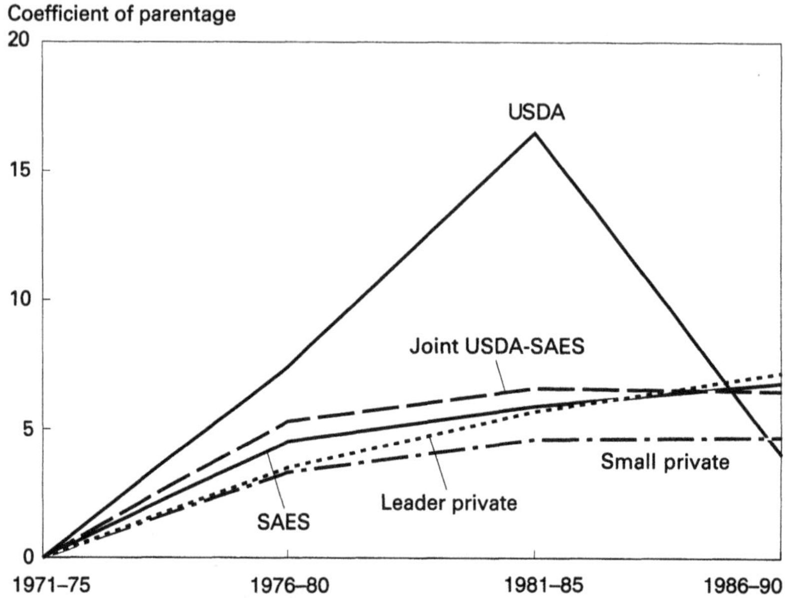

Figure 6.7. Sources of cotton germplasm for leader private breeder

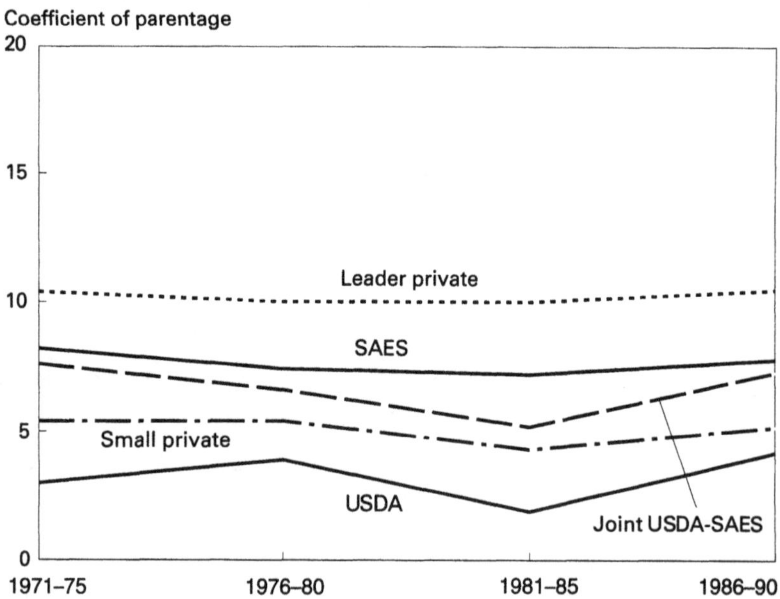

CONCLUSIONS AND DIRECTIONS FOR FUTURE RESEARCH

The four sectors involved in crop improvement research in the United States assume very different roles. The USDA effort is closely focused on germplasm enhancement research, a research area that is crucial to the overall germplasm improvement effort. Because the benefits of this research are difficult to appropriate, there is concern that it has become an area of acute underinvestment. In nearly all crops, the private sector dominates cultivar development research, but it has become the largest employer of germplasm enhancement scientists, and nearly equals the SAESs in the provision of basic plant breeding research.

Given these facts, the impact of public policies on private-sector incentives must be a major concern. Large amounts of capital are flowing into the sector, putting it (rather than the public sector) in a position to determine future technical changes in crop production. In particular, the emergence of practical biotechnology protocols for creating transgenic plants is inducing a rapid restructuring of the U.S. seed industry. As an example, over the past two years DuPont and Monsanto have invested more than $8 billion in germplasm firms. In the seed corn industry, DuPont now owns 20 percent of Pioneer, and Monsanto has taken over Holden's Foundation Seed Co. and DEKALB. Monsanto also owns the number two soybean and cotton seed companies. The amount these two companies have spent acquiring germplasm firms is more than the twice the sum all public research institutions spent on all commodities and all disciplines last year. Given this growing presence of the private sector in crop research, the USDA and SAESs must reevaluate their own portfolio of research investments, adjusting their role to compliment the new activities of the private sector.

The issues raised in this paper also suggest new avenues for future analytical research. The relationship between market structure and the distribution of rents induced by technical change in germplasm enhancement has received relatively little attention, although it is increasing in importance as seed markets become more concentrated (Falck-Zepeda and Traxler 1998). The cotton sector is an interesting case study, particularly in light of the appearance of biotechnology varieties such as *Bt* cotton from Delta & Pineland and BXN cotton from Stoneville. The relationship between the sustainability of yield increases and possible societal losses due to increased market imperfections introduced by IPR policy instruments is another important issue. The USDA must continue to provide the public goods research necessary to provide a complete system of seed technology generation for U.S. farmers. One vital USDA role will be to maintain the competitiveness of increasingly concentrated seed markets by providing free access for startup firms to quality public germplasm.

NOTES

1. This last term, inspired by Monsanto's recent restructuring, refers to biotech firms that supply genes to traditional seed companies, which then incorporate the genes into commercial varieties.

2. Frey (1996) uses the following definitions. *Plant-breeding research* involves genetics, methodologies, and biotechnology and is usually undertaken to provide basic information that will make plant breeding more efficient and productive. It does not include basic research on plant molecular biology. *Germplasm enhancement* is any activity that involves transferring genes using either sexual or asexual means from germplasm accessions and increasing the desirable genes in crop gene pools used to develop parents or cultivars. *Cultivar development* is any crossing, transformation, and/or selection (including marker-assisted selection) of plants with the direct purpose of releasing a crop variety.

3. This activity is subsumed under germplasm enhancement in Frey's (1996) classification.

4. In the seed corn industry for example, Dupont now owns a 20 percent share in Pioneer, and Monsanto's acquisitions (Holden's, Asgrow and DEKALB) have a share of more than 40 percent of the U.S. market.

5. Previous studies (Souza et al. 1994; Souza and Sorrells 1989; Murphy, Cox, and Rodgers 1986) have used the *coefficient of diversity* in different crops, which is defined as the coefficient of parentage subtracted from one, as indicated in the formula:

$$\text{Coefficient of diversity (CD)} = 1 - CP.$$

Their intention was to measure the degree of separation between germplasm, while we are interested in the contribution a specific germplasm makes to another germplasm.

6. The pedigrees were traced to four original ancestors: Eastern Green Seed Introductions (1622), Burling's Mexican (1806), Sea Island (1787), and Siam (1732).

7. Public-sector varieties include those released by the USDA/ARS program in Shafter, California (now closed) and the California Producing Cotton Seed Distributors (CPCSD). Public-sector varieties also include varieties released by state agricultural experiment stations and jointly produced varieties released by federal and state programs.

8. Delta & Pineland purchased the Paymaster program from Cargill in 1991, increasing the area planted by Delta & Pineland to 67 percent for the 1991–96 period.

REFERENCES

Alston, J.M., and P.G. Pardey. 1996. *Making Science Pay: The Economics of Agricultural R&D Policy.* Washington, DC: American Enterprise Institute.

Begemann, B.D. 1997. "Competitive Strategies of Biotechnology Firms: Implications for U.S. Agriculture." *Journal of Agricultural and Applied Economics* 29 (July): 117–22.

Bowman D.T., D.S. Calhoun, and O. L. May. 1995. "Coefficients of Parentage for 260 Cotton Cultivars Released Between 1970 and 1990." Technical Bulletin Number 80. Washington, DC: U.S. Department of Agriculture.

Butler, L.J., and B.W. Marion. 1985. *The Impacts of Plant Protection on The U.S. Seed Industry and Public Plant Breeding.* North-Central Regional Publication 304. College of Agricultural and Life Sciences, University of Wisconsin, Madison: University of Wisconsin Press.

Calhoun, D. S., D. T. Bowman, and O. L. May. 1994. "Pedigrees of Upland and Pima Cotton Cultivars Released Between 1970 and 1990." Mississippi Agriculture and Forestry Experiment Station Bulletin 1017.

Duvick, D. 1984. "Genetic Diversity in Major Farm Crops on the Farm and in Reserve." *Economic Botany* 38:161–78.

Falck-Zepeda, J.B., and G. Traxler. 1998. "Rent Creation and Distribution from Transgenic Cotton in the U.S." Paper presented at the symposium Intellectual Property Rights and Agricultural Research Impacts. Sponsored by North Central Region Cooperative Project NC-208 and International Center for the Development of Wheat and Maize (CIMMYT) Economics Program. El Batan, Mexico, March 5-7.

Frey, K. 1996. *National Plant Breeding Study I: Human and Financial Resources Devoted to Plant Breeding Research and Development in the United States in 1994.* Special Report 98. Ames: Iowa Agricultural and Home Economics Experiment Station, Iowa State University.

Fuglie, K., N. Ballenger, K. Day, C. Klotz, M. Ollinger, J. Reilly, U. Vasavada, and J. Yee. 1996. *Agricultural Research and Development: Public and Private Investments under Alternative Markets and Institutions.* Agricultural Economic Report Number 735. Washington, DC: Economic Research Service, U.S. Department of Agriculture.

Huffman, W., and R.E. Evenson. 1993. *Science for Agriculture: A Long-term Perspective.* Ames: Iowa State University Press.

Knudson, M.K., and C.E. Pray. 1991. "Plant Variety Protection, Private Funding, and Public Sector Research Priorities." *American Journal of Agricultural Economics* 73(August): 882–86.

May, O.L., D.T. Bowman, and D.S. Calhoun. 1995. "Genetic Diversity of U.S. Upland Cotton Cultivars Released between 1980–1990." *Crop Science* 35:1570–74.

Murphy, J. P., T. S. Cox, and D. M. Rodgers. 1986. "Cluster Analysis of Red Winter Wheat Cultivars Based on Coefficients of Parentage." *Crop Science* 26:672–76.

National Research Council. 1995. *Colleges of Agriculture at the Land Grant Universities: A Profile.* Washington, DC: National Academic Press.

Souza, E., and M.E. Sorrells. 1989. "Pedigree Analysis of North American Oat Cultivars Released from 1951 to 1985." *Crop Science* 29:595–601.

Souza, E., P. N. Fox, D. Byerlee, and B. Skovmand. 1994. "Spring Wheat Diversity in Irrigated Areas of Two Developing Countries." *Crop Science* 34:774–83.

Stallman, J.I. 1990. "Plant Patents and Public Research Priorities" *Choices* 3:8–11.

U.S. Department of Agriculture (USDA). 1996. *Cotton Varieties Planted.* Washington, DC: Agricultural Marketing Service (ISSN: 0498–1987, microfilm).

7

PUBLIC AND PRIVATE INVESTMENTS IN ANIMAL RESEARCH

KEITH O. FUGLIE, CLARE A. NARROD,
AND CATHERINE NEUMEYER

THE MODELS of agricultural research described by Fuglie in chapter 2 and Falck-Zepeda and Traxler in chapter 6 depict a system in which public and private research largely complement one another, with the public sector focusing on areas where private-sector incentives are weak. One implication of this model is that as private incentives change, public research priorities are also affected. The changes occurring in the organization of agricultural research in the United States affect not only crop research but animal research as well. Like plant seed technology, specialized animal breeding companies have become major developers and suppliers of improved genetic stock to farmers. The research incentives faced by these firms are being affected by the significant changes taking place in consumer demand for animal products and in the structure of animal production. These firms are also being influenced by new technological developments in animal reproduction and genetics. Thus, it is important to document private research expenditures and priorities and assess their implications for public research policy.

The purpose of this chapter is threefold. The first goal is to review the roles of the public and private sectors in animal research in light of major structural changes occurring in the animal sector. In particular, shifts in demand for animal products and the trend toward vertically coordinated animal production have changed incentives in the private sector for animal research. Second, the chapter examines levels and trends in animal research expenditures by the private and public sectors. New survey evidence is presented on private investment in animal breeding. Secondary sources are used to derive estimates for private research in animal health, nutrition, and mechanization. Trends over time in the allocation in public animal research resources are also examined to see how public priorities have responded to these developments. The third goal of the chapter is to describe some of the new institutional arrangements that have developed to strengthen research collaboration between the public and private sectors, including regional research consortiums and jointly funded research.

PUBLIC AND PRIVATE ROLES IN GENERATING NEW ANIMAL TECHNOLOGY

Farm animal industries in the United States have experienced remarkable productivity growth since the end of World War II. One of the most important factors behind this growth has been investments, both public and private, in animal research. These investments have provided farmers with a steady stream of new technologies that have significantly increased productivity. Table 7.1 shows trends in two partial productivity measures for livestock and poultry since 1955: output per animal and output per unit of farm labor. Improved breeds, nutrition, and disease control helped to significantly raise animal yields for most commodities. Growth in animal yield between 1955 and 1995 was greatest for dairy and showed steady improvements for all major animal commodities except beef. (In fact, beef yield is the exception to the trend: Much of the growth in beef output per animal since 1975, shown in table 7.1, is due to a decline in veal slaughter—so that a larger proportion of beef cattle are grown to maturity—rather than from improvements in tech-

Table 7.1. Changes in U.S. livestock and poultry productivity indicators, 1955–95

Animal yield

Year	Beef cattle (lb beef/cow)	Hogs (lb pork/sow)	Dairy (lb milk/cow)	Broilers (lb/bird)	Layers (eggs/layer/year)
1955	590	788	5,842	3.07	192
1965	591	1,022	8,304	3.48	218
1975	546	1,167	10,360	3.76	232
1985	680	1,310	12,994	4.21	247
1995	723	1,503	16,451	4.67	253

Beef cattle yield: pounds of beef produced divided by the number of cows and heifers that have calved.
Hog yield: pounds of pork produced divided by the number of farrowing sows.

Labor productivity

Year	All farm	All livestock	Meat animals	Milk cows	Poultry
1955	1.00	1.00	1.00	1.00	1.00
1965	1.90	1.92	1.47	2.05	2.73
1975	2.97	3.54	2.42	4.26	5.40
1985	5.17	7.29	4.53	12.84	11.53

Index of output per hour worked: 1955=1.00.
Labor productivity indices for specific commodities are no long published after 1986.

Source: *Agricultural Statistics*, annual issues.

nology.) New technology has also facilitated the substitution of capital for labor to increase output per unit of labor. Milk output per unit of labor increased more than 12-fold and poultry output per unit of labor 11-fold between 1955 and 1985. Animal productivity has been improved in other dimensions as well. For example, the efficiency of feed conversion in meat production (pounds of feed required to produce one pound of meat) has improved over time. In poultry, feed conversion efficiency was improved from around four in the 1940s to under two in the 1980s (Bugos 1992).

The U.S. system of animal research and technology development responsible for producing new technology is composed of federal and state research institutions (which comprise the public sector) and the private sector. The principal federal institution for animal research is the Agricultural Research Service of the USDA. At the state level, land-grant universities and their affiliated state agricultural experiment stations conduct research, training, and extension in animal production. In the private sector, specialized firms conduct breeding research for most major animal commodities, and farmers themselves are important sources of new innovations. Outside of the agricultural sector, the pharmaceutical, feed manufacturing, machinery, and electrical industries develop new technologies for use in animal production (Huffman and Evenson 1993).

Specializations in Animal Research

In general, the public and private sectors tend to support different kinds of research. The public sector is the source of most basic research (original investigation that advances scientific knowledge but with no immediate commercial application), while the private sector focuses on market-oriented applied and developmental research (Fuglie et al. 1996).

Table 7.2 shows the relative importance of the public and private sectors in five major technology areas for animal production. In breeding and genetic improvement, the private sector puts greatest emphasis on commodities where technology improvements can be most easily appropriated, such as hybrid poultry breeding. The public sector undertakes a wide range of breeding and genetics research, from basic studies on animal genome mapping and molecular biology, to applied farm record-keeping systems for livestock breed improvement. In disease protection and nutrition, both sectors make large research investments. Private research focuses on health and feed research to develop marketable products in which intellectual property can be protected and benefits from research realized. Public research emphasizes basic studies on animal diseases, pathology, physiology and metabolism. Public institutions also conduct applied research on technologies with a high information content, such as farm management methods, that may be less saleable as a marketable product. Mechanization research, especially of the labor-saving type, is almost entirely conducted in the private sector, whereas the public sector conducts almost all of the research on livestock economics and statistics. The public sector has also assumed a major responsibility for developing technology to reduce animal waste pollution and improve the safety of animal food products. Pollution and food safety problems are exacerbated by the failure of market forces to properly signal consumer demand to producers, since their full social costs of these problems are not reflected in market incentives. Thus, in the absence of regulations or other nonmarket incentives, the private sector is not likely to devote sufficient attention to these issues (Narrod and Pray 1995; Fuglie et al. 1996).

The stylized representation of animal research and development (R&D) in table 7.2 depicts a system in which the public and private sectors largely complement each other in research, with the public sector concentrating on areas

Table 7.2. Roles of the public and private sectors in animal research by technology area

Animal technology area	Private sector[a]	Public sector[a]
Breeding and genetics	** Applied breeding where technology can be protected through trade secrets (hybridization), market structure (e.g., large market share), or patents on genetically modified animals	*** —Basic studies in animal genetics, physiology, and biological efficiency —Applied breeding where private incentives are weak —Development of livestock genetic improvement schemes for producers
Disease protection and nutrition	*** —Applied research on products that can be patented or protected through trade secrets (e.g., veterinary pharmaceuticals and compound feeds) —In vertically integrated systems, applied research on management systems for disease control and nutrition management	*** —Basic research on animal pathology, nutrition, and metabolism —Applied research on management systems to control diseases and improve feed efficiency
Mechanization	*** Applied research on new equipment and machines that can be protected by patents and trade secrets	* Little unless related to other technology areas
Management and economics	* Little except in large integrated systems	** Farm management, market statistics and analysis
Externalities (pollution reduction and food safety)	* Little unless producers and/or processors are subject to regulation and fines	*** Basic, applied, and development research

[a] Stars indicate relative emphasis on research by sector:
*** major focus of research.
** important but secondary focus of research.
* little research conducted.

where private incentives are weak. One implication of this model is that changes in private incentives affect public research priorities. Therefore, a more complete understanding of the factors affecting private incentives for animal research and how these factors are changing over time have a direct bearing on public research priorities and policy.

Incentives for Private Animal Research

Pray and Fuglie in chapter 13 present a model in which private research is determined primarily by (1) market demand or size, (2) technology opportunities, and (3) the ability of a firm to capture the economic benefits from research, called "appropriability." Private research will tend to be biased toward those commodities or technologies that favor these determinants. Moreover, an increase in any one of these determinants will increase the incentive for the private sector to invest in research.

Trends in Market Demand

Demand for new production technology is directly derived from demand for the final product. Therefore, a shift in consumer demand for animal products will similarly shift the demand for new inputs and technology to produce those products. While an increase in consumer demand for a product increases private incentives to develop new and improved methods for producing that product, a fall in demand would similarly cause private research to decline.

In animal production, two major elements have been at work to shift the demand for new technology. First is the changing pattern of consumer demand for animal products in the United States. Over time, per capita consumption has tended to increase for poultry meat, remain stable for pork products, and decline for eggs, beef, and veal (figure 7.1). Demand for lamb, never a large item in the American diet, has also declined steadily over the past three decades, and is currently less than one pound per person per year. For dairy products, total consumption has been increasing slightly, although shifting from fresh milk in favor of processed products like cheese.

These changes in market demand are partly the result of past technical changes which affected the relative prices of close substitutes in food consumption. For example, more rapid technical changes in poultry production helped reduced retail prices of poultry relative to beef. This facilitated the substitution of chicken for beef in consumer diets (Brester, Schroeder, and Mintert 1997). The poultry and pork industries have also developed myriad new food products, including convenience food items, to bolster demand (Thurman 1987; Brester, Schroeder, and Mintert 1997). Health concerns have also contributed to shifts in consumption patterns. Consumers increasingly choose leaner meats in order to reduce fat intake, for instance. And increased awareness of the link between cholesterol consumption and heart disease may

Figure 7.1. Per capita consumption of animal products since 1970

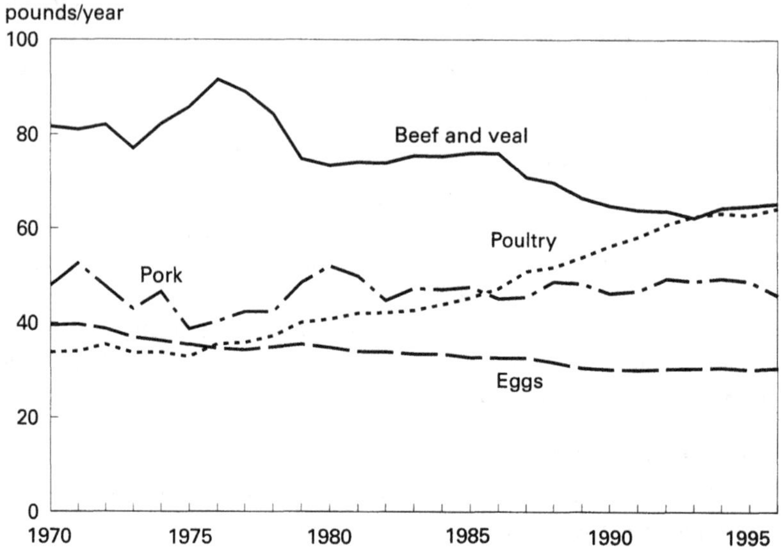

Source: Economic Research Service.

have contributed to the decline in egg consumption (Henneberry and Charlet 1992).

A second factor is the globalization of agricultural input markets (see chapter 13). Firms that develop new and improved animal breeds, health products, feed additives, and production equipment, for example, can market these inputs to producers not only in the United States but worldwide. Several of the major poultry and pork breeding firms are multinational companies with breeding and testing stations located in several countries. The increase in the size of the potential market strengthens incentives to invest in research to develop new animal production inputs.

New Technology Opportunities

Major advances have occurred in the past decade in applying the tools of biotechnology to enhance animal productivity. In breeding, technological advances are enhancing animal genetics and improving the efficiency of animal reproduction methods. In disease control, genetic engineering has opened new possibilities for developing new animal vaccines and other pharmaceutical products (Jordan 1990). Technological opportunities affect animal commodities unevenly, however.

Technical factors, especially gestation and fecundity, provide greater incentives to private breeding in poultry and swine than in cattle (Ward 1997).

It takes only five months from the time a chicken egg is fertilized for a hatched chick to become mature enough to breed. Moreover, chicks retained for breeding are only a small proportion of the production flock since one layer can produce 300 offspring per year. Swine require less than 12 months (from the time a sow is bred and farrowed) to reach breeding age. Fecundity in swine has been increasing: Most sows are currently farrowed at least twice a year and produce nearly nine pigs per litter. With cattle, however, the production cycle is considerably longer. Each cow can produce one offspring per year, and it takes about 24 months from the time a cow is bred for her calf to reach breeding age. Thus, the time it takes to develop new breeds and introduce them into the production flock or herd is relatively short for poultry, moderately long for swine, and very long for cattle. Since this factor directly influences the payback period for research investments, private companies are likely to invest more in poultry breeding than in breeding other animal commodities and, similarly, to prefer investments in swine breeding to those in cattle breeding.

Advances in reproductive technologies, such as artificial insemination, embryo transfer, and cloning (when it becomes commercially feasible) increase the incentives for private research in animal breeding.[1] These technologies can significantly expand the market for a successful breeder of improved parent stock. They speed up the process of genetic improvement, reduce the risk of disease transmission, and greatly expand the number of animals that can be bred from a superior parent. For example, a boar can fertilize up to 2,000 sows per year through artificial insemination but only 100–130 through natural mating (Estienne 1993).

Artificial insemination (AI) is more widely used with dairy cows than with other animals because it requires daily monitoring of animals to time ovulation. This is relatively easy for dairy farmers who are already in frequent contact with their cows for milking but is more costly for other livestock farmers. By 1991, 70 percent of U.S. dairy cows were being fertilized through AI (Johnson and Ruttan 1997). Nevertheless, the use of AI has been increasing in beef cattle and swine due to increasingly stringent sanitary restrictions on live animal trade designed to curb the spread of diseases. Artificial insemination was being used in an estimated 48 percent of sows in 1998 (Lawrence, Grimes, and Hayenga 1998) and 15–20 percent of beef heifers in 1996 (Keaton, cited in Johnson and Ruttan 1997). Embryo transfer is currently too expensive for general commercial use but is used by breeders to produce superior bulls.

Appropriating Benefits from New Technology

"Appropriability" is the ability of a firm to capture the economic benefits of its research. Generally, a firm needs to charge more for a newly developed input than the cost of manufacturing it in order to recoup the initial costs of

R&D. Technology that is easily duplicated by other firms once it is first sold is not very appropriable. A competing firm that merely copies a technology doesn't need to cover the sunk costs of R&D and therefore could offer the input to users more cheaply than the original developer. Thus, a firm investing in R&D will try to protect its intellectual property from rival firms in order to maintain a monopoly, at least for a time, on its proprietary technology.

Patenting. Several means may be available to a firm to protect the results of its research from rival firms. Many inventions may be patentable. A patent awards the inventor the sole right to use a technology for a specific period of time (usually 20 years) in exchange for publicly disclosing the nature and details of the invention. Patenting is widely used to protect mechanical inventions, especially those which may be easy to copy through "reverse engineering." Other technologies may be difficult or impossible to patent. Animal breeds and killed viruses used to produce animal vaccines, for example, are generally not patentable since they are products of nature. Genetically engineered breeds or vaccines, however, may be patentable (Lesser 1993).

Trade Secrets. Firms may also protect intellectual property by keeping key research findings out of the public domain. For example, a firm may keep secret the processes or formulas used in the manufacture of new technology. In the manufacture of animal vaccines using killed viruses, firms often protect intellectual property by keeping secret the processes used to rapidly multiply a virus, formulate the vaccine, and test for efficacy (Carl Pray, personal communication, 1998).

Trade secrets are commonly used to protect intellectual property in poultry breeding (see box 7.1). Poultry breeders develop superior inbred poultry lines and then cross them to produce hybrid chicks. While the hybrid chick exhibits desired characteristics and superior performance, its progeny loses genetic uniformity, performance, and yield. Thus, a producer needs to repurchase hybrid chicks each generation from a breeder with access to the parent inbred lines.

Hybridization has also been used to protect intellectual property in pig breeding. However, in higher mammalian species such as cattle, the hybridization process is limited by the difficulty of maintaining fertile inbred parent lines since animals bred with close relatives tend to lose reproductive potential and become sterile (Huffman and Evenson 1993). Instead, breeders may use crossbreeding or select among superior performing animals within a breed.

Market Structure. Unlike a small firm which relies on sales of inputs and technology to producers to recoup its research investments, large companies

BOX 7.1. THE USE OF HYBRIDS IN ANIMAL BREEDING

Poultry breeders have developed a highly successful way of protecting their intellectual property investment in superior breeds by exploiting heterosis, or hybrid vigor (Bugos 1992). Hybrid vigor is the yield advantage obtained when two or more pure inbred lines are crossed in a breeding scheme. While the offspring of this cross exhibit some superior yield performance, the yield advantage steadily declines if the offspring themselves are bred. Thus, by restricting access to the pure parent line stock (a form of trade secret) a breeder remains the sole supplier of the hybrid. Farmers then need to continue purchasing new stock from the breeder in order to maintain high yields.

The original experiments with hybridization in poultry were done in 1908 by Bateson, who demonstrated that combs, crests, and feather color followed the Mendelian laws of genetics. Heterosis was produced through a series of crossings that enabled hybrids to out-produce the average yield of their parents. Two companies that had already successfully used hybridization in corn seed production, Pioneer Seed and DEKALB AgResearch, were largely responsible for developing this method commercially for poultry. Beginning in the 1940s, multipurpose breeds (used for both meat and egg production) were replaced with specialized layer and broiler breeds. Currently, virtually all commercial poultry flocks are produced from hybrid baby chicks purchased from commercial hatcheries that have access to the parent poultry breeds. Breeding and selection are entirely under the control of scientists employed by poultry genetics companies. Vertically integrated companies, including companies such as Perdue (which has its own in-house breeding program), Tyson Foods, and Charoen-Pokphand (which owns Cobb and Avian Farms), use this system even with contract production.

Private breeders have also had some success exploiting hybrid vigor in swine. Producing hybrids for beef or dairy cattle is not really possible. Close inbreeding of parent lines (the process that produces maximum hybrid vigor) reduces the reproductive fertility and health of inbred offspring (Warwick and Legates 1979). Beef cattle, for example, have very little excess reproductive capacity—cows normally give birth to a single calf—and even a small deterioration results in infertility of the inbred line. Thus, in livestock, commercial breeding is normally restricted to unrelated animals, called crossbreeding or outbreeding.

(continued on next page)

(box 7.1 continued)

Crossbreeding, which mates animals of different breeds, is used in other animal production industries. Swine producers began using crossbreeding widely in the 1950s and 1960s, and beef cattle ranchers in the 1970s (Huffman and Evenson 1993). Unlike poultry farmers, livestock farmers often maintain their own breeding herds and retain crossbred sows and beef cows for breeding. By the mid-1980s, at least 80 percent of all marketed hogs were being crossbred (Huffman and Evenson 1993), although the use of hybrid pigs was growing (Johnson and Ruttan 1997). By 1996, beef cattle producers were crossbreeding some 87 percent of cows, and 49 percent from three or more breeds (APHIS 1997). In dairy, however, the dominance of the Holstein breed in low-fat milk production has caused breeding to focus on outcrossing (pure-line breeding of unrelated animals) rather than crossbreeding. In dairy, yield gains are more easily obtained through genetic improvement with breeds rather than with hybrid vigor (Warwick and Legates 1979). Commercial dairy herds are improved by mating farmers' best cows with superior bulls of the same breed supplied by commercial breeders.

can capture benefits from research by exerting market power or by applying the technology directly within the firm. Large, vertically integrated operations now dominate the broiler and layer industries and are gaining market share in swine, lamb, and beef (figure 7.2). In vertically integrated operations, an "integrator" firm may either wholly own breeding flocks, hatcheries, feed manufacturing, animal production, processing, and distribution or coordinate them through contractual arrangements (Martinez and Reed 1996). Long-term contractual arrangements between breeding companies and integrator firms strengthen private incentives for research. Furthermore, in the animal breeding industry, concentration has risen over time, and in some cases only a few firms supply most of the genetic stock to producers (box 7.2).

PRIVATE INVESTMENT IN ANIMAL RESEARCH

No recent, comprehensive estimates of private investment in animal production research exist, although some estimates are available from previous surveys. A survey by Malstead (cited in Ruttan 1982, 185) reported that in 1978–79 the private sector spent an average per year of $52 million on animal breeding, $99 million for animal health (mostly veterinary drugs), and $32 million for animal feed and feed ingredients, for a total of $183 million per year in animal-related research. In a 1985 survey of private agricultural

Figure 7.2. Vertical coordination in animal production

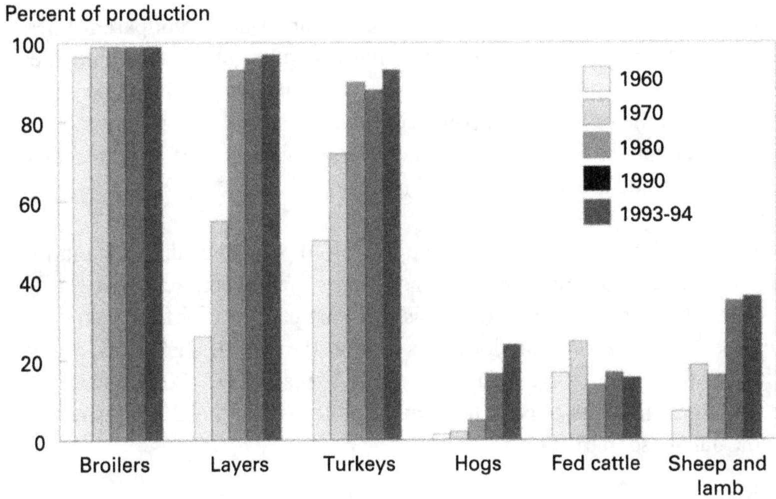

Bars show production under vertical integration, production contracts, or forward contracts.

Source: Economic Research Service.

BOX 7.2. STRUCTURE OF THE ANIMAL BREEDING INDUSTRY

Poultry broilers. The layer and broiler genetics industries emerged as two separate subsectors following the development of specialized poultry breeds in the 1940s and 1950s (Bugos 1992). Currently, each subsector has fewer than 20 companies conducting poultry breeding programs worldwide which dominate the research, development, and production of poultry breeds (Rose 1997). Seven companies conduct most of the breeding for broilers: Avian Farms, Arbor Acres, Cobb-Vantress, Hubbard (ISA), Peterson Farms, Perdue, and Ross. All of Perdue breeding stock is used in-house. Cobb-Vantress and Avian Farms are both owned by integrator firms but are operated independently and provide chicks to both in-house and outside producers (Henry and Rothwell 1995). Peterson Farms produces stock for its integrated broiler operation, Crystal Lake, but most of its breeding stock is produced for external use. The other companies all serve international markets and sell

(continued on next page)

(box 7.2 continued)

their breeds as fertilized eggs, chicks, or parent and grandparent lines. Although transferring grandparent lines to other affiliated companies reduces the company's control over its intellectual property, in some cases it is the only way breeding companies can sell in foreign markets. Some countries have restricted such imports in an effort to foster local breeding industries (Narrod 1997).

Poultry layers. Just two firms supply 85–90 percent of the U.S. market for layer breeds. Hy-Line, originally part of Pioneer, accounts for 70 percent of the U.S. market and has a unique program for field testing its genetic stock in various locations around the world. DEKALB, which holds the second largest share of the U.S. market (15–20 percent), has its largest markets outside the United States (including a large share of the Japanese market).

Swine. The swine-breeding industry in the United States is divided into two groups. One consists of companies that specialize in breeding and supplying hybrid boars and gilts to farms. The companies in the first group, formed in the 1960s and 1970s, resemble the large international poultry-breeding companies (Johnson and Ruttan 1997). They have developed superior pureline breeds (which exploit heterosis for improved yield and quality) for producing hybrid crosses. They provide both male and female crossbred parent lines to hog farms for terminal crossbreeding. The second group is composed of small breeders who provide purebred boars to farmers for rotational crossbreeding. In the last several decades the specialized breeding companies have gradually increased their market share (McLaren 1992) and in 1989 accounted for 28 percent of all boars and 14 percent of gilts bought by U.S. farmers (Johnson and Ruttan 1997). The companies with the largest market share in 1996 were Dalgety PIC, DEKALB Swine Breeders, Newsham Hybrids, Cotswold USA, Seghers, Babcock Swine, and Danbred USA. DEKALB recently acquired the National Pig Development Company (NPD) and JSR.

Artificial insemination for beef and dairy. The number of companies providing artificial insemination for dairy and beef cattle declined from 200 in 1950 to approximately 20 in the 1980s. The top five suppliers are ABS Global, Accelerated Genetics, Alta Genetics, Select Sires, and SirePower, who together have about 80 percent of the market for dairy

(continued on next page)

(box 7.2 continued)

and beef cattle insemination services in the United States (Charles Sattler, personal communication, 1998). Several of these companies also sell internationally. As the technology for storing and transporting frozen semen improves and the market expands, the cost of artificial insemination has been falling (Funk 1996).

research, Crosby (1987) estimated that the private sector spent $44.1 million for research on animal health products and $42.5 million for animal nutrition and feeds. The Crosby survey did not provide any estimates for animal breeding, and neither survey estimated research on animal machinery. In addition to these specialized surveys, the Pharmaceutical Research and Manufacturers of America (PhRMA) reports annual estimates of research on animal health products by its member companies in the United States, although these estimates include research for both animal agriculture and household pets. Klotz, Fuglie, and Pray (1995) reported these estimates for 1960 to 1992, and they showed that private research on animal health products in the United States was $331 million in 1992.

In this section we provide new and updated estimates of private research in animal breeding, animal health, animal feed and nutrition, and animal machinery. The most detailed estimates are provided for animal breeding, and these are drawn from a survey by the authors of private-sector breeding programs for broilers, layers, swine, beef, and dairy in the United States. The PhRMA (1998) data provide updated estimates of private research on animal health. Indirect methods are used to develop estimates of private research on animal feed and animal machinery.

Private Research in Animal Breeding

In 1997 and 1998 we conducted a phone survey of the companies listed in table 7.3 to determine their investment in animal breeding and molecular genetics. Companies were asked to give the number of full-time equivalent scientists (i.e., a "scientist-year," or SY) at the Ph.D. and M.S. level employed in the United States for research in applied animal breeding (quantitative genetics) and basic breeding research (animal molecular biology). We also asked the companies to estimate their annual expenditures for animal breeding research in the United States. For a comparison, we collected data on public sector SYs allocated to animal breeding and molecular biology research at USDA and SAES laboratories from USDA's Current Research Information System (CRIS).

Many of the animal breeding companies are multinational. Some large international firms with substantial research investments in other countries do

Table 7.3. Companies with breeding programs for poultry and livestock in the United States, 1996

Broilers	Arbor Acres (Booker PLC, UK) Avian Farms (Charoen Phophand, Thailand) Cobb-Vantress (Tyson Foods, US) Hubbard ISA (includes Hubbard, ISA, and Shaver; Merial, UK)[a] Perdue (US) Peterson Farmers (US) Ross Breeders (Hillesdown Holdings, UK)
Layers	DEKALB (Toshoku, Japan) Hubbard ISA (includes Hubbard, Shaver, ISA, Babcock; Merial, UK)[a] Hy-Line International (Lohmann-Wesjohann, Germany)
Swine	Babcock Swine (US) Hartland (Cotswold, UK) DanBred (Denmark) DEKALB Swine Breeders (Monsanto, US) Farmers Hybrid (US) Genetic Improvement Service (US) Genetic Porc (Canada) Nesham Hybrids (US) Pig Improvement Company - PIC (London-Dalgety PLC, UK) Seghers Hybrid (Belgium)
Dairy and beef[b]	ABS Global Accelerated Genetics ALTA Genetics Cooperative Resources International Select Sires Sire Power

[a]Merial is a London-based company that by 1997 had acquired Hubbard, ISA, and Shaver for broiler breeding and Hubbard, Shaver, ISA, and Babcock for layer breeding.
[b]Mainly suppliers of artificial insemination.

Source: Various industry sources.

not currently breed animals in the United States but nevertheless market their products to U.S. farmers. Other firms, such as the European swine-breeding firms Dalgety Pig Improvement Company (PIC) and Cotswold, conduct breeding and genetics research in the United States and abroad. The survey figures on private research report only in-house breeding research conducted in the United States by these firms. They may also conduct other types of animal research and contract out some research to consultants or universities.

For the layer and broiler industries, the companies interviewed provide virtually all the breeding stock for these commodities, so the estimates account for most (if not all) private-sector breeding in the United States. The layer-breeding industry reported 13 SYs (9 at the Ph.D. level) and $16.7 million in research for 1996 (table 7.4). In the broiler industry, there were 45.25 SYs (29.25 at the Ph.D. level) and $64 million in research investments.

The estimates reported in table 7.4 for swine breeding only include research by swine-breeding companies and do not include research by indepen-

Table 7.4. Public and private research in animal breeding and genetics, 1996

	Layers	Broilers	Swine	Beef and dairy	*Total*
Private sector					
Ph.D. (SYs)	9.0	29.25	18.6	16.5	73.35
M.S. (SYs)	4.0	16.0	14.25	20.0	54.25
Molecular biology (SYs)	4.0	4.5	8.6	5.0	22.1
Breeding and genetics (SYs)	9.0	40.75	24.25	31.5	105.5
Total (SYs)	13.0	45.25	32.85	36.5	127.6
Expenditure ($ million)	16.7	64.0	19.1	18.6	118.4
USDA					
Molecular biology (SYs)	2.3	2.3	11.0	11.0	26.6
Breeding and genetics (SYs)	1.2	2.0	3.1	14.1	20.4
Total (SYs)	3.5	4.3	14.1	25.1	47.0
SAES					
Molecular biology (SYs)	2.7	4.2	18.1	49.5	74.5
Breeding and genetics (SYs)	2.1	4.9	10.4	58.6	76.0
Total (SYs)	4.8	9.1	28.5	108.1	150.5
USDA and SAES					
Molecular biology (SYs)	5.0	6.5	29.1	60.5	101.1
Breeding and genetics (SYs)	3.3	6.9	13.5	72.7	96.4
Total (SYs)	8.3	13.4	42.6	133.2	197.5

Figures in scientist-years (SYs) except where otherwise noted.

Sources: Private-sector research from authors' survey. Public-sector research derived from USDA's *Current Research Information System* (1998).

dent breeders. Independent breeders provide a substantial but declining share of the market for swine breeds, especially purebreds for local and regional markets (Johnson and Ruttan 1997). Swine-breeding companies produce mainly hybrid pigs, often for large, integrated hog farms. These breeding companies reported a total of 32.85 SYs for swine breeding and an annual investment of $19.1 million.

For both the dairy and beef industries, the figures reported in table 7.4 account only for private-sector research by companies providing artificial insemination. Because most dairy cows are artificially inseminated, these figures probably account for most of the research in the dairy industry. The use of artificial insemination is growing in the beef industry but is not yet as extensive, so the figures do not capture all breeding research investment in beef. The industries together reported a total of 36.5 SYs and $18.6 million in research investment.

Compared with the public sector, the private sector had fewer resources devoted to animal breeding overall: 127.6 SYs in the private sector compared with 197.5 SYs at public institutions. However, the figures indicate a division of labor between public and private institutions in basic and applied breeding research for each commodity except large cattle (beef and dairy). Public breeding was heavily oriented toward basic research in molecular biology, while private research emphasized applied breeding and quantitative genetics. The private sector had more applied breeders for layers, broilers, and swine than the public sector, while the public sector had more SYs devoted to molecular biology. But in beef and dairy, where long gestation and low fecundity limit private incentives for research, the public sector dominated both basic and applied breeding.

Compared with state agricultural experiment stations (SAESs), USDA laboratories allocated a somewhat larger percentage of their total breeding and genetics resources to upstream research on molecular genetics than to applied breeding. On the basic-applied research spectrum, then, federal agencies appear to concentrate most heavily on basic research, private research is mostly applied, and research at state institutions falls somewhere between these two poles. These divisions of labor in animal research are similar to those for crop breeding research reported by Frey in chapter 5.

Private Research in Animal Health

The demand for animal health products has increased markedly in recent years because of the growth in global animal production and in confined animal production systems, especially for poultry and swine. The dense animal populations produced in large, confined systems intensify disease problems, especially the risk of epidemics among herds and flocks (Schmit 1987). Veterinary pharmaceuticals and medicated feeds are widely used to help control the incidence and spread of animal diseases in these systems.

Globally, the market for animal health products was estimated to be $14,370 million in 1995 (table 7.5). Veterinary pharmaceuticals represented slightly under one-half this market, and nutritional feed additives accounted for around one-third. By commodity, cattle represented the largest market for animal health products, especially for pharmaceuticals and nutritional feed

Table 7.5. Global sales of animal health products, 1995

Animal species	Nutritional feed additives	Medicinal feed additives	Biologicals	Pharmaceuticals	*Total*
Cattle	1,025	440	610	2,475	4,550
Pigs	1,100	730	285	1,120	3,235
Sheep	130	95	145	485	855
Poultry	1,065	765	500	240	2,570
Pets/other	470	60	600	2,030	3,160
Total	3,790	2,090	2,140	6,350	14,370

Figures are for 1995 in millions of U.S. dollars.

Source: Wood Mackenzie cited in James (1997).

additives. Pigs were the second largest market, followed by household pets/other. Poultry had the most sales in feed additives, but sales of poultry pharmaceuticals were relatively small. Pharmaceuticals are more important for cattle and household pets because the animals live longer and rely more heavily on curatives than on the preventatives provided in feed.

According to industry survey data published by PhRMA, private-sector research expenditures for animal health products was $324 million. To derive an estimate of private research on animal health for agricultural commodities, we assume that health research is distributed across commodities in proportion to sales of animal health products. From table 7.5, about 78 percent of global sales of all animal health products in 1995 were for agricultural commodities and about 22 percent for household pets and other animals. Thus, we estimate that about $252.7 million out of the total of $324 million was spent by private industry on animal health research for agricultural commodities.

Private Research in Other Sectors: Animal Nutrition and Animal Machinery

Direct estimates of research spending by private firms for other areas of animal production are unavailable. However, it is possible to derive estimates for private research on animal feed and animal machinery using indirect methods. For animal feed research, previous studies (the Malstead and Crosby surveys cited above) found that private research spending amounted to 0.17–0.22 percent of the value of annual feed sales. Assuming an average of 0.19 percent of feed sales are allocated to research on animal feeds, an estimate of private research spending can be derived for other years by multiplying this percentage with data on total feed sales. For 1996, feed purchased by farmers amounted to $25,235 million (Economic Research Service 1998), implying about $47.9 million was spent on research.

Research spending on machinery for animal production is more difficult to determine because machinery innovations applied to animal production may arise from a large number of manufacturing firms, many of which may not specialize in agriculture. To derive an estimate of research spending on animal machinery, the total number of animal machinery patents awarded in the United States was multiplied by an estimate of average research spending per patent. The number of machinery patents is taken from data developed by Johnson and Evenson that links patents from industry of manufacture (IOM) to a sector of use (SOU). For 1995, the latest year for which sufficient data are available, Johnson and Evenson (1998) identified 120 patents out of a total of 18,063 patents from machinery industries that were applied to animal agriculture. Also for 1995, the National Science Foundation (1996) reported that the machinery industry in the United States spent $5,041 million on research, or $279,000 per patent obtained by the industry. Applying this figure to the animal machinery patents gives an estimate of $33.5 million for animal machinery research by private industry.

THE ALLOCATION OF PUBLIC ANIMAL RESEARCH RESOURCES

The USDA's Current Research Information System (CRIS) provides a detailed catalog of public agricultural research by institution, commodity, technology area, field of science, and source of funding. Using CRIS and research definitions developed by Huffman and Evenson (1993), we analyzed trends in the allocation of animal research resources from 1970 to 1996. We first examined trends in annual research spending by commodity over this period. We then disaggregated research spending for three years (1970, 1984, and 1996) into technological areas and fields of science. We examined the levels of animal research relative to commodity value by institution (USDA or SAES) and by source of funds (federal, state, and nongovernment sources). These measures provide information on changing priorities in public-sector animal research and on the relative importance of private support for animal research at public institutions.

Research Spending by Commodity

The allocation of public research resources among animal commodities has adjusted slowly over time, although rankings by commodity in order of importance have not changed in the past 25 years (figure 7.3). Between 1970 and 1996, the largest allocation of public research funds for animal commodities went to beef cattle, followed by dairy cattle (figure 7.3). However, annual spending in constant dollar terms for beef research fell between 1980 and 1996. Swine research, on the other hand, increased by 66 percent in real terms between 1970 and 1996. In poultry research, there was a marked shift in emphasis from layers to broilers, and research on layers declined steadily in constant dollar terms.

The changes in the allocation of research resources among commodities may reflect economic as well as other factors. A starting point for examining the allocation of research resources among commodities is to compare the level of research expenditure relative to the value of commodity production. The *research intensity ratio* is calculated by dividing the level of research by the value of commodity production (or the value averaged over a period of years, to account for short-run fluctuations in price and yield) and multiplying by 100 percent. Whether research intensity ratios should be the same for all commodities depends on two key assumptions: First, the possibilities for technological advance for a given level of research must be the same across commodities; and second, the value of a scientific or technical innovation must be proportional to the current value of the commodity (Bredahl and Peterson 1976; Ruttan 1982).

There are several reasons why a public research administrator may want to depart from a constant research intensity ratio among commodities in alloca-

Figure 7.3. Public research expenditures for animal commodities, 1970–1996

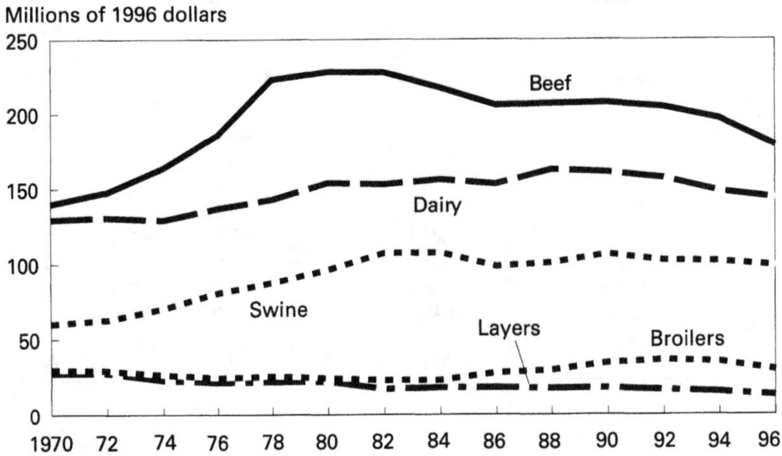

Annual research expenditures deflated by cost-of research price index (Klotz, Fuglie, and Pray 1995).

Source: *Current Research Information System*, USDA.

tion of research resources. First, consumer demand for a commodity may be changing. In such cases, the current production value may not be directly proportional to the long-run economic value of enhancing production through research. Second, technical spillovers may be large. For example, research on one commodity may provide information that can benefit other commodities and may be less costly to conduct than research on the other commodities it affects. And there may be noneconomic objectives, such as the desire to promote production in low-income areas or to help certain groups of producers, such as small farmers.

Table 7.6 reports research intensity ratios for animal commodities for 1970, 1984, and 1996. For total animal production, research intensity increased over this period, from 0.43 in 1970 to 0.81 in 1996. Research for sheep and wool is a clear outlier in the data: In 1996, $7 in research was spent for every $100 in production, compared with $0.90 or less per $100 of production for all other animal commodities. Another result that stands out is that the differences in research intensity ratios among commodities widened over this period. Excluding sheep and wool, the ratios ranged from a low of 0.25 for beef cattle to a high of 0.49 for broilers in 1970. By 1996, they ranged from 0.24 for broilers to 0.90 for swine. While research intensity ratios for poultry commodities declined, they rose for beef, dairy, swine, and sheep. The ratio for swine grew the most rapidly, tripling from $0.33 to $0.90 per $100

Table 7.6. Public-sector animal research relative to commodity value in 1970, 1984, and 1996

1970

Commodity	SYs	Research expenditure ($ thousand)	Production ($ million)	Research intensity ($ of research/$ of production) × 100%
Layers	122.1	6,719	2,034	0.33
Broilers	132.2	7,287	1,496	0.49
Turkeys	39.3	2,583	485	0.53
All poultry	403.7	22,577	4,192	0.54
Beef	510.9	33,840	13,730	0.25
Dairy	515.8	31,345	6,512	0.48
Swine	197.8	14,593	4,444	0.33
Sheep and wool	163.5	9,963	386	2.58
Total	2,032.6	126,931	29,528	0.43

1984

Commodity	SYs	Research expenditure ($ thousand)	Production ($ million)	Research intensity ($ of research/$ of production) × 100%
Layers	61.8	10,677	3,608	0.30
Broilers	85.1	13,707	5,520	0.25
Turkeys	46.5	8,566	1,581	0.54
All poultry	336.9	56,789	11,145	0.51
Beef	614.1	129,822	29,484	0.44
Dairy	461.5	92,974	18,244	0.51
Swine	286.1	64,069	9,509	0.91
Sheep & wool	127.0	23,805	464	5.13
Total	2,310.9	453,281	70,774	0.64

1996

Commodity	SYs	Research expenditure ($ thousand)	Production ($ million)	Research intensity ($ of research/$ of production) × 100%
Layers	40.1	12,989	4,139	0.31
Broilers	95.4	29,988	12,347	0.24
Turkeys	42.5	12,991	2,841	0.46
All poultry	306.1	91,915	19,947	0.46
Beef	467.7	178,383	33,846	0.53
Dairy	381.8	143,802	20,888	0.69
Swine	269.9	98,850	10,930	0.90
Sheep and wool	121.7	42,634	608	7.01
Total	2,108.0	720,919	89,359	0.81

SYs = scientist-years.
Production = a three-year moving average of annual cash sales.

Sources: SYs and research expenditures derived from USDA's *Current Research Information System* (1998); Value of commodity production from Economic Research Service, USDA.

of production between 1970 and 1996. The widening of research intensity ratios may reflect several factors: a decision of the public sector to concentrate on the commodities where private incentives are relatively weak, such as beef and dairy; a sluggish response in public research to changing market conditions; or the influence of noneconomic factors, such as a desire to help certain types of farms.

Research Spending by Technology Area

Between 1970 and 1996, more than 70 percent of public animal research could be categorized as either improving biological efficiency or protection and maintenance (table 7.7). Improving biological efficiency includes basic and applied research on breeding and animal genetics. Protection and maintenance includes disease control, nutrition management and feed efficiency. The rest of public animal research was allocated among other technology areas, including mechanization, management, postharvest uses, food safety, and pollution control issues. With the exception of food safety research on poultry, which saw an increase between 1984 and 1996, relatively little work was done in these other areas.

The emphasis on yield-increasing technology (biological efficiency and protection and maintenance) rather than on labor-saving technology (mechanization) may reflect policy objectives to develop scale-neutral technologies and concentrate on areas where private research incentives are relative weak. Ruttan (1982) suggests that research on agricultural mechanization should be left primarily to the private sector, for two reasons. First, mechanical innovations are more easily protected by patents than biological innovations, and second, improvements in mechanization may involve issues of scale that favor large farms. However, in some cases innovations in biological efficiency and disease protection may also entail economies of scale. Improvements in disease control, for instance, were an important factor in the development of large, confined poultry production facilities (Peterson 1967), and larger farms have tended to adopt new livestock breeding technologies more rapidly than small producers (Johnson and Ruttan 1997).

Basic and Applied Research

The growth in private agricultural research may allow the public sector to shift resources from applied and near-market research (where most private research is concentrated) to more basic agricultural sciences (see chapter 2). Public animal research in basic scientific fields did increase between 1984 and 1996: The share of SYs devoted to the five basic science fields increased from 33 to 41 percent of total SYs in animal research (table 7.8). The increase occurred both at USDA research agencies (from 32 to 40 percent) and at SAESs (from 35 to 41 percent). Scientist-years in biotechnology fields have increased rapidly since the early 1980s, rising from 165 in 1984 to 338 in 1996, even as total animal SYs declined from 2,311 to 2,109. This increase reflects a rise in

Table 7.7. Public-sector animal research by technology area

Commodity	Biological efficiency	Protection-maintenance	Mechanization	Management	Post-harvest	Food safety	Pollution alleviation	*Total*
	(percent of total research expenditures on commodity)							
All poultry								
1970	41.7	33.9	0.4	7.7	10.8	4.1	1.4	100
1984	45.8	37.7	1.2	4.3	7.0	3.2	0.8	100
1996	35.5	38.5	0.3	2.8	8.6	13.3	1.0	100
Layers								
1970	53.7	27.5	0.5	3.8	6.0	7.1	1.5	100
1984	56.8	33.8	0.7	4.2	3.7	0.2	0.7	100
1996	39.7	50.7	1.0	2.8	4.0	1.0	0.8	100
Broilers								
1970	39.4	33.1	0.8	6.7	16.0	3.3	0.7	100
1984	48.0	28.8	3.6	9.2	7.0	2.7	0.7	100
1996	33.4	36.6	0.5	4.8	7.9	16.1	0.8	100
Beef								
1970	50.4	24.8	0.2	9.1	12.8	1.5	1.2	100
1984	52.2	30.5	0.1	8.1	6.9	1.6	0.6	100
1996	44.0	37.1	0.0	8.5	6.3	3.3	0.9	100
Dairy								
1970	44.1	23.9	0.6	8.9	16.5	4.7	1.2	100
1984	50.0	29.5	0.4	7.9	9.7	1.9	0.5	100
1996	42.5	34.7	0.1	6.0	13.6	2.2	0.9	100
Swine								
1970	45.7	31.9	0.8	6.5	12.2	2.4	0.4	100
1984	50.0	36.3	0.7	3.6	5.4	3.4	0.6	100
1996	44.1	39.6	0.2	3.4	8.3	3.7	0.8	100
Other								
1970	40.3	30.6	0.1	7.9	16.6	4.3	0.3	100
1984	30.3	57.3	0.8	3.8	3.3	3.7	0.7	100
1996	31.4	60.1	0.2	2.6	2.7	2.3	0.7	100
Total								
1970	45.0	28.1	0.4	8.3	13.9	3.4	1.0	100
1984	45.9	37.7	0.5	6.0	6.5	2.6	0.6	100
1996	39.4	43.0	0.1	5.0	7.5	4.2	0.8	100

Research expenditures for these technology areas are determined by combining subsets of CRIS Research Problem Area codes according to the definitions developed by Huffman and Evenson (1993, p. 114, table 4.9), except for food safety and animal pollution, categories we have added. Another area of research is animal welfare (not shown in the chart). Approximately $7 million per year was allocated to animal welfare research in the 1990s.

Source: Derived from USDA's *Current Research Information System* (1998).

the application of biotechnology research methods to animal research at public agricultural research laboratories.

Another trend is evident from table 7.8—a steady shift in responsibility for animal research from the federal to the state level. Between 1970 and 1996, the USDA reduced the number of SYs in animal research from 631 to 377, a reduction of 40 percent. The number of animal scientists at SAESs, however, increased between 1970 and 1984 and then remained steady at about 1,730 through 1996. Thus, an increasing share of public animal research is being performed at SAESs rather than at the USDA.

Research Spending by Source

Public agricultural research institutions receive financial support from both government and private sources. The federal government funds the great majority of the research performed at USDA labs. But SAESs receive support from a number of sources, including federal agencies, state governments, private companies, producer groups, and product sales (National Research Council 1995; Fuglie et al. 1996).

Between 1970 and 1996, federal funds for animal research at public research institutions (including USDA and SAES institutions) declined while funding from nongovernment sources increased (table 7.9). The share of public animal research funded by the federal government fell from 54 percent in 1970 to 42 percent in 1996. Federal funding declined in real terms for each

Table 7.8. Public-sector animal research in basic and applied sciences and biotechnology at USDA and SAESs

Field of science	1970 USDA	1970 SAESs	1970 *Total*	1984 USDA	1984 SAESs	1984 *Total*	1996 USDA	1996 SAESs	1996 *Total*
Biochemistry and biophysics	72.5	131.9	204.2	32.5	130.1	162.6	26.6	109.2	135.8
Molecular biology	3.1	4.8	7.9	6.8	22.5	29.2	30.3	137.9	168.2
Animal genetics and breeding	32.3	141.8	174.2	30.9	131.0	162.2	24.3	115.3	139.7
Microbiology	67.1	71.3	138.5	39.7	96.0	135.7	39.3	130.7	169.9
Physiology	25.5	141.9	167.3	49.5	219.5	269.3	29.6	222.8	252.5
Basic biological sciences	200.5	491.7	692.2	159.0	599.1	759.0	150.1	715.9	866.1
All other science fields	430.8	909.8	1,340.4	378.2	1,174.1	1,552.4	226.7	1,015.8	1,242.4
Total basic and all other sciences	631.3	1,401.5	2,032.6	537.6	1,773.2	2,311.4	376.8	1,731.7	2,108.5
Basic biological sciences (%)	31.8	35.1	34.1	29.7	33.8	32.8	39.8	41.3	41.1
SYs in biotechnology fields	70.2	76.1	146.5	46.5	118.5	164.9	69.6	268.6	338.1

Figures in scientist-years (SYs).

Using the definitions developed by Huffman and Evenson (1993), we classify the basic animal science research areas as biochemistry and biophysics, molecular biology, animal genetics and breeding, microbiology, and physiology. We classify all other animal science fields in the CRIS system as applied animal research. Research utilizing new biotechnology tools and methods is probably most closely represented by molecular biology and microbiology research.

Source: Derived from USDA's *Current Research Information System* (1998).

commodity between 1970 and 1996, except for swine (but federal research funding for swine also declined after 1984). Between 1970 and 1996, non-government funds for public animal research increased from 14 to 20 percent. The largest allocations of private-sector funding for animal research went to beef and dairy (more than 20 percent of total funds in 1996) and the smallest to poultry (only 11 percent of layer and broiler research). Nongovernment sources make up an even larger share of animal research in SAES budgets, since most of the nongovernment funds reported in table 7.9 were directed to SAES rather than to USDA laboratories.

Table 7.9. Public-sector animal research by sources of funds

Commodity	Federal	State	Non-gov.	Federal	State	Non-gov.
	(thousands of 1996 dollars)			(percentages)		
All poultry						
1970	86,327	32,092	8,752	56.7	34.0	9.3
1984	58,524	44,120	9,159	45.7	45.0	9.3
1996	47,472	40,125	12,299	43.0	43.7	13.4
Layers						
1970	23,027	10,322	2,931	52.8	36.8	10.4
1984	10,659	9,166	2,044	39.2	49.7	11.1
1996	7,749	5,734	1,502	44.3	44.1	11.6
Broilers						
1970	26,288	9,332	3,027	59.4	30.7	9.9
1984	13,829	11,062	2,213	43.9	46.7	9.3
1996	17,849	10,803	3,332	52.9	36.0	11.1
Beef						
1970	80,160	42,426	26,935	50.9	30.0	19.1
1984	102,960	83,128	41,587	44.4	37.1	18.5
1996	74,677	68,505	37,198	40.7	38.4	20.9
Dairy						
1970	80,147	41,820	17,140	55.0	32.0	13.1
1984	73,344	63,926	26,756	43.5	39.8	16.7
1996	55,642	57,048	33,108	37.3	39.7	23.0
Swine						
1970	42,844	17,461	8,848	56.8	28.7	14.5
1984	58,227	38,774	17,096	49.5	35.0	15.4
1996	45,895	38,532	16,418	44.4	39.0	16.6
Sheep						
1970	35,215	9,653	4,952	64.9	23.2	11.9
1984	25,234	15,519	3,790	53.0	37.7	9.2
1996	21,632	15,674	7,325	46.1	36.8	17.2
Other						
1970	51,817	27,461	6,409	44.5	45.0	10.5
1984	90,499	45,628	22,396	54.1	30.8	15.1
1996	77,687	56,151	37,485	43.4	34.0	22.7
Total						
1970	376,510	170,914	73,035	54.0	32.2	13.8
1984	408,788	291,099	120,784	47.4	37.2	15.4
1996	323,005	276,035	143,832	41.8	38.3	20.0

Source: Derived from USDA's *Current Research Information System* (1998).

Maximizing Economic Returns to Animal Research

One way to assess the economic performance of research is to examine empirical studies on its rate of return in the United States. These studies use past investments in research to explain differences in productivity and productivity growth among states or regions. One limitation of some of these studies is that they may ignore or underestimate the importance of technology spillovers across regions. Interstate research spillovers can be significant because improved animal technology is often directly transferable across geographic boundaries, especially for commodities produced in confined facilities, such as poultry, swine, and dairy. Another limitation is that most of the studies focused exclusively on public-sector research and did not take private-sector contributions into account. By not considering private research, they may tend to overstate the benefits attributable to public research (Fuglie et al. 1996).

The studies offer a wide range of estimates of the annual internal rate of return to research on animal commodities (table 7.10). Estimates of the return to poultry research vary from 21 to 61 percent; to dairy research, from 25 to 43 percent; and to other livestock commodities (beef, swine, and sheep combined), from 22 to 66 percent. These estimates do not favor one particular animal commodity over another, and even the lower bound of the estimates indicates a high rate of return relative to the opportunity cost of investment capital (Fuglie et al. 1996). Given the rates of return—20 percent or higher—

Table 7.10. Rates of return to animal research in the United States

Investigator	Year published	Commodity	Period of study	Annual internal rate of return (%)
Peterson	1967	Poultry	1915–60	21
Bredahl and Peterson	1976	Poultry	1969	37
		Dairy	1969	43
		Beef, swine, and sheep	1969	47
Norton	1981	Dairy	1974	33
		Beef, swine, and sheep	1974	66
Smith, Norton, and Havlicek Jr.	1983	Poultry	1978	61
		Dairy	1978	25
		Beef, swine, and sheep	1978	22
Norton and Ortiz	1991	Poultry	1950–82	46
		Beef, swine, and sheep	1950–82	55
Huffman and Evenson	1993	Public livestock research (all)	1950–82	negative
		Public livestock research (basic)	1950–82	83
		Private livestock research	1950–82	86

These studies used econometric models to relate investments in research to changes in animal productivity on farms. Studies using time series data evaluated changes in animal productivity over time. Studies using data from a single period compared investment in research to differences in animal productivity among states.

it seems that the U.S. economy would do well to increase public investments in research on these commodities.

A different picture emerges from the most recent and probably most comprehensive study (Huffman and Evenson 1993). While Huffman and Evenson do not estimate rates of return for individual commodities, theirs is the only model to take private-sector contributions explicitly into account. They also attempted to include in their model the possibility of spillovers of technology across regions. In addition, they consider two different types of public research: research in the basic animal sciences (see table 7.8) and applied animal research. While the study finds very high rates of return to public investments in basic animal sciences and to private sector investments in animal research, it finds a negative return to applied animal research by the public sector.

Huffman and Evenson suggest that one possible reason for the poor performance of applied public animal research is that too many public resources have been concentrated on areas that are declining in importance, such as small dairy farms. Political pressures may be responsible for this trend. Meanwhile, private-sector research has been more responsive to expanding sectors, such as large integrated poultry and swine operations. Another reason for the low estimated returns may be that applied public animal research is increasingly focused on nonmarket issues, such as environmental and food safety concerns. Social and economic benefits from the development and diffusion of technologies in these areas are likely not to show up in traditional measures of output from animal production and therefore not be captured in the estimates of the rate of return. Nevertheless, these areas offer weak incentives for private-sector investment and will require continued public-sector support for applied research.

The studies reviewed in table 7.10 carry implications for public policy toward animal research. The decline in government funding for animal research is inconsistent with the general high rate of return to this research. Continuation of this trend would imply that a significant amount of economic growth will be foregone in future decades. Second, the shift in responsibility for animal research from the federal to the state level is inconsistent with the finding of large regional spillovers in animal technology. States may become increasingly reluctant to fund animal research adequately when other states can derive substantial benefits from that research without sharing the cost. In fact, state support for animal research fell 5 percent in real terms between 1984 to 1996, while federal support was off 21 percent.

Third, the finding that public-sector investment in basic animal sciences had a higher rate of return than applied sciences is consistent with the finding that public resources have been shifting toward basic sciences since 1984. In the long run, basic animal research is likely to generate greater economic benefits than applied research, because the results can be applied to a larger set of commodities in a wider geographic area over a longer period of time. In order for the public sector to improve the performance of its applied animal research, it may need to become more responsive to changing market and demand conditions.

INSTITUTIONS FOR PUBLIC-PRIVATE COLLABORATION IN ANIMAL RESEARCH

In recent years, the increased capacity of private-sector research and the fiscal constraints on public-sector research have led to greater collaboration in agricultural research, including in the animal research community. As we have seen, private-sector funding of research on animal commodities has become increasingly important at public institutions (table 7.9). This section discusses two institutional approaches to public-private cooperation in animal research. The first is the regional consortium model, which is used to achieve economies of scale in public research and education programs and to increase private support for these programs. The second is the Cooperative Development and Research Agreement (CRADA) model, in which private companies and USDA laboratories collaborate on R&D to develop and commercialize specific technologies (see chapter 8 for a general discussion of the CRADA model).

Regional Consortiums for Animal Research and Graduate Training at Universities

Agricultural land-grant universities and colleges have traditionally been the primary suppliers of skilled technical and scientific personnel for the agricultural industries. But a decline in student enrollment in animal science departments has forced some animal science programs to close or consolidate. For example, since 1950 the number of dairy and poultry science departments at U.S. universities has fallen by more than 30 percent, and by 1992 only four Ph.D. programs remained in dairy science and only six in poultry science (National Research Council 1995, 49).[2]

Because of the close relationship between teaching and research, reductions in student enrollments affect research programs (Ruttan 1982). The falling number of students enrolled in animal science programs has raised industry concerns about possible shortages of trained labor and the loss of publicly supported research. In an effort to generate support for animal science programs, public universities and private entities have been working together to establish regional consortiums (Havenstein 1998). These consortiums coordinate research, teaching, and extension among universities in neighboring states and work to win more support from private industry for university programs. This type of cooperative effort helps to avoid the free-rider problem in research (where one region benefits from, but does not help pay for, research conducted in a neighboring region). It also reduces research redundancies that may occur when universities and/or private companies undertake the same type of research.

The poultry sector has probably had the most experience with regional cooperation, but with mixed results. Some efforts have failed due to political and administrative problems, inadequate funding and support, insufficient numbers of students, and a lack of mechanisms to reward faculty for out-of-state teaching, research, and extension (Havenstein 1998). Experiments with regional consortiums are beginning to show some success, however. One of the most successful is the Midwest Poultry Consortium, which was formed with the cooperation of 24 private companies and 13 universities. In 1993, the consortium used approximately $1 million in private-sector contributions to establish the Midwest Poultry Science Undergraduate Center at the University of Wisconsin. The principal goal of the center is to attract students to poultry science (Havenstein 1998). Private-sector members also use the consortium to channel money to universities for poultry research.

Public and private cooperation in research and education is also working to address environmental problems associated with animal production. This development is of special interest, because private incentives to fund environmental research are particularly weak (due to the externality problem discussed above). In 1994, North Carolina State University organized the Animal and Poultry Waste Management Center. The new entity—a partnership among seven universities, two commodity groups (swine and poultry), and eight private companies and nonprofit groups—seeks to find new ways to improve livestock and poultry waste management. Each member of the consortium makes an annual contribution to a research fund and has a voice in how the funds are allocated. The consortium expected to raise about $1 million for research during its first three years.

Cooperative Research and Development Agreements

In 1986 Congress authorized cooperative research and development agreements, or CRADAs, to enhance research and technology transfer between federal laboratories and the private sector. Between 1987 and 1997, 161 CRADAs were established between the USDA's Agricultural Research Service (ARS) and private companies for animal-related technology (table 7.11). These CRADAs brought in approximately $9 million from the private sector to help finance animal research at ARS laboratories during this period. In addition to CRADAs, ARS also receives private funds for research through private trusts and reimbursables.[3] From 1990 to 1997, the ARS received over $15 million in private-sector funds for animal research—about half of which was generated by CRADAs and half from trusts or reimbursables.

The CRADAs have had some significant technological successes. One is the development of a more efficient method of vaccinating broilers and layers. The INOVOJECT machine allows producers to vaccinate 20,000–50,000 eggs per hour for Marek's and Gumboro disease, eliminating the need for manual postharvest injection. The ARS has patented this technology and licensed it to a private partner, Embrex, for commercialization. Approximately 80 percent of the North American market uses the procedure, and the market is expanding worldwide. Another is a technique for sorting cattle sperm by sex. This technology was developed with collaboration of Colorado State University and has been licensed to the XY Corporation.

Table 7.11. Number of CRADAs established and incoming funds from CRADAs and other sources for animal research at ARS/USDA, 1987–97

Year	New CRADAs established	Incoming CRADA funds[a] ($ thousand)	Other incoming funds[b] ($ thousand)
1987	2	58.3	
1988	14	803.7	
1989	10	347.0	
1990	14	396.6	138.0
1991	19	676.1	648.0
1992	12	634.9	1,195.8
1993	14	1,714.0	2,961.8
1994	25	868.2	3,731.5
1995	17	2,103.2	2,451.3
1996	16	505.3	1,917.0
1997	18	881.1	2,345.1
Total	161	8,988.4	15,388.5

[a] Scheduled to be received by USDA from the private sector over the life of the CRADA.
[b] Other sources of soft money comes from trusts (research grants) and reimbursables (compensation for materials).

Source: Compiled by Agricultural Research Service, U.S. Department of Agriculture for this volume.

To be successful, collaborative research needs to benefit both public and private partners. The ARS and private companies often share in basic research that leads to new products. For the government laboratory, cooperative research can speed up the commercialization process for new technologies and lower the R&D costs paid by taxpayers. This collaboration benefits the private partner by reducing the costs and risks associated with such research. As companies look to reduce internal research costs, developing partnerships with federal laboratories to conduct basic research can be an attractive alternative. In addition, collaboration can provide a company with access to services available only in government laboratories. For example, a company wishing to assess the efficacy of a new vaccine or drug using ARS's unique library of animal viruses can set up a cooperative agreement with ARS to conduct the assessment.

Formal, cooperative ventures between the public and private sectors in animal research technology development are fairly new. The specific institutional arrangements establishing how research is financed and performed, the ownership of intellectual property, and the sharing of economic benefits among the different participants are still being modified and refined. To date, little formal analysis has been done on the performance of these institutions relative to the performance of alternative arrangements for technology development and transfer. However, it is likely that leveraging public and private resources benefits both parties. These ventures promise to be a fruitful area for further study.

SUMMARY AND CONCLUSIONS

The private sector is an increasingly important part of the research system that develops new technology for animal production in the United States. Private incentives for animal research are strongest where markets for improved technology are large, technical advances can be made relatively easily and quickly, and intellectual property can be protected. Private animal R&D tends to concentrate on projects that are likely to result in market applications in the near future.

Table 7.12 provides a summary of public and private expenditures for animal research in the United States. We estimate that the private sector spent about $453 million for research on animal commodities in 1996, compared with $667 million spent by public agricultural research institutions. More than half of the private-sector research investment was for animal health. The private sector is the principal source of new veterinary pharmaceuticals and medicated feeds for animal producers.

In animal breeding, several multinational companies have emerged over the past several years to become major developers of improved animal breeds to producers. While private-sector animal breeding expenditures were less than public-sector breeding expenditures, the private sector was a major investor in basic poultry research and was responsible for most of the applied breeding research for broilers, layers, and swine. The willingness of the private sector to make investments in basic poultry research may be due to the high degree of market concentration and vertical coordination in these industries, which reduces technology spillovers. For large animals (dairy and beef), the public sector predominates in both basic and applied breeding research. Long gestation, low fecundity, and an inability to produce high-performing hybrid animals limit private research incentives for these commodities.

Research for new machinery for animal production was conducted almost entirely in the private sector. Intellectual property in this kind of innovation is relatively easy to protect through patenting. Research on animal economics

Table 7.12. Summary of public and private investments in animal research in the United States, 1996

<table>
<tr><th>Technology area</th><th>Private</th><th>Public</th></tr>
<tr><td>Breeding and biological efficiency</td><td>118.4</td><td>284.0</td></tr>
<tr><td>Health and protection</td><td>252.8</td><td rowspan="2">310.0</td></tr>
<tr><td>Feed, nutrition, and maintenance</td><td>48.5</td></tr>
<tr><td>Machinery</td><td>33.5</td><td>0.7</td></tr>
<tr><td>Management and economics</td><td>?</td><td>36.0</td></tr>
<tr><td>Externalities (pollution control and food safety)</td><td>—</td><td>36.0</td></tr>
<tr><td>Total</td><td>453.2</td><td>666.7</td></tr>
</table>

Sources: Estimates for the private sector are from authors' survey and other sources (see text). Estimates for the public sector are derived from USDA's *Current Research Information System* (1998).

and production externalities (pollution control and food safety) were primarily public-sector responsibilities because the private sector finds it difficult to appropriate research benefits in these areas.

In public research, there was an increasing degree of divergence in the value of research relative to production among animal commodities between 1970 and 1996. The widening of research intensity ratios may reflect several factors: a decision of the public sector to concentrate on the commodities where private incentives are relatively weak, such as beef and dairy; a sluggish response in public research to changing market conditions; or the influence of noneconomic factors, such as a desire to help certain types of farms.

Another trend in public animal research is a reallocation of some resources toward more basic animal sciences, especially in biotechnology fields such as animal molecular biology. This trend appears to have been underway since the mid-1980s. The increasing proportion of public resources spent on basic animal science complements the growth of private-sector capacity in applied animal research. It is also consistent with the results of economic research which showed a higher rate of return to research in more basic agricultural sciences. There remain important areas in applied science, however, where private research is likely to remain inadequate. Applied breeding for large animals and R&D for environmental and food safety issues related to animal production, in particular, are likely to require continued support from the public sector. In fact, since the mid-1980s there has been a substantial increase in public research resources devoted to food safety issues for poultry.

Within the public agricultural research system, there appears to have been a tendency for the federal government to reduce its support for animal research relative to state governments and the private sector. However, individual states may become increasingly reluctant to fund research when a large share of the benefits flow to producers in other states. The large regional and interstate spillovers in animal production technology would imply that a larger role for federal support for animal research may be necessary to overcome the "free rider" problem.

One response of state universities to declining support for animal science programs has been to form regional consortiums. Successful regional consortiums, such as the Midwest Poultry Consortium and the Animal and Poultry Waste Management Center have been able to garner significant support from private firms and commodity associations to support research and educational programs. The regionalization of state university programs faces considerable administrative and political challenges, however, and will require continuing innovation and experimentation in management and institutional structures.

Since the 1980s several changes have taken place in the way public research institutions transfer new science and technology into commercial use. The USDA and land-grant universities have made greater use of patenting new technologies and licensing them to private companies. USDA laboratories have also made extensive use of CRADAs, a type of joint research venture, with private firms. Through CRADAs, private resources can augment public research and help bring new technology to the marketplace at less cost to taxpayers. These ventures have already resulted in the commercialization of important new technologies for animal production. Institutional innovations such as these can lead to greater coordination and complementarity among public and private investments in agricultural science and technology, although their full implications for social welfare need further study.

ACKNOWLEDGMENTS

The authors would like to sincerely thank all those who helped us with information for parts of the paper and in reviewing the draft. Special thanks goes to Derek Byerlee, Basil Eastwood, Bill Phelps, Ray Gamble, Cornelius de Haan, Wallace Huffman, Cassandra Klotz-Ingram, John Lawrence, Ken Matthews, George Norton, Willis Peterson, Carl Pray, Richard Reynells, Michael Rouff, Charles Sattler, Tjaart Schillhorn Van Veen, Dave Schimmelpfennig, Dennis Unglesbee, and Terry Wing. In addition we would like to thank the numerous representatives from each of the poultry and livestock breeding companies who provided us with much of our insight as to what was occurring in terms of animal research in the private sector.

NOTES

1. Artificial insemination places sperm from selected males in the female reproductive tract by mechanical means. In embryo transfer, high-performance females are superovulated; the eggs are removed and fertilized with sperm from su-

perior males and then transferred to host mothers. Both techniques are reproductive technologies rather than breeding methods per se and can be applied either with outcrossing or crossbreeding. Cloning is also being investigated as a means of reproducing animals.

2. Industry changes are partly responsible for the decline in poultry programs. Poultry production is now concentrated in certain regions, and states that have lost production facilities have eliminated their university programs in poultry science.

3. Trusts are research grants which do not involve the transfer of intellectual property rights to the private funder. Reimbursables are funds received for research materials provided to companies or other institutions. As a matter of policy, the ARS does not do contract research and only accepts support from industry when the privately funded research complements on-going ARS research projects (June Blalock, personal communication, 1998).

REFERENCES

Agricultural Statistics, National Agricultural Statistics Service, Washington DC: U.S. Department of Agriculture, NTIS: NASS-ZAG-99INU.

APHIS. 1997. *Beef Cow-Calf Production Management and Disease Control.* National Animal Health Monitoring System, Veterinary Service. Washington, DC: Animal and Plant Health Inspection Service, U.S. Department of Agriculture.

Bredahl, M., and W. Peterson. 1976. "The Productivity and Allocation of Research: U.S. Agricultural Experiment Stations." *American Journal of Agricultural Economics* 58:684–92.

Brester, G., T. Schroeder, and J. Mintert. 1997. "Challenges to the Beef Industry." *Choices* 12(4): 20–25.

Bugos, G. 1992. "Intellectual Property Protection in the American Chicken-Breeding Industry." *Business History Review* 66 (Spring): 127–68.

Crosby, E. 1987. "Private Sector Agricultural Research in the U.S." In V.W. Ruttan and C.E. Pray, eds. *Policy for Agricultural Research.* Boulder, CO: Westview Press.

Economic Research Service. 1998. *Economic Indicators of the Farm Sector.* Washington, DC: Economic Research Service, U.S. Department of Agriculture.

Estienne, M. 1993. "Swine Artificial Insemination Workshops." University of Maryland Swine Report. University/Extension Factsheets and Research Report. Princess Anne, MD: Department of Agriculture, University of Maryland Eastern Shore.

Fuglie, K., N. Ballenger, K. Day, C. Klotz, M. Ollinger, J. Reilly, U. Vasavada, and J. Yee. 1996. *Agricultural Research and Development: Public and Private Investments under Alternative Markets and Institutions.* Agricultural Economic Report 735. Washington, DC: Economic Research Service, U.S. Department of Agriculture.

Funk, D. 1996. "Genetic Technologies in the 1990s." Proceedings of the National Dairy Congress, Venado Tuerto, Sante Fe, Argentina, September 4–5, 1992.

Havenstein, G. 1998. "Mid-Atlantic Poultry Consortium: A Department Head's Viewpoint on Regionalization." *Poultry Science* 77:201–07.

Henneberry, S., and B. Charlet. 1992. "A Profile of Food Consumption Trends in the United States." *Journal of Food Products Marketing* 1:3–23.

Henry, R., and G. Rothwell. 1995. *The World Poultry Industry.* Washington, DC: International Finance Corporation.

Huffman, W.E., and R.E. Evenson. 1993. *Science for Agriculture: A Long Term Perspective.* Ames: Iowa State University Press.

James, C. 1997. "Progressing Public-Private Sector Partnerships in International Agricultural Research and Development." ISAAA Briefs 4. Ithaca, NY: International Service for the Acquisition of Agribiotech Applications, Cornell University.

Johnson, D.K.N., and R.E. Evenson. 1998. *Johnson-Evenson Patent Set.* Available from internet at djohnson@wellesley.edu.

Johnson, N., and V.W. Ruttan. 1997. "The Diffusion of Livestock Breeding Technology in the U.S.: Observations on the Relationship Between Technical Change and Industry Structure." *Journal of Agribusiness* 15 (Spring): 19–35.

Jordan, F. 1990. *Poultry Diseases.* 3rd ed. London: Baillière Tindall.

Klotz, C., K. Fuglie, and C. Pray. 1995. "Private-Sector Agricultural Research Expenditures in the United States, 1960–92." AGES 9525. Washington, DC: Economic Research Service, U. S. Department of Agriculture.

Lawrence, J., G. Grimes, and M. Hayenga. 1998. "U.S. Pork Industry Structure." Staff paper. Department of Economics, Iowa State University, Ames.

Lesser, W. 1993. "Royalty Collection for Patented Livestock." AER 93-10. Ithaca, NY: Department of Agricultural Economics, Cornell University.

Martinez, S., and A. Reed. 1996. *From Farmers to Consumers: Vertical Coordination in the Food Industry.* Agricultural Information Bulletin 720. Washington, DC: Economic Research Service, U. S. Department of Agriculture.

McLaren, D. 1992. "Biotechnology Transfer: A Pig Breeding Company Perspective." *Animal Biotechnology* 3:37–54.

Narrod, C. 1997. *Technology Transfer in the Poultry Industry: An Examination of Supply Factors and Externalities Associated with Increased Production.* Ph.D. dissertation, University of Pennsylvania, Philadelphia.

Narrod, C., and C. Pray. 1995. "Technology Transfer in the Poultry Industry: Factors and Externalities Associated with Increased Production." In Kenneth Steel, ed. *Animal Wastes and the Land-Water Interface.* Boca Raton, FL: CRC Press.

National Research Council (NRC). 1995. *Colleges of Agriculture at the Land Grant Universities: A Profile.* Washington, DC: National Academy Press.

National Science Foundation (NSF).1996. *Research and Development in Industry.* Washington, DC: National Science Foundation.

Norton, G. 1981. "The Productivity and Allocation of Research: U.S. Agricultural Experiment Stations Revisited." *North Central Journal of Agricultural Economics* 3 (January): 1–12.

Norton, G., and J. Ortiz. 1991. "Reaping Returns to Research." Staff Paper 91-6. Blacksburg, VA: Department of Agricultural Economics, Virginia Polytechnic Institute and State University.

Peterson, W. 1967. "Returns to Poultry Research in the United States." *Journal of Farm Economics* 49 (August): 656–69.

PhRMA. 1998. Internet homepage www.phrma.org. Pharmacuetical Research and Manufacturers of America.

Rogers, G. 1979. "Poultry and Eggs." In L. Schertz, ed. *Another Revolution in U.S. Farming*. Washington, DC: U. S. Department of Agriculture.

Rose, S. 1997. *Principles of Poultry Science.* New York: CAB International.

Ruttan, V.W. 1982. *Agricultural Research Policy.* Minneapolis: University of Minnesota Press.

Schmit, B. 1987. "Disease Prevention Crucial in Intensive Livestock Production." *Zootecnica International* (July): 49–51.

Smith, B., G. Norton, and J. Havlicek, Jr. 1983. "Impacts of Public Research Expenditures on Agricultural Value-Added in the U.S. and the Northeast." *Journal of the Northeastern Agricultural Economic Council* 12 (Fall): 109–14.

Thurman, W. 1987. "The Poultry Market: Demand Stability and Industry Structure." *American Journal of Agricultural Economics* 69 (February): 30–37.

U.S. Department of Agriculture (USDA). 1998. *Current Research Information System.* Washington, DC: Cooperative State Research, Extension, and Education Service.

Ward, C. 1997. "Vertical Integration Comparison: Beef, Pork, and Poultry." Selected Papers of the 1997 Annual Meeting of the Western Agriculture Economics Association, Reno, Nevada, July 13–16.

Warwick, E., and J. Legates. 1979. *Breeding and Improvements of Farm Animals.* 7th ed. New York: McGraw-Hill.

III. Institutions for Public-Private Collaboration in Agricultural Research

8

THE CRADA MODEL FOR PUBLIC-PRIVATE RESEARCH AND TECHNOLOGY TRANSFER IN AGRICULTURE

KELLY DAY-RUBENSTEIN AND
KEITH O. FUGLIE

THE 1980s witnessed a heightened interest in strengthening the links between federally funded and private-sector research. Several pieces of federal legislation were enacted to encourage the transfer of technology between public research laboratories and the private sector. These laws affected ownership rights to new technology developed with federal funds and established new mechanisms for collaboration between public- and private-sector researchers. The primary policy goal was to increase the economic returns to federal research and development (R&D) by disseminating those federal technology resources that have commercial possibilities and moving the technologies more rapidly into the marketplace (Congressional Research Service 1991).

Although agriculture was rarely the direct focus of the technology transfer initiatives, the U.S. Department of Agriculture (USDA) responded to the new laws by strengthening its ties to the private sector. Historically, the USDA has been actively engaged in technology transfer through its agricultural extension service, which works directly with end users (mainly farmers, but also agribusiness firms, households, and communities) to promote the adoption of new technology and services. The atomistic structure of the agricultural sector, which includes many small, competitive firms, initially limited the incentives for private agricultural research (Alston and Pardey 1995). However, the past few decades have seen a rapid expansion of private-sector capacity in agricultural R&D by specialized firms, creating new opportunities for public-private partnerships in technology development (Fuglie et al. 1996).

The Cooperative Research and Development Agreement (CRADA) is a principal means of fostering technology transfer. (Other means of encouraging technology transfer include patent licensing, direct research grants, and research consortia.) The distinguishing characteristics of the various forms of research collaboration are the way resources are shared and intellectual property rights assigned among participating institutions. A CRADA is a partnership undertaken by a federal laboratory and a private company with the goal of developing and commercializing technology. It specifies the activities and responsibilities of each partner, defines how data are to be shared and released, and stipulates how intellectual property rights will be assigned. In principle, this type of agreement allows the public sector to focus on generic or precommercial aspects of technology development while the private sector concentrates on downstream activities such as developing manufacturing techniques and testing and promoting new products.

The CRADAs have resulted in several economically important technologies (GAO 1994). The anti-cancer drug Taxol was developed via a CRADA (Day and Frisvold 1993). The CRADAs have also helped commercialize useful agricultural technologies, including significant pest and disease controls. Some of these CRADAs have produced technologies with additional social benefits, such as environmental protection (Stanley 1998; Becker 1997; Harden 1997).

Despite their success, however, CRADAs have been criticized on several grounds. Some critics have charged that these arrangements are not particularly efficient at commercializing technology (Bozeman and Coker 1992). There is also concern that CRADAs may have diverted public research from its central research mission (National Research Council 1995). And finally, successful CRADAs may have created unfair competitive advantages for specific companies (Cohen and Noll 1995). But previous studies of CRADAs and other technology transfer mechanisms have relied heavily on case studies. Either they have not evaluated the overall performance of CRADAs or have not considered the USDA's experience with technology transfer in depth. This paper provides the first in-depth study of CRADAs at the USDA, using the entire set of CRADAs.

The paper is organized in the following format: It first presents an overview of recent technology transfer legislation and several models for public-private research and technology transfer. The discussion then turns to alternative models of public-private technology transfer employed by the USDA. Using the USDA's unique research classification systems (set out in the appendix), we examine funding at USDA research laboratories, comparing the shares of public, public-private, and private research. In addition, we attempt to determine how USDA, CRADA, and other types of public-sector research compare with private-sector in-house research, using data from studies funded by the Small Business Innovation Research (SBIR) program. The final section discusses some implications for agricultural research and technology transfer policy.

THE CHANGING ENVIRONMENT FOR TECHNOLOGY TRANSFER

There was widespread concern during the 1970s and 1980s about both the declining rate of productivity growth in the U.S. economy and the loss of international competitiveness among U.S. companies. Some believed that constraints to the transfer of technology between the public and private sectors were in large part responsible for this situation. Traditional methods of transferring technology to U.S. firms (publications, education and training, consultancies, and personnel exchanges) were viewed as inadequate; research innovations were not being commercialized quickly enough, if at all. The result was legislation easing the transfer of new technology between public research institutions and the private sector (Congressional Research Service 1991; National Academy of Sciences 1992).

Table 8.1 lists key technology transfer legislation enacted since 1980. One of the first major changes dealt with patent policy. While public institutions had for some time had the right to patent, the federal government nevertheless assumed ownership of any inventions resulting from federally funded research. Technologies described in patents are often far from commercial viability, and without exclusive licenses companies are often unwilling to make the investments necessary to commercialize them. Responding to this constraint, the Government Patent Policy Act of 1980 (the Bayh-Dole Act) gave institutions what is called "certainty of title" for inventions resulting from federally funded research. The legislation allowed universities and other cooperating institutions receiving federal funds for research to obtain the patents on any inventions they discovered. It also expanded the right of federal laboratories to issue licenses for patents of their inventions. Previously, these laboratories had been able to grant only nonexclusive or open licenses, but the new act allowed them to grant exclusive licenses. These changes to patent policy were designed to encourage scientists at universities and other public institutions to seek more patents and to provide private companies with incentives to work with these patents.

Other legislation sought to promote greater collaboration between federal laboratories and the private sector. The 1980 Stevenson-Wydler Technology Innovation Act mandated that each federal research agency develop specific mechanisms for disseminating government innovations. Prior to this act, federal technology transfer activities had been voluntary, and each agency had used its own method of disseminating information on new research discoveries and technology. The 1986 Technology Transfer Act gave government agencies additional means to foster technology transfer by authorizing CRADAs. Prior to the 1986 act, federal researchers had not been permitted to

Table 8.1. New federal legislation to promote technology transfer

Year	Legislation	Action
1980	Bayh-Dole Act	Authorized federal agencies to grant exclusive licenses for federally owned patents and allowed institutions to own patents on technologies developed from federally funded research
1980	Stevenson-Wydler Act	Encouraged federal laboratories to increase cooperation with private sector in technology development and transfer
1982	Small Business Innovation Development Act (amended 1993)	Earmarked 1.5 percent of federal extramural research funds to small businesses (increased to 2.5 percent in 1993)
1984	National Cooperative Research Act	Provided antitrust exemptions for private participation in research consortia
1986	Federal Technology Transfer Act	Established rules for federal-private Cooperative Research and Development Agreements (CRADAs), gave authority for exclusive licenses
1995	National Technology Transfer and Advancement Act of 1995	Granted CRADA cooperators the first right of negotiation for exclusive licensing of one use of any invention created through a CRADA

Sources: National Academy of Sciences (1992).

collaborate directly with the private sector (Congressional Research Service 1991). The new act allowed federal laboratories to enter into collaborative research agreements with private companies as well as with universities and state and local government entities. This act was followed by the 1995 National Technology Transfer and Advancement Act, which clarified ownership of intellectual property in CRADAs. Prior to the passage of the act, ownership was granted to the government. The act gave private-sector cooperators first right to negotiate for exclusive licensing of at least one use of any invention created through a CRADA; it also gave the federal government rights to royalties from the invention.

Related initiatives were enacted in 1982 and 1984. The Small Business Innovation Act of 1982 required federal agencies to earmark a portion of extramural research funds to private companies with 500 or fewer employees through the Small Business Innovation Research (SBIR) grant program.[1] The act originally required federal agencies to earmark 1.5 percent of their extramural R&D funds for small business, but this figure was increased to 2.5 percent in 1993. The 1984 Cooperative Research Act encouraged joint research among private companies (with or without public partners) by limiting the antitrust liability of companies collaborating in precommercial R&D.

Models of Technology Transfer

The legislation just described governs the exchange of information and resources between public and private collaborating institutions and, in general terms, states how rights to new inventions are to be owned and benefits shared. Figure 8.1 presents a stylized model of the various research and technology development mechanisms currently available to federal research agencies for cooperating with the private sector and other nonfederal institutions.

Figure 8.1. Models of public-private research collaboration and technology transfer

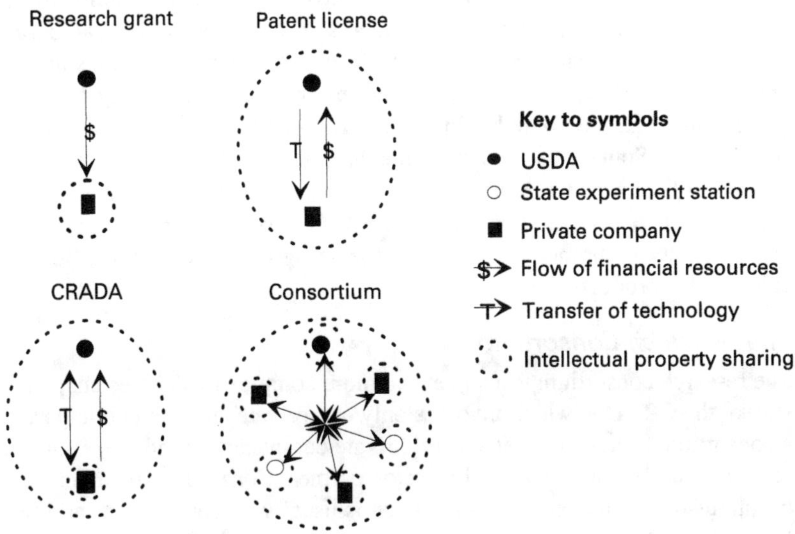

The Research Grant Model

With this model, the simplest mechanism for collaborative research, the government provides the monetary resources for private in-house research. In this situation there is no formal research collaboration between the federal lab and the nonfederal partner, and the grant recipient has sole ownership over any patentable technology. This type of arrangement characterizes the SBIR program and the Advanced Technology Program run by the Department of Commerce.

The Patent Licensing Model

Here, a public research institution develops and patents a technology and then assigns the rights to use the patented technology to nonfederal institutions or private companies. The rights may be exclusive, partially exclusive, or nonexclusive. Exclusive patent licenses are awarded when they are necessary to promote private commercialization—for example, when a company must make significant investments in product and market development, or when substantial commercial risk is involved. Patent licenses usually include a royalty payment that returns either a fixed fee or a percentage of revenues to the public institution.

The CRADA Joint-Venture Model

With this model, both sides contribute in-house resources to R&D, and the nonfederal collaborator may provide the federal laboratory with some research funds. Federal laboratories may provide personnel, equipment, and laboratory privileges, but not financial resources, to a nonfederal partner (Congressional Research Service 1991). The partner may give funds directly to the federal lab, however. Patents resulting from a CRADA may be jointly owned, although the nonfederal partner has first right to negotiate an exclusive license. Some data also may not be publicly disclosed for a certain amount of time. The CRADAs or other types of joint public-private ventures are appropriate in cases where there is a well-developed private sector supplying inputs to farmers or processing raw agricultural commodities into value-added products.

The Research Consortium

The research consortium is a somewhat more complex model (see chapter 9). Unlike the CRADA, which involves only one private and one public partner, a consortium brings together several private companies to undertake joint research, with or without a public-sector partner. Consortium members contribute resources for the research, which is usually precommercial, and have first rights to technologies developed by the consortium. Companies can protect spin-off technologies through trade secrets or new exclusive patents. Research consortiums have proven useful for increasing support for strategic research and research that is considered to be long term and high risk (see chapter 9). Additional applied and adaptive research is often required, however, to develop and diffuse technology to farmers. Thus, consortiums often rely on the in-house research capacity of their members (either private companies or national agricultural research programs) to develop specific applications from the more generic results of consortium-sponsored research.

Public and Private Goods

Ideally, the new mechanisms and models permit the allocation of public resources to research that has the characteristics of a public good. Such research has value to society, but is unlikely to be undertaken by for-profit firms because it is nonrival and nonexcludable (Samuelson 1954). A good is *nonrival* if one person can use it without reducing what is available to others. A lighthouse is a nonrival good because it serves all passing ships; so is information, the main product of research, because many persons can use it without reducing its availability to others.

A good is *nonexcludable* if others cannot be prevented from using it after it first becomes available. Information can be replicated through photocopies at negligible costs, for instance, although a firm may have invested heavily in developing the information. Patents and other protections can make the results of some research partially excludable by restricting the use of information for a limited time, but patenting is difficult with more basic research that may not have specific commercial applications. Important areas of applied research, such as technologies designed to address environmental externalities or improve food safety and nutrition, may also lack commercial interest. In these cases, the externalities reduce end-user demand. But when adopted, these technologies provide important social and economic benefits.

Figure 8.2 illustrates some of the different research mechanisms along a spectrum, moving from purely public to purely private goods.

TECHNOLOGY TRANSFER AT THE USDA

The USDA has also had a long and unique history of close collaboration with private agricultural industries, in part because public agricultural research is often more applied in nature than other types of public research (Fuglie et al. 1996). The USDA has transferred many technologies to the private sector and maintains close ties to farmers and suppliers of agricultural technologies. Yet the new mechanisms for public-private collaboration in research have had a significant impact at the agency. The USDA's research agencies have signifi-

Figure 8.2. The public-private research continuum

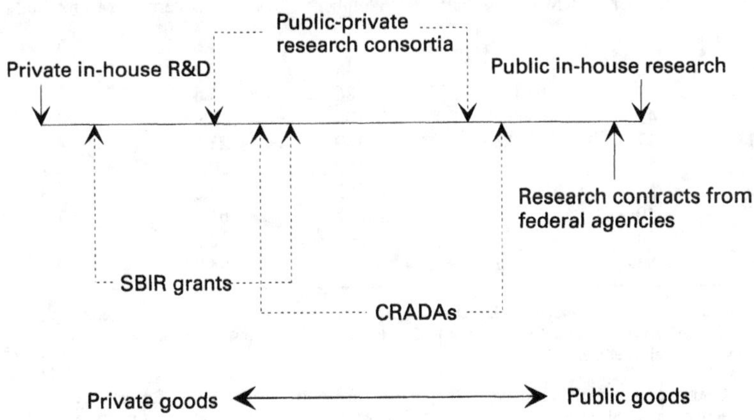

cantly increased technology transfer to the private sector in the last decade (table 8.2).

Statistics showing the number of CRADAs established and patents granted, plus the growth in the SBIR program, provide a measure of increased USDA activity in technology transfer. In 1996 the Agricultural Research Service (ARS), the USDA's primary in-house research agency, was participating in 258 active CRADAs involving nearly $100 million of research resources.[2] In total, the USDA has established more than 700 CRADAs since 1987. The relative contributions of the USDA and outside collaborators show that the private sector plays a substantial role in the technology transfer process. Financial data from 366 CRADAs entered into by the ARS between 1987 and 1995 show that 64 percent of CRADA resources come from collaborators and only 36 percent from the USDA.[3]

The changes in patent policy have increased incentives for USDA scientists to seek patents for the results of their research and for the USDA to license patents to the private sector. Royalties from patent licenses have risen steadily since 1987, indicating increased licensing activity. In 1996 the USDA received more than $2 million in patent royalties (a fraction of ARS's annual research budget of some $700 million.)

The growth of the SBIR program is tied to appropriations for USDA extramural research funding as a whole and is managed by the agency's Cooperative State Research, Education, and Extension Service. In 1996, SBIR grants totaled $9.1 million of the USDA's $364 million budget for extramural research.

Table 8.2. USDA technology transfer activities

Year	Patents awarded	Patent license royalties	Active CRADAs[a]	Value of CRADAs[b]	Small business research grants
	Number	Million dollars	Number	Million dollars	Million dollars
1987	34	0.09	9	1.6	3.5
1988	28	0.10	48	8.7	3.6
1989	47	0.42	86	15.6	3.7
1990	42	0.57	104	18.9	4.1
1991	57	0.83	139	25.6	4.5
1992	56	1.0	160	30.0	4.4
1993	57	1.5	185	34.0	6.9
1994	40	1.4	212	61.3	7.2
1995	38	1.6	227	80.1	9.4
1996	53	2.1	258	98.9	9.1

[a]Number of Cooperative Research and Development Agreements (CRADAs) with the private sector.
[b]Value of CRADAs includes the total value of USDA and private-sector resources committed to active CRADAs over their lifetime.

Source: Compiled by Agricultural Research Service and Cooperative State Research, Education, and Extension Service for this volume.

Federal regulatory guidelines for the new technology transfer mechanisms leave federal agencies with considerable discretion to formulate policies for collaborating with the private sector. As technology transfer activities have increased, the USDA has instituted a number of administrative procedures to help ensure that CRADAs, patent licensing, and other collaborative efforts achieve USDA policy objectives. These guidelines, which govern the selection of private-sector collaborators and the sharing of research costs and benefits, aim to assure that all parties fulfill their commitments.

The CRADA Program at the USDA

The USDA has a relatively high degree of experience in establishing research partnerships. This is evidenced by the fact that the USDA's CRADA program has been operating longer and has had more agreements per appropriated dollar than similar programs at other federal agencies (June Blalock, personal communication, 1997).

The CRADAs are generally initiated by ARS scientists (Willard Phelps, personal communication, 1997). According to USDA technology transfer officials, the guidelines for these arrangements are that the research must be consistent with the agency's mission, that there be no conflicts of interest, and that fairness be shown to potential cooperators (June Blalock, personal communication, 1997). Once scientists determine that an innovation has market potential, they begin seeking outside collaborators. Sometimes industry representatives suggest a collaboration based on interest in an ARS scientist's work—for example, research findings presented at a professional meeting. The ARS has a staff of technology transfer coordinators to assist in the process of setting up CRADAs. The USDA policy has always been to grant cooperating firms the option of negotiating an exclusive license to any patented inventions resulting from a CRADA, even though this option was not required by law until 1995 (USDA 1992).

The ARS informs potential cooperators of research advances through announcements at workshops and conferences, advertisements in the Federal Register, and electronic postings. It also maintains a database called TEKTRAN, which reports research findings that have been peer reviewed and cleared by ARS management.

Examples of important CRADAs begin with the first CRADA between a private company, Embrex, and a government lab, the ARS. This CRADA developed a method to vaccinate chicken embryos inside the eggshell against Marek's disease (a significant and potentially fatal poultry disease which attacks birds' nervous systems). The injections are expected to be used on more than 2.5 billion chickens a year (Becker 1997).

The CRADAs have also advanced the commercialization of biopesticides. Biopesticides may control pests with fewer negative effects on the environment. One example is a micronutrient called Keyplex, which uses plants' natural defense mechanisms and has fungicidal and insecticidal properties (developed with Morse Enterprises Limited). Because Keyplex is no more toxic than fertilizer, it does not even require registration as a pesticide (Stanley 1998). Other important biopesticides developed via CRADAs are derived from the neem tree. Neem oil has important fungicidal and insecticidal properties, yet is more environmentally benign than many alternatives.

Another example of a successful CRADA was a joint effort with Cytec Industries focused on the agricultural amendment polyacrylamide (PAM). Adding PAM to irrigation water can virtually halt soil erosion caused by furrow irrigation. In 1995, the first year of PAM's commercialization, farmers used the product on 50,000 acres, saving an estimated one million tons of soil (Harden 1997).

MEASURING THE EFFECT OF TECHNOLOGY TRANSFER ON THE ALLOCATION OF RESEARCH RESOURCES

The USDA employs a unique classification system for its in-house and extramural programs that makes it possible to examine the pattern of federal funding for agricultural research in considerable detail (see appendix). Under this system, agricultural research can be grouped into five main areas: (1) natural resources, (2) plant production and protection, (3) animal production and health, (4) postharvest use of agricultural commodities, and (5) human health and nutrition. A sixth category captures all other research. The research can be broken down even further, although with finer distinctions the limitations of the classification process are likely to introduce more measurement errors.

The estimates of resources devoted to CRADAs are from an ARS database of CRADA projects undertaken between 1987 and 1995. During this period the ARS entered into 528 CRADAs with nonfederal partners, the great majority of which were for-profit companies. The research and financial data for each of these CRADAs have been compiled and aggregated into the research categories mentioned above. In addition, public and "private" contributions to CRADA research are tabulated separately for each research category.[4] One limitation of the analysis is that detailed financial data are available for only 366 of the 528 projects. Therefore, the research allocation by category has been calculated as both the share of CRADA projects and the share of financial resources for those projects with complete data.

To capture market demand for agricultural innovations, we use the allocation of SBIR grants between 1987 and 1995. Although these grants represent a very small component of private agricultural research, the composition of successful grant proposals helps to determine the private sector's research preferences. More aggregate estimates of private-sector research serve as another indicator, although these data are not easily classified into comparable categories (Klotz, Fuglie, and Pray 1995).

Hypotheses and Assumptions

We begin with the assumption that joint collaboration between public and private researchers through CRADAs involves, in some cases, sharing economic benefits as well as R&D costs. From the standpoint of many cooperators, CRADAs can reduce not only R&D costs but also the risk inherent in certain types of research. By reducing research costs and giving private companies the opportunity to obtain exclusive rights to the results of research, CRADAs may increase the range of R&D alternatives explored by private firms.

Besides specific information about a particular technology or line of development, CRADAs offer cooperating firms access to the human scientific capital found within the government. Compared with private industry, the USDA's scientific knowledge is often more grounded in basic research, so that USDA scientists offer expertise to which private industries otherwise might not have access. The research effort itself can also give the cooperator useful information for future R&D efforts. Even if a technology is not marketable, the results of "negative" research can be valuable because they give companies information about potential technologies and the development process. The CRADAs benefit the USDA by giving government scientists not only insight into market demand for new technology and into industry's needs and resources, but also information about certain lines of research inquiry. The USDA also benefits through shared license fees and royalties, and individual USDA researchers earn a share of patent royalties.

We hypothesize that public in-house research will focus on research with the characteristics of a public good. While we hypothesize that the allocation of resources for joint projects will reflect a middle ground between each partner's priorities, we expect that the public share of R&D costs will be greater for research that has a larger public good component. The CRADA priorities, for instance, will be relatively close to the public-sector agenda, because the initial R&D effort is determined by the government. However, private interest in CRADAs will depend on market considerations, so that private goods will also be on the agenda. We expect SBIR grants, however, to be closer to purely private goods, since these projects originate from the private sector. But SBIR proposals are awarded competitively by government-selected peer reviewers, potentially affecting the type of projects chosen. By examining the public good content of these research categories, then, we form four hypotheses on the priorities for public and private research.[5]

First, public research will concentrate more heavily on natural resources and human nutrition than private research, since the social returns in these areas are greater than the private benefits. Second, animal research, which is likely to provide relatively high economic returns, will receive more emphasis from the private sector. Market concentration in the livestock industry has already increased the private sector's ability to appropriate research benefits, particularly for swine and poultry research (Johnson and Ruttan 1997). Third, plant and postharvest research will receive approximately the same amount of attention from public and private researchers. Although stronger intellectual property protection is now available for new crop varieties and genetically engineered plants than in the past (Fuglie et al. 1996), private incentives for several important components of crop improvement research remain weak (see chapter 6).

Fourth, postharvest research places particular emphasis on developing new uses for agricultural commodities. While new products and processes can often be protected (through patents and trademarks), the R&D required to bring them to market often entails substantial commercial risk. For example, large investments in manufacturing scale-up may be required before the commercial viability of a product or process can be determined, and a new product that is unfamiliar to consumers may need to be widely advertised. But once these investments are made, they significantly lower the hurdle for other firms to enter the market. The public good nature of the initial investments thus reduces the incentive for private research in new uses for agricultural commodities, so this type of research can be expected to interest both the public and private sectors.

What the Results Show

Public research priorities are shown by the share of total ARS research resources allocated to each of the five categories. The CRADA data are used to indicate public-private activities and the SBIR data to represent private-sector priorities. Table 8.3 shows the percent of total research allocated to each category for ARS, CRADAs, and SBIR grants.

The results show that the public sector emphasized nutrition research more heavily than the private sector or public-private collaborations. While human nutrition research constituted 9 percent of ARS research, it accounted for only 3 percent of CRADA resources and less than 1 percent of SBIR grants. The apparent challenge in finding private-sector partners for nutrition research is further illustrated by the relative public and outside-collaborator contributions to CRADA research (table 8.4). While on average the ARS contributed 36 percent of the resources in CRADA research, it contributed 52 percent of research resources for CRADA projects on nutrition.

Research on natural resources shows a similar though less pronounced pattern. The ARS devoted a greater share of its research budget to natural resources research than either CRADAs or SBIR programs. However, CRADAs may be underutilized as a technology transfer mechanism for natural resources. Private-sector demand revealed through SBIR grants seems to indicate a larger interest in this type of technology than CRADA activity suggests.

The data support our hypothesis that animal research has a stronger private good component compared with the other categories. While 23 percent of

SBIR grants were allocated to animal research, this category received only about 18 percent of ARS research. The CRADA activity appears to reflect public research priorities, with 17 percent of CRADA resources devoted to animal research, although 23 percent of all CRADA projects were for livestock research.

The public sector places a somewhat higher priority on plant research. The

Table 8.3. Agricultural research allocation by public, public-private, and private research institutions by technology area

Research category	Public (ARS)	Public-private (CRADAs)[a] By no. of agreements	By resources	Private (SBIR)
		(percent of total)		
Natural resources	13.8	4.4	6.3	9.5
Plants	36.9	32.2	36.5	25.7
Animals	17.6	23.1	17.2	23.0
Postharvest utilization of agricultural commodities	20.8	37.6	34.6	27.0
Human nutrition and well-being	9.3	1.1	2.7	0.8
General	1.6	1.5	2.7	4.4
Rural issues				9.6
Total	100.0	100.0	100.0	100.0

ARS: USDA Agricultural Research Service (USDA in-house research).
CRADA: Cooperative Research and Development Agreement between ARS and an outside cooperator.
SBIR: Small Business Innovation Research grants (private-sector initiated research).
[a]Based on data from 528 CRADAs initiated by ARS between 1987 and 1995. Value of research resources available for only 366 projects.

Source: Calculated from ARS databases and USDA Current Research Information System *Inventory of Agricultural Research* (1998).

Table 8.4. Public and private contributions to CRADAs by technology area

Research category	Public	Private (outside cooperators)
	(percent of total)	
Natural resources	40.2	59.8
Plants	33.4	66.6
Animals	36.5	63.5
Postharvest utilization of agricultural commodities	36.6	63.4
Human nutrition and well-being	52.0	48.0
General	37.1	62.9
All	36.1	63.9

Contribution based on the value of resources contributed to 366 CRADA agreements between USDA and outside cooperators between 1987 and 1995 (69 percent of all CRADAs during this period).

ARS and CRADAs devote roughly one-third of their research resources to plants, the SBIR only one-quarter. What these data may mask is a significant difference within this category in the responsibilities each sector assumes. A recent comprehensive survey of plant-breeding research by the public and private sectors shows that the ARS concentrates most of its research on long-term, prebreeding activities, while the private sector devotes most of its resources to short-term variety development (Frey 1996). The ARS has terminated most of its research on variety development, preferring to concentrate on areas of research not pursued as intensely by the private sector.

Postharvest research on new uses for agricultural commodities diverges somewhat from the expected pattern. This research constitutes a much larger share of CRADA research (35 percent) than it does of either ARS research (21 percent) or private-sector research (27 percent). However, the allocation of SBIR grants may understate private-sector interest in research for new products and processes, since the grants are restricted to small businesses. The food industry also comprises many large companies that are ineligible for the SBIR program, so their preferences are not reflected in these data. Klotz, Fuglie, and Pray (1995) estimate that private-sector in-house research on food and related products makes up about 30 percent of all private-sector food and agricultural research. This estimate also suggests, however, that a disproportionate share of CRADA resources may be allocated to research on new uses.

Finally, research data can be used to address the question of whether CRADA activities have diverted ARS researchers from the USDA's central research mission. Shares are compared for 1986 (before the CRADA program began) and 1995 to see whether public priorities have shifted over time in response to private interests. The results suggest that public priorities in agricultural research have remained relatively constant over time (table 8.5). However, two areas with strong public good components grew. The ARS research allocations for natural resources research increased from 11 percent of the ARS's research budget to 13 percent between 1986 and 1995. Human nutrition research also showed a slight increase over time. At the same time, animal research, a field with strong private-sector demand, fell by about 5 percent of total resources. The growth of public-private collaboration in agricultural research may therefore have allowed public research to shift more resources to areas where incentives for private research are likely to be relatively weak rather than to areas of greater private-sector interest. This finding differs from that of Ulrich, Furtan, and Schmitz (1986), who found that closer ties between public research and the private sector shifted public priorities toward private-sector interests.

Table 8.5. Allocation of ARS/USDA research funds by technology area

Research category	1986	1995
	(percent of total)	
Natural resources	11.0	13.4
Plant	38.4	37.7
Animal	22.5	17.3
Postharvest utilization of agricultural commodities	19.4	21.4
Human nutrition and well-being	8.0	9.1
General	0.7	1.1
Total	100.0	100.0

Source: Adapted from USDA Current Research Information System *Inventory of Agricultural Research* (various years).

SOME CONCLUSIONS AND IMPLICATIONS FOR FUTURE RESEARCH

The evidence presented in this paper suggests that closer cooperation between the USDA and the private sector in agricultural R&D may have enhanced research efficiency by enabling the public sector to focus resources on areas where private incentives are relatively weak. While the overall pattern of research allocation by the USDA has remained relatively stable over time, USDA research concentrates most heavily on technologies that serve important societal needs and areas where market failures are most evident. Specifically, public research concentrates more on natural resources and human nutrition, areas where environmental externalities and information asymmetries are likely to reduce private incentives to develop new technology. Private research concentrates more heavily on livestock, a sector of agricultural production in which a consolidated industrial structure has made it easier for private firms to appropriate gains from research. Public and private research devotes roughly the same resources to plant research and research on new uses for agricultural commodities. However, even here the public sector appears to concentrate on basic, precommercial research rather than applied R&D, which the private sector emphasizes.

Resource allocation for joint research between the USDA and the private sector appears to reflect a middle ground between public and private priorities. While private-sector participants in CRADAs contribute, on average, around two-thirds of the financial resources, the public share is higher for areas where private incentives are lower. The pattern of resource allocation among CRADAs also suggests that the USDA may be underutilizing these partnerships for natural resources and overutilizing them for R&D in postharvest uses of agricultural commodities. However, our indicator for private research priorities (SBIR grants) may not fully reflect private-sector demand for public science and technology, since this indicator includes only research projects involving small companies.

A key administrative issue with both CRADAs and patent licensing is the selection process for private-sector partners. Public-private joint endeavors may give participating companies a competitive advantage. The USDA's strategy to assure fairness in the licensing selection process has been to publicize its research program widely and give preference to the first company that demonstrates a credible commitment to R&D and a sound commercialization timetable. This approach appears to rely heavily on the experience and judgment of public-sector technology transfer officers. An alternative approach for identifying the most efficient private-sector partner may be a more explicit competitive process, such as competitive bidding. But while a competitive selection process may help identify the cooperator able to commercialize a technology at least cost, the process may be confounded if higher commercialization costs are associated with better quality proposals which are more likely to succeed.

An important area for future research is the quantitative assessment of the economic returns to alternative models for technology transfer. As with many federal research programs, the economic performance of CRADAs is difficult to evaluate on a case-by-case basis. A common measure of project performance is whether a patent is awarded or a license granted. However, patent and license data fail to account for several important factors. First, private companies are often slow to license or patent CRADA technologies, because the exclusivity granted by a CRADA means that there is less urgency to seek intellectual property protection immediately (J. Blalock, personal communication, 1997). Second, R&D is a risky undertaking, and not all projects succeed. However, negative research contributes to future successful innovations by narrowing the path of inquiry and should be accounted for in economic measures of success. Finally, the value to a private company of a CRADA may not lie solely in a specific technology. Collaborators often cite strong benefits from the interaction with government scientists, which can increase their exposure to a wide range of emerging technologies (Beardsley 1993). Nonetheless, as the duration of technology transfer activities exceeds a decade, better information continues to become available about these programs. Future research will attempt to better measure and compare economic outcomes among these alternative mechanisms for transforming new knowledge into improved goods and services.

APPENDIX: THE RESEARCH CLASSIFICATION SYSTEM

The detailed classification system developed by the USDA to track research allocation is unique within the federal government. Since the mid-1970s, each USDA research project has been assigned one or more codes under the Current Research Inventory System (CRIS). This system makes it possible to examine the allocation of research by detailed topical areas (for example, by scientific field, commodity, or resource problem areas). In-house agricultural research conducted by the ARS, as well as SBIR grants, are all assigned CRIS codes. In addition to the CRIS system, the ARS maintains its own internal system for classifying research based on its strategic plan using what are called strategic planning codes (STPs). All CRADA research, for example, is assigned STPs. To match and compare research allocated to the ARS, CRADAs, and private companies through SBIR grants, we developed a correspondence, or crosswalk, between the CRIS and the STP classification system. This technique is straightforward for broad research categories such as the ones used in this study, but using more defined research areas increases the risk of introducing measurement errors when moving between the two classification systems.

In this paper, we report the allocation of research for the ARS, CRADAs, and SBIR grants according to the five main STP categories (resource systems, plant systems, animal systems, postharvest systems, and nutrition systems). The ARS also has two other STP categories, a systems integration research category and a small, general classification category (general systems technologies). Since the subcategories under systems integration correspond to the other main categories, we allocated systems research to the other categories (table 8.A.1). SBIR grants contain an additional category, rural issues research, which appears below the other categories on our tables.

The ARS maintains a database for each CRADA that describes the nature of the research, the timetable, and a budget that shows the resources the public and private partner have agreed to contribute. The budget includes the value of monetary and in-kind resources devoted to the project over its lifetime (usually two to four years). (Note that CRADA agreements occasionally address more than one project, and some projects involve more than one agreement. We merged the data by projects.) The ARS gave the authors a copy of the database, which describes 528 CRADAs initiated between 1987 and 1995. In the database, most projects were assigned an STP (those without an STP were assigned one by the authors). Detailed financial data are available for only 366 (69 percent) of the CRADAs. Table 8.3 shows the distribution of CRADA research among the five principal areas by the number of projects and by financial resources for the subset of projects with these data. Overall, the two methods show similar results, although the share for animal research

Table 8.A.1. Classification of agricultural research into major technology areas

Technology area	Assignment of STP[a] codes used to classify CRADA research	Assignment of RPA[b] codes classify ARS and SBIR research
Natural resources	1.1 Atmosphere and climate 1.2 Soil 1.3 Water 6.1 Resource management: systems and models	101, 102[c], 103, 105, 106, 107, 109, 110[c], 306, 901
Plants	2.1 Plant production 2.2 Plant protection 6.2 Plant production and protection systems	102[c], 111, 112, 201, 202, 204, 205, 206, 207, 208, 209, 214[c], 301, 302, 304, 305, 307, 308, 309, 314, 318[c], 706[c], 905, 906
Animals	3.1 Animal production 3.2 Protection of animal and human health 6.3 Animal production systems	110[c], 210, 211, 212, 213, 214[c], 310, 311, 312, 313, 317, 318[c], 706[c], 904
Postharvest utilization of agricultural commodities	4.1 New uses, new products 4.2 Safety 4.3 Postharvest losses and quality changes 4.4 Quality definition, grading and assessment 6.4 Systems and models for conversion and delivery	401, 402, 403, 404, 405, 406, 407, 408, 409, 410, 411, 412, 501, 503, 512, 604, 701, 702, 704[c], 707
Human nutrition and well-being	5.1 Human nutritional requirements 5.2 Food composition and bioavailability 5.3 Nutritional status and well-being 6.5 Human nutrition systems and models	703, 704[c], 708, 709, 802, 805
Rural issues		801, 804, 907, 908
General	6.6 General systems technologies 8 General	104, 113, 114, 203, 315, 316, 502, 506, 508, 509, 601, 603, 902, 903

[a]STP: planning codes defined by the Agricultural Research Service (ARS).
[b]RPA: Resource Problem Areas, defined by USDA's Current Research Information System (CRIS).
[c]RPA: falls into more than one major technology area. In these cases, research resources in the RPA were split (usually 50/50) between the technology areas.

is somewhat higher in terms of number of projects (23 percent) than in terms of financial resources (17 percent). When these shares are compared with the allocation of ARS and SBIR research, each method of determining the share of CRADA research shows a similar ranking. An advantage of the budget data is that it allows us to identify the contributions of both the USDA and the outside collaborator.

NOTES

1. At the USDA, SBIR grants are allocated competitively to small businesses for research on food and agricultural concerns.

2. These figures represent the value of research resources in CRADAs over the life of the projects, which typically last about three years. Annual resources devoted to CRADAs are around $30–35 million.

3. Collaborator resources were estimated by the collaborators themselves, not by the USDA.

4. "Private" research is represented by the research of outside cooperators. Most of these cooperators are private-sector firms.

5. Federally funded research contracted to outside scientists may also be expected to conform to the public research agenda. Private-sector in-house research is likely to emphasize private goods, although some precommercial research (which cannot easily be appropriated) may be funded through research consortia or by very large firms that can internalize these benefits.

REFERENCES

Alston, J., and P. Pardey. 1995. *Making Science Pay: The Economics of Agricultural R&D Policy.* Washington, DC: American Enterprise Institute.

Beardsley, T. 1993. "CRADA Mania, Will Joint Research Make U.S. Industry Competitive?" *Scientific American* 268:104–06.

Becker, H. 1997. "Researchers' Egg Vaccines Protect Billions of Chickens." *Agricultural Research Service News.* USDA. December 17.

Bozeman, B., and K. Coker. 1992. "Assessing the Effectiveness of Technology Transfer from U.S. Government R&D Laboratories: The Impact of Market Orientation." *Technovation* 12:239–55.

Cohen, L., and R. G. Noll. 1995. "The Future of the National Research Labs." Paper presented at the National Academy of Sciences Colloquium on Science, Technology, and the Economy, Irvine, CA, October 20–22.

Congressional Research Service. 1991. *Transfer of Technology from Publicly Funded Research Institutions to the Private Sector.* A Report Prepared by the Congressional Research Service for the Use of the Subcommittee on Oversight and Investigations of the Committee on Energy and Commerce, U.S. House of Representatives. Washington, DC: U.S. Government Printing Office.

Day, Kelly A., and George B. Frisvold. 1993. "Medical Research and Genetics Resources Management: The Case of Taxol." *Contemporary Policy Issues* 11:1–11.

Frey, K. J. 1996. *National Plant Breeding Study I: Human and Financial Resources Devoted to Plant Breeding Research in the United States in 1994.* Special Report 98. Ames: Iowa Agriculture and Home Economics Experiment Station, Iowa State University.

Fuglie, K., N. Ballenger, K. Day, C. Klotz, M. Ollinger, J. Reilly, U. Vasavada and J. Yee. 1996. *Agricultural Research and Development: Public and Private Investments Under Alternative Markets and Institutions.* Agricultural Economics Report 735. Washington, DC: Economic Research Service, U.S. Department of Agriculture.

GAO (Government Accounting Office). 1994. *Technology Transfers: Benefits of Cooperative R&D Agreements.* GAO/RCED-95-52. Report to the Vice Chairman, Joint Economic Committee, U.S. Congress.

Harden, B. 1997. "USDA Scientists Honored for Moving Research to Market." Agricultural Research Service Press Release. April 10.

Johnson, N. L., and V. W. Ruttan. 1997. "The Diffusion of Livestock Breeding Technology in the U.S.: Observations on the Relationship Between Technical Change and Industry Structure." *Journal of Agribusiness* 15:19–35.

Klotz, C., K. Fuglie, and C. Pray. 1995. *Private-Sector Agricultural Research Expenditures in the United States,* 1960–1992. Staff Paper 9525. Washington, DC: Economic Research Service, U.S. Department of Agriculture.

National Academy of Sciences. 1992. *The Government Role in Civilian Technology: Building a New Alliance.* Washington, DC: National Academy Press.

National Research Council (NRC). 1995. *Allocating Federal Funds for Science and Technology.* Washington DC: National Academy Press.

Samuelson, Paul. 1954. "The Pure Theory of Public Expenditure." *Review of Economics and Statistics* 36:387–389.

Stanley, D. 1998. "Exploring Plants' Protective Proteins." *Agricultural Research.* 46:8–10.

Ulrich, A., H. Furtan, and A. Schmitz. 1986. "Public and Private Returns from Joint Venture Research: An Example from Agriculture." *Quarterly Journal of Economics* 10:103–29.

U.S. Department of Agriculture (USDA). 1987–1998. Current Research Information System, *Inventory of Agricultural Research.* Washington, DC: Cooperative State Research, Extension, and Education Service.

———. 1992. Technology Transfer Agreements with the Agricultural Research Service. Washington, DC: Agricultural Research Service.

9

THE RESEARCH CONSORTIUM MODEL FOR AGRICULTURAL RESEARCH

MARY K. KNUDSON[1]

In the 1970s and 1980s, U.S. laws governing intellectual property rights (IPRs) and technology transfers underwent significant changes. These changes have had profound effects on agricultural research and development (R&D). By providing limited monopoly control through patents and patent-like protections, for instance, IPRs increased incentives for the private sector to invest in R&D. They also opened up new possibilities for cooperation between the private and public sectors in plant breeding research and development (R&D). Given the continuing decline in public-sector funding for agricultural R&D, these developments have been particularly important.

New federal legislation enacted in the 1980s and 1990s further facilitated scientific cooperation between public and private R&D laboratories (see chapter 8). One such institutional structure is the research consortium model, in which private firms and public agencies collaborate on basic and applied research, have joint ownership of the products of their collaborative work, and sometimes share any profits. This structure has presented the research community with a viable alternative to public funding for basic R&D.

However, a number of concerns have been raised about the possible adverse effects of these private-public partnerships on the traditional mission of public-sector research. The public sector tends to focus on basic research that produces results with the attributes of a public good.[2] Do the new partnerships support enough high-risk, basic research to assure long-run technological progress? Do consortia invest in applied research in cases where private incentives are weak but social value may be high—for instance, research on the environment? Is the private sector setting the public R&D agenda?

This paper presents case studies of two relatively recent public-private research consortia in agriculture: the Germplasm Enhancement of Maize (GEM) project and the Biotechnology Research and Development Corporation (BRDC). Both exist because of the changes that have taken place in IPRs and technology transfer laws. The GEM consortium is examined first, and in some detail, with an eye to determining if the consortium produces scientific and technological advances that meet the twin policy goals of strengthening research on public goods and developing commercial applications. The paper shows that GEM does appear to be achieving both goals. The consortium mainly pursues research that the private sector is reluctant to undertake on its own but that demonstrates significant potential for long-term commercial applications with high social returns. Furthermore, GEM's unique arrangement helps private-sector partners share the risk when conducting relatively basic research. Based on this case study, the paper identifies the factors that help a public-private research consortium succeed, including the important role of the federal government in management and information sharing.

A briefer study of the BRDC considers whether the factors identified as key to GEM's success hold true for other research consortia. Because the BRDC is older than GEM, it offers the opportunity to examine a consortium that has had time to bring technological successes closer to commercialization. The study identifies a set of institutional design factors in the BRDC that are similar to those in GEM, strengthening the argument that a viable research consortium model of public-private R&D collaboration not only exists but has proven successful in more than one instance.

The paper is arranged as follows: First, it reviews the changes in IPRs and technology transfer laws that gave rise to the new public-private initiatives in agricultural research. It then sets out the history and structure of GEM and presents a list of the factors responsible for the consortium's success. A brief analysis of the BRDC follows. The paper concludes by considering the policy implications of public-private R&D arrangements for federal resources and leadership.

A BRIEF LOOK AT PLANT BREEDING RESEARCH AND DEVELOPMENT

Agricultural R&D in the United States has three principal components: federal and state programs, which together comprise the public sector, and the private sector. The federal government conducts both intramural and extramural agricultural research through the U.S. Department of Agriculture (USDA). Intramural agricultural research is carried out by the USDA's Agricultural Research Service (ARS), the Forest Service, and the Economic Research Service. Extramural research is carried out through state agricultural experiment stations (SAESs) housed at universities and is also supported by state government and private funds. The USDA's Cooperative State Research, Education, and Extension Service administers most of USDA's funds for extramural research.

The USDA's intramural R&D serves a number of purposes:

- It provides the in-house science expertise that is essential to maintaining a position of national (and international) leadership in agricultural science;
- It accommodates high-risk, long-term research that the private sector is reluctant to support (see chapter 2);
- It addresses national and regional problems with high social payoffs but few incentives for investment at the local level;
- It maintains expensive research infrastructure and laboratory capacity;
- It supports regulatory agencies and the development of science-based regulations and policy;
- It offers a means of cooperating in multinational agricultural research partnerships; and
- It facilitates technology transfer and commercialization (Fuglie et al. 1996).

In the past several years, the ARS has concentrated most of its resources for plant breeding on basic research because the private sector has significantly increased its investment in applied breeding (Frey 1997; Fuglie et al. 1996). However, there is some concern that applied programs for minor crops, which offer limited economic but high social returns, are being underfunded (Frey 1997).

States typically direct their research efforts toward resource and production constraints that are important locally (Fuglie et al. 1996; Ruttan 1982). State-supported R&D may therefore be more applied than federal research. Nevertheless, like federal R&D programs, state programs still concentrate on basic research that will result in knowledge with the characteristics of a public good. States also carry out long-term, risky R&D. Some research may result in products for which there is little or no market (such as varieties that conserve environmental resources or enhance genetic diversity). In other cases, states may undertake research on which private investors cannot adequately capture returns (such as pureline varieties that farmers can reproduce from year to year without having to purchase new seed) (Knudson 1990). States are still the major source of new pureline varieties in wheat, rice, barley, oats, rye, and triticale, for instance (Fuglie et al. 1996).

The private sector focuses more intensively on applied or commercially oriented products, such as breeding cultivars for higher yield. However, private firms have little incentive to do either basic research (for example, genome mapping, germplasm development, conservation, or characterization) or research that improves public and consumer decision making (such as basic and applied research on water quality, global climate change, and other environmental and health concerns). While the private sector's investments in plant breeding have been growing, private firms still depend on the public sector for advances in basic research (Fuglie et al. 1996).

Neither the public nor the private sector has played a static role in agricultural R&D. Both the amount and kind of research each sector undertakes have changed. The most significant change has been the growing importance of private-sector R&D in plant breeding. Private-sector investments in plant breeding and genetics have grown dramatically in recent years, rising from $5.6 million in 1960 to $400 million in 1992 (Klotz, Fuglie, and Pray 1995). Private-sector investments have been particularly strong in soybeans and wheat (Perrin, Kunnings, and Ihnen 1983; Butler and Marion 1985). In 1992 federal and state contributions toward *all* crop research, including plant breeding, totaled approximately $1 billion (Fuglie et al. 1996). Despite the growth in private-sector investment in plant breeding, however, funding for the USDA and SAESs has barely kept pace with the rising costs of conducting research (Frey 1997). While the private sector's investments in plant breeding have been growing, it has become an increasingly important source of funding for agricultural research at SAESs (Fuglie et al. 1996).

A 1994 survey of the plant breeding industry verifies the importance of the private sector in R&D (see chapter 5). Of the 2,241 scientist-years (SYs, or the scientific input of one person working full time for a year) devoted to the plant breeding industry in 1994, 1,499 were in the private sector, 529 at SAESs, 177 at the ARS, and 36 at the USDA's Plant Materials Centers. The figure for the private sector represents an increase of 160 SYs between 1990 and 1994. The figures for the public sector, however, show a decline of 12.5 SYs during the same period.

Frey (1997) also finds that the type of R&D each sector conducted differed. Some 88 percent of the SYs at the ARS and 59 percent of those at the SAESs were devoted to either basic plant breeding or germplasm enhancement that will help build a more genetically diverse parental pool—long-term, relatively risky research with uncertain returns. But the private sector dedicated a full 80 percent of its SYs to *applied* research in crop development, leaving only 20 percent for basic research (Frey 1996). The basic R&D conducted by the private sector is initiated with short-term goals in mind. For example, the private sector has been using germplasm enhancement in gene transformation research in an effort to locate specific genes for which companies can obtain the IPRs (Frey 1998).

Given these figures, it appears that the basic research in agricultural sciences typically undertaken by the public sector is not being carried out at a level that will guarantee future growth, much less support the kinds of research the private sector is unlikely to consider. Had the climate surrounding property rights in the United States (which discouraged public-private collaboration in research, as the next section shows) continued unmodified, the chances of changing this situation would be slim. But over the last three decades, revisions to the laws governing public and private R&D have created new opportunities for public-private cooperation.

CHANGES IN INTELLECTUAL PROPERTY RIGHTS AND TECHNOLOGY TRANSFER LAWS

The legal framework for protecting intellectual property in biological inventions has undergone significant changes since the 1970s (table 9.1). The first major change came with the passage of the 1970 Plant Variety Protection Act. This act, which was amended in 1980 and 1994, gives plant breeders plant variety protection certificates (PVPCs) for new varieties of seed-generated crops. The certificates provide patent-like protection for a period of 20 years, subject to two exemptions. The *research exemption* allows researchers to use protected material in research programs. This exemption was partly curtailed by the 1994 amendments, which addressed concerns regarding the potential misdirection of R&D funds by withdrawing the act's protections for minimal (cosmetic) alterations in plants. The second exemption, the *farmer* or *crop exemption,* originally allowed farmers to sell harvested seed to other farmers if the sellers received less than 50 percent of their income from such sales. The 1994 amendments also curtailed this exemption, prohibiting farmers from selling seed derived from any variety released since April 1995 without a license from the owner. A 1995 Supreme Court decision, *Asgrow vs. Winterboer,* further reduced the scope of the crop exemption, allowing farmers to save seed for their own replanting but barring them from selling protected seed.

Table 9.1. Laws affecting plant breeding research and technology transfer

Law	Date	Description
Plant Variety Protection Act	1970	Amended in 1980 and 1994; gives 20 years of patent-like Protection Act protection for new crop varieties; includes a limited research (PVPA) exemption; no longer permits farmers to sell or exchange protected seed, but farmers can keep seed for own use.
Bayh-Dole Act/ Patent Policy Act	1980	Allows nonfederal institutions to own patents resulting from federally funded research.
Stevenson-Wydler Act/ Technology Innovation Act	1980	Allows federal agencies to grant exclusive licenses for federally owned patents and for federal labs to receive research money from the private sector. Amended by the Federal Technology Act of 1986 which established framework for public-private Cooperative Research and Development Agreements (CRADAs).
Ex parte Hibberd	1985	Patent office ruling which extended utility patent protection for plants and plant parts; there is no research exemption.
Asgrow vs. Winterboer	1995	Supreme Court decision prohibiting farmers from selling or exchanging seed protected by the PVPA.

In a 1980 case involving a genetically engineered microbe *(Diamond vs. Chakrabarty)*, the Supreme Court ruled that new forms of living material are patentable. In 1985 the U.S. Patent Office extended the concept of living material to include plants or plant parts, including genes *(Ex parte Hibberd)*. The advantage for a plant breeder of a utility patent instead of a certificate is that a utility patent has no crop or research exemption clause.

The 1980s saw other changes in the agricultural R&D environment as well. The Bayh-Dole Patent Policy Act of 1980 allowed scientists conducting research with federal funds to own patents to their discoveries. The Stevenson-Wydler Technology Innovation Act of 1980 and its amendment, the Technology Transfer Act of 1986, allowed federal agencies to grant exclusive licenses for federally owned patents and to accept research funds from private sources. The 1986 act also created a new vehicle for joint research between federal laboratories and private companies by authorizing Cooperative Research and Development Agreements, or CRADAs (see chapter 8).

The changes in IPRs and technology transfer laws were aimed at increasing private-sector investments in agricultural research. As we have seen, the laws appear to have succeeded for plant breeding. Between 1970 and 1994, a total of 3,306 PVPCs for plant breeding were issued, of which 87 percent were granted to the private sector (Fuglie et al. 1996). A survey by Knudson and Pray (1991) showed that the original Plant Variety Protection Act may also have increased public research in corn, cotton, sorghum, and wheat.

Similar trends have occurred in the use of utility patents. Between 1985 and 1994, 324 patents were granted for multicellular organisms. Of this number, 286 were for plant parts or new plants, primarily in corn (Fuglie et al. 1996). The private sector received most of these patents, with 204 going to U.S. companies. The public sector (including state and federal governments as well as American universities) obtained 48 patents. Future trends should see increasing levels of public-sector involvement. In a 1989 survey of SAES directors, many reported that they planned to increase their use of patents, particularly in wheat, corn, and soybeans (Brooks 1989).

The growth of private sector involvement in plant breeding R&D raises several issues about the direction of future research in the area. In pursuit of funds in a time of diminishing resources, the public sector may begin to focus on research that generates products eligible for IPRs. If it does, the government will find itself competing directly with the private sector. And if the level of basic research the public sector typically supports declines, applied research will eventually suffer as the pool of potential new technological advances is exhausted (Fuglie et al. 1996). Further, there is concern that the availability of utility patents for new plants and plant parts could restrict the flow of research materials and crop germplasm among researchers. A 1989 survey of SAES directors offers some support for this concern: 84 percent thought patents would adversely affect the flow of plant germplasm between public and private breeders in the United States; another 73 percent extended their concern to the flow of international sources of germplasm (Brooks 1989).

Pray, Knudson, and Masse (1993) found in a 1990 survey of private plant breeders that U.S. firms had mixed views about the effects of the Plant Variety Protection Act and utility patents on access to new technology. Most of the 90 firms surveyed said that the act had had no impact or a positive impact on information and germplasm spillover from both the public and private sectors. However, 35 percent thought that utility patents had had a negative impact on the exchange of germplasm. But 25 percent responded that germplasm exchange was likely to increase, since with patent protection firms were less likely to rely on trade secrets to protect their intellectual property.

A central objective of the new and revised laws has been to strengthen investments in basic research and accelerate the rate of technology transfers from federal laboratories and universities to the private sector. In the area of plant breeding, the research consortium model has provided an innovative means of encouraging cooperation between the public and private sectors on basic research. By spreading the costs of research among the partners, a consortium arrangement can significantly improve the climate of support for basic research that is too costly for any one member to undertake alone. Further, it provides a ready means of sharing data and information among public and private scientists. An examination of two of these consortia demonstrates the ways in which new institutional arrangements have altered the relative balance between basic and applied research in plant breeding.

THE GERMPLASM ENHANCEMENT OF MAIZE CONSORTIUM: CASE HISTORY

By looking at the Germplasm Enhancement of Maize consortium (GEM), it is possible to more carefully assess the way in which new institutional arrangements have altered the relative balance between basic and applied research in plant breeding. The GEM project grew out of the Latin American Maize Project (LAMP), an international effort dedicated to promoting greater biodiversity in corn (maize) cultivars. Developed in 1994, GEM was designed to satisfy the need for both basic and applied agricultural research. It would carry out basic research by testing the most promising lines or accessions (an accession is a distinct plant genotype, such as a variety) from LAMP as well as other exotic accessions, adapting them to growing conditions in the United States, and introducing them into more-advanced breeding programs. To satisfy the need for applied research, GEM would seek commercial applications for the new materials.

The GEM consortium has already successfully introduced 51 exotic materials from LAMP. The consortium's institutional arrangement spreads the risk associated with basic research on genepool enhancement among all members. Futhermore, until the material becomes the property of a firm, it is available to all members and to nonmembers, giving it the added dimensions of a public good. The consortium has also had some laboratory successes that show potential for commercial success, and private firms appear optimistic that GEM material will result in economically valuable traits in hybrid varieties. In the fall of 1996, private-sector members increased their in-kind support by 25 percent, from $450,000 to $562,000 (GEM 1996).

The following sections present the history, development, and structure of GEM and discuss key issues surrounding this innovative institution.

The Latin American Maize Project

LAMP was initiated in 1987 as a cooperative effort among the United States, 11 Latin American countries (Argentina, Bolivia, Brazil, Chile, Colombia, Guatemala, Mexico, Paraguay, Peru, Uruguay, and Venezuela), and the Mexico-based International Maize and Wheat Improvement Center. Plant breeders from national agricultural research programs sought to preserve and evaluate the 14,000 corn germplasm accessions in germplasm banks in their countries. The ultimate goal was to utilize these accessions more fully in corn breeding programs (for more information on LAMP, see Salhuana 1988; Salhuana, Jones, and Sevilla 1991).

The impetus for this organization was provided by Dr. William Brown, at that time the CEO of Pioneer Hi-Bred International, the leading private-sector producer of hybrid corn. Brown had two main goals. First, he wanted to increase biodiversity in agriculture through the approximately 50,000 accessions of corn found in gene banks around the world (Knudson 1994). For these accessions to be useful, they needed to be *characterized.* (Characterization provides plant breeders with vital information on growth conditions and agronomic and plant breeding traits. Without this information, plant breeders do not know which accessions to use and, in many cases, will simply not use any uncharacterized material.) Second, Brown wanted to ensure that the accessions were regenerated so that the seed would remain viable. To meet these two goals, he envisioned a collaborative effort among nations.

Pioneer donated $1.5 million dollars for LAMP activities. The ARS provided administrative support, corn accessions, and overall coordination. Participating countries contributed land, laboratory facilities for growing out and evaluating their native accessions, a project leader, and scientific, technical, and support staff to assess the collections (Salhuana 1995).

LAMP was divided into five one-year stages. Each stage increased the amount of testing, selection, and exchange of materials among countries.[3] Stage IV marked the end of the regional cooperation phase.[4] Each country within LAMP was free to use any LAMP materials, especially those that performed best locally, in research programs. Ultimately, the United States earmarked 51 accessions to start the GEM project, although more accessions from LAMP would eventually be used (Linda Pollak, personal communication, July 1997). The GEM project also used an additional seven tropical accessions that DEKALB donated.

Developing the Germplasm Enhancement of Maize Project

The public sector usually undertakes research involving the enhancement of exotic germplasm because the commercial worth of this material is very uncertain. Even assuming that a commercial product can be obtained from such accessions, it typically takes at least 15 years for promising new traits to be incorporated into commercial hybrids. The private sector has a greater interest in more profitable short-term projects, and individual companies probably could not afford to invest in adapting these accessions on their own.

At the time GEM was developed, public programs in the United States also lacked the resources to carry out the necessary adaptations. One important constraint was the distribution of expertise. A great deal of plant breeding effort is expended on corn; in fact, more than 25 percent, or 545, of the SYs used annually in plant breeding R&D in the United States are in corn. But this high aggregate level disguises the relatively asymmetric relationship between the public and private sectors. Most of the research funds for corn are in the hands of the private sector. For example, in the maize plant breeding industry in 1994, the private sector employed 509.05, or 64.7 percent of SYs. State agricultural experiment stations and the ARS employed only 27.1 and 8.2 SYs (Frey 1996). The 35.3 SYs used in the public sector in 1994 represent a huge decrease from the 125 SYs Huffman (1984) reported 10 years before. Further, the overwhelming majority of private-sector research involves downstream varietal development. (The assumption is that the private corn industry devotes 80 percent of its SYs to applied R&D [Frey 1996]).

The recipe for resolving this impasse was public and private interaction. Two teams of public and private corn breeders each wrote a proposal that would involve public and private interaction in conducting germplasm enhancement of earmarked LAMP accessions (basic breeding research). The adapted accessions would eventually be used in U.S. corn breeding programs (applied breeding research).The first team's proposal focused on adapting temperate accessions for the United States, and the second on tropical accessions for the South.[5] These proposals first defined GEM.

The proposals caught the attention of the Corn-Sorghum Basic Research Committee of the American Seed Trade Association (ASTA) at ASTA's December 1992 meeting. Members of ASTA come mainly from the private sector, and they saw in these proposals a response to the widespread concern in the corn seed industry about the diminishing role of the public sector in breeding. This issue had been emphasized in a report the Corn-Sorghum Basic Research Committee presented at the meeting that detailed a significant decline in public-sector scientific and financial resources for corn breeding (David Harper, personal interview, 1996).

The ASTA meeting in December 1992 had two important outcomes. First, the two teams decided to combine their proposals. In March 1993, the combined proposal was distributed to the ASTA membership to gauge levels of interest and potential in-kind support. The response was largely positive, with most firms supporting the proposal and agreeing to give in-kind support in the form of yield trial plots, nursery rows, and disease observation rows. Second, the Corn-Sorghum Basic Research Committee decided to launch a major lobbying effort to increase public funding for basic research.

While industry sources cite the importance of lobbying efforts in building political support for GEM, it is probably the case that lobbying, while necessary for public support, was almost certainly not sufficient. In the absence of lobbying efforts, the political incentives of legislators operating in a world of scarce resources guarantee that funding will not be forthcoming. But lobbying, by itself, is not enough to guarantee federal support. Legislators also had to find a compelling case for funding, and in the case of GEM such a case could be made. Legislators were convinced that the historical yield increases of two bushels per acre would not be sustainable after two decades, and that without further efforts the current genetic base may at that point be used to its capacity (David Harper, personal communication, 1996). In addition, the willingness of the private sector to provide significant levels of in-kind support made it possible for the government to gain political benefits while having costs subsidized. This suggests that public-private partnerships are unlikely to be sustained solely due to either industry lobbying efforts on the one hand or especially persuasive public good demands on the other.

The lobbying efforts that began later that year won significant support from Midwestern senators. Congressional members were impressed by the \$1.5 million Pioneer had contributed to LAMP and by the private companies' in-kind contributions, which were valued at approximately \$450,000.[6] The lobbyists argued that funding the new consortium would help the United States enhance its international competitiveness in corn.[7]

In 1994, ASTA sought an initial \$1 million from Congress to initiate GEM. This request was not included in the original House and Senate agricultural appropriations bills, but \$500,000 was added in the House-Senate conference committee. In 1995, 1996, and 1997 GEM received \$500,000 annually from the federal government.

The Structure of GEM

In December 1993, ASTA's Corn-Sorghum Basic Research Committee established the Technical Steering Group to oversee the project's execution (GEM 1994). The group, which is made up of representatives from each member organization, reflects one of the consortium's primary concerns: establishing a balance between the interests of public and private members. At least 5 of the group's 9 to 11 members must come from the private sector; the others can come from the public or private sector. The ARS provides two ex officio representatives (GEM 1995). Members are elected for three-year terms.

The Technical Steering Group established a protocol for research and membership in the GEM consortium for both private and public entities. Private sector members supply in-kind support annually in the form of nursery and yield trial plots, cross the accessions sent to them by the coordinator with one of their own elite lines, and share breeding crosses and relevant data with the coordinator. In return, members have access to GEM material. The protocol also draws a distinct line between material that is in the public domain and material that can be owned by an individual firm (see box 9.1). With these guidelines clearly spelled out, GEM is able to achieve its two main goals: to strengthen support for long-term basic research in genetic enhancement by sharing the cost of such research among its members and to provide incentives for the commercial exploitation of promising new genetic traits.

Public-sector members of GEM do not have to provide in-kind support when they sign the protocol. But public-sector members receiving GEM funding for experiments must provide technical support, supplies, or test plots. In fact, many of the public signatories do not have the resources to provide in-kind support because of the diminishing resources available to their programs. The total amount of in-kind support that GEM's public-sector members could provide is far less than what the private sector supplies. Nevertheless, the public sector's involvement in GEM shows support for the consortium's goals and signals to Congress the sector's belief in the project. Public-sector signatories also have access to GEM material.

The GEM consortium has a coordinator who directs research and data management and analysis, manages and releases enhanced germplasm, and provides a breeding site. The consortium finances scientists and laboratory equipment. The GEM laboratory at Ames, Iowa, for instance, has three full-time scientists and has been furnished with equipment for assessing value-added traits. The GEM consortium also provides research grants to ARS or SAES corn scientists. In 1995, it funded nine projects at SAESs and four at ARS, for a total of $65,775. In 1996, $77,500 was allocated to 11 projects at SAESs and 4 at the ARS. In 1997, collaborative research money to public cooperators totaled $77,500 (GEM Annual Reports).

Protecting Public and Private Interests

One concern with consortium research arrangements is that information sharing will endanger the competitive position of private firms. The GEM project protocols and the trust members place in the project coordinator have minimized this concern. To evaluate and adapt GEM material, firms must provide their own elite inbred lines for crossing the GEM accessions.

BOX 9.1. GEM RESEARCH PROTOCOLS

A key to successful public-private partnerships in research is to clearly specify the responsibilities and obligations of each participant, particularly concerning ownership of intellectual property on new discoveries. The central goal of GEM is to strengthen support of basic research in germplasm enhancement of corn so that important new traits can eventually be incorporated into cultivated varieties. GEM achieves this goal by sharing the costs of research that no one partner could afford to undertake alone.

Private partners need assurances that they will be able to successfully exploit new scientific developments commercially—that is, an agreement stating when they can claim private ownership rights for new discoveries. These rights, and the companies' obligations, are spelled out in the GEM research protocol. The protocol follows the standard breeding procedures used in corn breeding, which allow for the proper evaluation of yield and other desirable traits, as well as crossing ability with other elite inbred lines.

The research protocol is defined in four steps. The first step begins with the initial adaptation of the exotic material, and the last step ends with the adapted material becoming the property of the private or public cooperator. The steps occur chronologically, so each additional step entails additional adaptation and testing of the exotic material.

Step 1: First Hybrid Cross

(1.1) The coordinator sends an exotic LAMP accession to a private cooperator (private cooperator #1) to be crossed with an elite inbred line, producing a hybrid cross. This hybrid progeny, which contains 50 percent elite line and 50 percent exotic line, is a *50:50 hybrid*. The elite line selected for the cross needs to be genetically similar to the exotic accession so that eventually a parental line from the 50:50 hybrid can be crossed with a parental line from a different breeding pool to produce a commercially successful hybrid. The resulting hybrid will be the commercial seed a farmer plants.

(1.2) The 50:50 hybrid is sent back to GEM. At this time, the GEM staff at Iowa State University and other public institutions test the hybrid for disease resistance and value-added traits.

(*continued on next page*)

(box 9.1 continued)

Step 2: Second Hybrid Cross

(2.1) The coordinator sends the 50:50 hybrid to a different private cooperator (private cooperator #2). Cooperator #2 does not know the source of the elite line used in the first cross. Cooperator #2 crosses the 50:50 hybrid with one of its own elite lines. The progeny from this cross is 75 percent elite line and 25 percent exotic, or a *75:25 hybrid.* For the reasons given above, the elite line used in this step is genetically similar to the 50:50 hybrid with which it is crossed.

(2.2) This 75:25 hybrid seed is sent back to the coordinator. Again, public sector GEM members assess the disease resistance and value-added traits of the cross.

Step 3: Yield Testing of Hybrids

(3.1) The coordinator now sends hybrid seed to cooperator #3 in the form of either a 75:25 hybrid or a 50:50 hybrid. Cooperator #3 crosses these hybrids with a higher-yielding elite line. How well the progeny of this cross yields indicates the yield ability of the 75:25 or 50:50 hybrid. The best crosses are advanced to step 4 to develop into inbred parent lines.

Step 4: Selfing, Synthetics, and Assignment of Property Rights

(4.1) The private and public cooperators self-fertilize the hybrid for two generations in a process known as *selfing*. The first generation of selfing is referred to as S1, the second generation as S2.

(4.2) After the first generation of selfing (when the S1 is formed), the private and public cooperators each select the top 20 percent of the S1 lines and self these selections again to produce an S2. As in step 3.1, all S2s are tested for yield ability, and the best ones are advanced to step 4.3.

(4.3) The final stage involves the allocation of property rights. This determination depends on whether the cooperator is private or public, as follows:

> **(4.3.1)** A private cooperator selects the 10 best S2s and returns them to the GEM coordinator in the form of an S2 synthetic to be evalu

(continued on next page)

(*box 9.1 continued*)

ated for important traits.[1] The S2 synthetic is then placed in the public domain. With the return of the S2 synthetic, the firm's obligation to GEM with regard to this particular accession is over. The firm is now free to take any of the materials (including those lines not forwarded to GEM) for use in its own commercial breeding program. For example, this procedure allows the firm to protect an S2 line as a trade secret for use in its own breeding program while making key genetic traits in the S2 synthetic available to other breeders. The firm may also seek other IPR proteciton.

(4.3.2) A public cooperator must return its 10 best S2s (not synthetics) to the GEM coordinator. These lines will be selfed for one more generation, creating an S3, and evaluated for resistance to disease and insects and for value-added traits. A public cooperator can eventually release material derived from S1 and S2 lines and claim ownership of them. However, the public cooperator must wait until the GEM coordinator releases the 10 best S2s to the rest of the consortium. This waiting period is prescribed to allow GEM to obtain some credit for the consortium's work.

---

1. An S2 line is one that has undergone two generations of self-fertilization to increase its genetic purity. An S2 synethetic is a mixture of such lines, each produced by two generations of self-fertilization. By mixing the lines, the identity of individual S2 lines can be kept confidential by the firm that produced it. At the same time, the genetic traits contained in the lines are made available to the rest of the consortium members to be exploited through subsequent research.

In most cases, companies will already have used these lines successfully to produce commercial hybrids (Douglas Tiffany, personal communication, July 1996). While the firm is required to share the results of the cross with the consortium, only the firm knows what line it used. (In the hybrid seed industry, companies restrict outsiders' access to elite inbred parent lines.) The project coordinator knows only the name of the company that owns the line, not the specific line. Other companies with access to these crosses know neither the name of the company nor the specific line. An added measure of security is provided by the difficulty of extracting a pure elite line once it has been hybridized. These measures have provided enough assurance for many corn-breeding firms: 21 companies had signed the protocol and joined GEM by 1996.

The institutional structure of GEM also guarantees that public interests in enhancing genetic diversity are protected. The Technical Steering Group decided early on that public lines should not be crossed with the accessions, since the best public lines are already widely used and could limit the genetic diversity found in commercial hybrid seed. Instead, the group sought to introduce new material into the parental gene pools in order to develop new inbred lines that could be used to produce hybrids. Public-sector members of GEM therefore provide resources other than their elite lines, such as providing backstopping research. For example, in 1995, the Technical Steering Group asked public members to conduct research on resistance and tolerance to the corn root worm, the spider mite, the European corn borer, the fall armyworm, and other common pests (GEM 1995). This division of resources offers a good balance between private- and public-sector considerations.

The GEM material, whether from private or public sources, is eventually released to the North Central Regional Plant Introduction Station and made available to all research institutions. The release of new material is announced in the pages of *Crop Science,* a premiere plant breeding journal, which publishes a joint statement by the GEM project and the organization involved in developing the material. The statement serves a dual purpose: GEM receives recognition for its work in coordinating the project and in developing and testing breeding material, and the breeder receives the recognition essential for obtaining new funding and furthering scientific careers. However, the protocols specify that anything a firm or public institution develops further using GEM material can be protected as a trade secret or IPR. The decision to acknowledge GEM's role is up to the proprietary organization.

GEM Results

In 1996, 18 private and 5 public member organizations tested the yields of 564 hybrids. Fifteen crosses were selected for advancement in the 1996–97 winter nursery based on yield and value-added traits; more crosses would have been advanced if there had been more nursery room (Linda Pollack, personal communication, 1996; Duvick, 1996). Several scientific successes have emerged from this research:[8]

- Two breeding crosses have shown total protein levels of more than 16.5 percent, with a range of 36 breeding crosses having 8.7–16.5 percent. Currently the average protein level for commercial varieties in the corn belt is 10 percent. The new high-protein material may be useful for both food and feed.
- Four breeding crosses show a total saturated fatty acid content of more than 16 percent, with a range of 50 crosses having 12.2–18 percent. The current corn belt average is 12 percent. These crosses could be used in a solid non-hydrogenated margarine product.

- Two breeding crosses show total oil levels of more than 5.6 percent, with a range of 3.3–5.6 percent. These levels compare with the current corn belt average of 4 percent oil, making the new crosses potentially useful for feed.

It is expected to be at least another 7–10 years before GEM material is successfully incorporated into commercial hybrid varieties. But high-yielding lines are being developed from the two high-protein breeding crosses discussed above (GEM 1996). The commercial potential of GEM's efforts may be even greater in the next few years, when the breeding program introduces more LAMP accessions.

Factors in the Success of GEM

In balancing the interests of the public and private sectors, GEM provides both social and economic returns. It encourages long-term genepool enrichment, which tends not to be a profitable investment for individual companies. It provides adequate private returns by protecting intellectual property, so that the private sector has an incentive to participate. The success of GEM in these areas makes identifying the attributes that have contributed to its success, and determining whether those attributes can be applied to other public-private research consortia, important to future policy for agricultural R&D.

The factors responsible for GEM's success include the following:

Federal leadership. The size of many projects, plus the need to maintain both a regional and a national focus, make the federal government's involvement of prime importance. Federal support also provides a measure of funding stability that would be difficult to achieve with, for example, only university leadership. These are the same reasons the USDA engages in intramural research (Fuglie et al. 1996). In addition, the government has provided protection for intellectual property—a key incentive for private collaborators, and one that the private sector itself cannot guarantee (Eric Wellin, personal communication, July 1996).

Information sharing. Concern has been voiced that increased protection for IPRs will mean a decreased exchange of information, whether formal or informal (Schmid 1985). The organizational structure of GEM allows for this informal exchange. In fact, some plant breeders claim that the informal exchange of ideas with other breeders from different institutions has been invaluable to creative problem solving (Knudson 1988). Ultimately, all GEM material and information becomes public, except for proprietary materials that may be protected as trade secrets or IPRs.

In-kind support from private partners. One challenge for policy makers and public-sector research administrators is deciding which research projects are likely to have the greatest scientific and economic success. By providing in-kind support, private firms signal their confidence in a project's potential. In-kind contributions also increase the likelihood that private firms will actively participate in a project, since the firms are anxious to see that their resources are used efficiently. And risk sharing among public- and private-sector participants encourages private firms to make long-term, potentially risky investments in basic research that promises large social but uncertain economic returns. Currently, for example, researchers are working to increase the genetic diversity of commercial corn varieties grown in the United States. The GEM consortium has made an invaluable contribution to this research. Private member firms have contributed their elite lines to cross with the exotic accessions from LAMP, so that researchers need not rely solely on public elite lines. The result is a more diverse and elite set of germplasm in the field.

Good communication among members. The Technical Steering Group meets four times a year to discuss research plans, results, and protocol changes. In addition, GEM holds a meeting for all members at the December ASTA meetings, distributes a quarterly newsletter, maintains a web site, and holds two field days in the late summer and early fall seasons.

Well-defined IPRs. The private sector knows it can appropriate some returns from its work because of the safeguards established in the protocols. At the same time, firms have the satisfaction of helping provide research results that have the attributes of a public good.

Trust. The GEM consortium's institutional features have helped bolster the private sector's confidence in GEM. The key risk to private firms from exchanging germplasm is that their elite pure lines will be made available to their competitors. By having firms submit a synthetic (that is, a mix as opposed to the identifiable pure lines required of public cooperators), GEM makes it difficult to determine which seeds are from the same pure line and can thus be bred (selfed) to recover the proprietary pure line. In addition, the individuals who were instrumental in organizing GEM (primarily from the ASTA) knew each other, communicated well, and had credibility within the field of agricultural R&D. For these reasons, the industry tended to trust the organizers and coordinator. For example, private firms have trusted the coordinator to keep the identity of their elite lines secret.

To determine whether other research consortia share these characteristics, the paper now turns to the case history of another agricultural research consortium, the Biotechnology Research and Development Corporation (BRDC).

THE BIOTECHNOLOGY RESEARCH AND DEVELOPMENT CORPORATION: CASE HISTORY

Like GEM, the BRDC is structured as a research consortium that includes the federal government, private companies, and public players. The BRDC's goal is to facilitate the transfer of technology from federal and university labs to the private sector so that basic biotechnology research can be developed into commercial products. The consortium funds and manages biotechnology research in plant science and molecular biology, bioprocessing, fermentation, and animal health and production.

Development of the BRDC

The BRDC was initially the vision of a group of businessmen in Peoria, Illinois. The group had two goals: to enable the private sector to use the information generated by the public sector, and to encourage the public sector to conduct research that responds to the needs of the marketplace. The group particularly wanted to facilitate the use of private-sector funds to support research carried out at public-sector laboratories (The BRDC Story 1997).

In 1985 this group and the ARS conducted a feasibility study to determine if such a collaboration could take place. At first the answer seemed to be no: Private firms feared that any research produced by the public sector would be legally restricted to be in the public domain, preventing investors from capturing the economic returns from their investments. Private firms would have no incentive to conduct research if they could not obtain exclusive licenses to commercialize the outcomes. Overcoming this difficulty was critical to the subsequent success of the BRDC.

The businessmen turned to the U.S. Congress, throwing their support behind the Technology Transfer Act of 1986, which would establish new incentives for federal research laboratories to form partnerships with private industry (The BRDC Story 1997). Like GEM, the BRDC needed federal leadership in order to become a viable research vehicle.

Once the Technology Transfer Act was passed, several private companies promised money to a public-private consortium that would focus on agricultural applications for biotechnology. As it had with GEM, the commitment of private funds impressed and reassured the government, and further funds were forthcoming at the federal level and from the Illinois state legislature. As a result, the BRDC was able to incorporate in February 1988 as a for-profit corporation. It began funding research projects in July 1988.

The Structure of the BRDC

The BRDC has a Scientific Advisory Board that evaluates project proposals. The board consists of two voting representatives from each of the stockholder companies and the ARS, plus representatives from the University of Illinois and Iowa State University. As with GEM, the participatory nature of the BRDC promotes good communication and enhances the flow of information.

The BRDC conducts long- and short-term research projects—all targeted for the marketplace—that take one to three years to complete. Eighty percent of the consortium's projects are long-term basic research projects with some likelihood of being commercially useful. In fact, the criteria for funding research are the marketability of the results and the project's scientific merit (The BRDC Story 1997).

The BRDC receives funding the from federal government, the state of Illinois, and private member companies.[9] Stockholders pay an annual fee of $200,000. They share technology developed with BRDC funding exclusively among themselves, and private-sector researchers can work with federal and academic scientists in conducting the research (The BRDC Story 1997). This is similar to the public good nature of sharing research between public and private cooperators that was described previously in GEM.

The BRDC's annual budget has ranged from $1 million to $4.5 million, for a total R&D investment of $25 million over nine years. Of the $25 million, $14.5 million came from the federal government and $10.5 million from private shareholders and the state of Illinois (J. Grant Brewen, personal communication, March 1997).

Protecting Public and Private Interests

The BRDC and the institution conducting the research retain ownership of IPRs for technologies. The technologies are then licensed to members, and royalties are allocated to the institutions that conducted the research. If the government conducts the research, the BRDC holds the exclusive right to commercialize the technology and to decide who will receive the license. However, any income earned from the technology is divided equally between the holder of the license and the government.

If the research is conducted by a university, two contracts with different funding and equity arrangements are offered to the research institution. Under the *Illinois contract,* the BRDC receives 75 percent of the royalties and exclusive rights to license the technology and bears 50 percent of the indirect costs. The university receives 25 percent of the royalties, bears 50 percent of the indirect costs, and retains the right to use the technology for education or research purposes. The *Purdue contract* shifts more of the risk to the university in exchange for more potential income. The BRDC receives 50 percent of the royalties and exclusive rights to license the technology but bears only 25 percent of the indirect costs. The university retains the right to use the technology for education or research, receives 50 percent of the royalties, and bears 75 percent of the indirect costs.

BRDC Results

Because the BRDC has been in existence longer than GEM, it has had more time to establish R&D programs that have resulted in commercial success. As of 1997, the consortium had been granted 11 patents and applied for 26 more. It has issued 9–12 commercial licenses and is negotiating four more. (J. Grant Brewen, personal communication, March 1997).

Two new BRDC technologies stand out as especially promising. The first involves a cornstarch that, when linked to a polysaccharide, makes foam and plastics biodegradable. Dow Chemical developed the cornstarch, and the BRDC and ARS found a way to blend it with the polysaccharide. The fast food industry has expressed interest in using this technology for its packaging needs. It may also be successful in Europe, where environmental laws are stricter.

The second technology is the Gelvin promoter, which was developed at Purdue University. The biotechnology industry uses promoters for recombinant DNA procedures. The Gelvin promoter is an improvement over an earlier promoter made by Monsanto, which currently holds a monopoly in the field. Twelve companies are seeking licenses for this technology.

Lessons from the BRDC

Most of the factors that have been instrumental in GEM's success can also be seen in the BRDC. Federal leadership, for example, was crucial in providing the framework for the development of both GEM and the BRDC. Both consortia have the support of the private sector, although this support takes somewhat different forms. Both have well-defined IPRs that ensure private firms a good return on their investments, and both encourage good communication among their members.

Exceptionally important to both consortia is trust: In both cases it was essential for private firms to be certain they could capture the economic benefits of their research investments and remain immune from potential predatory behavior by competitors. Trust is the most difficult of the factors to replicate, however. The differences in the two cases allow some analysis of this point. What is important in both cases is that private firms feel confident revealing information to the consortia. These firms know that their competitiveness will not be affected, since they are placing information in the hands of individuals who have no incentives to leak it to rivals. The GEM consortium, for example, has been concerned primarily with diffusing LAMP germplasm, and this diffusion would stop if firms no longer felt assured that their material would be protected. Similarly, the potential for profit in the BRDC produces incentives to acquire and maintain a reputation for trustworthiness. Since the BRDC funds outside research, its ability to make a profit is dependent on its outside contacts. If the BRDC is unable to secure proprietary information, these contacts will be withdrawn.

Pursuing this argument, it might be possible to identify situations that would not foster a climate of trust. For example, if a research consortium is controlled by a for-profit firm that engages in its own research, assuring outside firms that their contributions will be protected will be difficult. This theory suggests that Pioneer's reluctance to fund the entire GEM project may have made obtaining elite lines from private firms easier. Similarly, trust may be difficult to establish in a not-for-profit research consortium overseen by government officials, at least if those officials have ambitions independent of the consortium.

CONCLUSION

These two case studies suggest that the research consortium offers a feasible model for public-private collaboration in agricultural R&D, but only as long as certain criteria are met. The research consortium model respects the traditional mission of the public sector in plant breeding—to discover and disseminate information. By spreading risk among members and providing IPRs, research consortia are able to pursue basic as well as applied research. Given declining public support for basic research, this model may be an important new alternative for maintaining adequate levels of basic research. Private firms, as well, benefit from the research consortium arrangement, which gives them access to new information and materials that enable them to make the technological advances necessary to stay competitive. Finally, the research consortium provides social benefits, such as enhanced genetic diversity in commercial cultivars. Even at this early date, GEM's breeding efforts have produced lines with higher yields, more disease resistance, and better nutritional content than older varieties.

Furthermore, there is no reason to think that the GEM consortium is an unusual case. The BRDC, for example, shares many of GEM's institutional strengths. Both consortia foster close interaction between the public and private sectors, with the goal of transferring knowledge and sharing expertise between the two. In both arrangements, the private sector provides either direct or in-kind support. Other common attributes are good communication, which is essential to sound R&D planning among so many players; well-defined IPRs; and trust among the participants that shared trade secrets will be honored. All these factors help drive the success of the consortia and thus provide a model for future efforts along these lines.

In terms of policy, federal leadership is clearly important to the success of public-private research consortia. From the definition of property rights to the allocation of initial funding, the federal government can play a major role in assuring that these consortia have the institutional prerequisites necessary for success. Given the constraints facing public funding for agricultural research, consortia promise a means of targeting federal attention and money to research that promises high rates of both social and economic returns.

NOTES

1. I would like to acknowledge Linda Pollak, Wilfredo Salhuana, David Harper, Doug Dion, Doug Tiffany, Susan Duvick, Eric Wellin, J. Grant Brewen,

Carl Pray, and Keith Fuglie for their invaluable comments and insights. I would also like to thank the Economic Research Service for funding this project. The views and findings of this chapter are the author's only and do not necessarily reflect those of the Economic Research Service.

2. A "public good" is characterized by a lack of rivalry (that is, each individual act of consuming the good does not diminish others' ability to consume it) and exclusivity (no one can prevent other people from consuming the good). Basic scientific knowledge has this characteristic; for instance, if I use Mendelian genetics in my research, it does not diminish other researchers' ability to use it. Private goods are proprietary products. While the private sector has an economic incentive to produce private goods, the creation of public goods often requires government support.

3. Five homologous areas (HAs) were defined based primarily on altitude and latitude (HAs are areas with similiar agricultural and environmental characteristics). The HAs would be used to help evaluate accessions under similar environmental conditions to determine the degree of adaptability. The five areas were

- HA1—Below 1200 meters altitude and 25° latitude (north or south)
- HA2—1200 to 1800 meters altitude and below 25° latitude (north or south)
- HA3—1800 to 2400 meters altitude and below 25° latitude (north or south)
- HA4—Above 2400 meters altitude and below 25° latitude (north or south)
- HA5—Below 1100 meters altitude and above 25° latitude (north or south)

Stage I accessions were evaluated in each region within a particular HA. With these results, the best germplasm was selected for further characterization during Stage II. In Stage III, selected accessions were exchanged among regions in the same HA and crossed with the best accession from each region. Stage IV evaluated the combining ability of these accessions with the local accession (Wilfredo Salhuana, personal communication, July 1997). (Combining ability measures how well a plant line hybridizes with other lines. Some plant lines have poor combining ability, as can happen, for example, when deleterious genes are linked to good genes.) Identifying plant lines with good combining ability is essential to future plant breeding efforts (Salhuana, Jones, and Sevilla, 1991).

4. There has been some interest from LAMP participants in an international GEM program. Financing has been the primary obstacle to realizing this vision.

5. The first team consisted of Linda Pollak (an ARS corn breeder stationed at Iowa State University at Ames, Iowa, and the U.S. project leader for LAMP), Wilfredo Salhuana (Pioneer Hi-Bred International and senior advisor to LAMP), and Douglas Tiffany (Pioneer Hi-Bred International). Members of the second team were Randy Holley (Northrup King) and Major Goodman (North Carolina State University).

6. This figure probably underestimates the actual value of the contributions, since it is based solely on nursery rows and yield plot contributions and does not include the value of elite lines, expert knowledge, and overhead costs (David Harper, personal communication, July 1996).

7. Even though currently less than 5 percent of all the corn germplasm available in the world is used in commercial corn breeding, many corn breeders believe there is enough diversity within this gene pool to continue to increase corn yields

by two bushels per acre for another 20 years. At that point, however, the current genetic base will likely be exploited to its maximum capacity. To assure that increases in corn productivity can be sustained well into the next century, efforts must be initiated today to expand the gene pool available to commercial breeders: Identifying and adapting exotic germplasm into commercial corn hybrids is expected to take 20 years or longer.

With an enhanced germplasm base, plant breeders will have a better chance of finding genes that will increase yields, promote disease resistance, increase tolerance to environmental stresses, and enable the creation of value-added traits. Value-added traits have economic value—for instance, a higher starch content will enhance the use of corn in the production of ethanol. New genetic material may also produce stronger corn kernels less likely to crack during transportation and storage, thus reducing postharvest losses, particularly in exports (Linda Pollak and David Harper, personal communication, July 1996).

8. These hybrids are the 50:50 or 75:25 hybrids crossed to an elite line. See the research procedures discussed in the protocol.

9. As of 1997, BRDC stockholder companies were Alexion Pharmaceuticals, American Cyanamid, Dalgety, Dow Chemical, Hewlett-Packard, Mallinckrodt Group, and McDonald's. Past members have included Amoco Technology, Cilcorp Ventures, ECOGEN, and International Minerals & Chemicals.

REFERENCES

The BDRC Story. 1997. Unpublished mimeo. Biotechnology Research and Development Corporation, Peoria, IL.

Brooks, H.J. 1989. "Questionnaire on Agricultural Experiment Station Policies on Release of Plant Materials." Agricultural Research Service, U.S. Department of Agriculture, Beltsville, MD. Unpublished mimeo.

Butler, L.J., and B.W. Marion. 1985. *The Impacts of Patent Protection on the U.S. Seed Industry and Public Plant Breeding.* North Central Regional Research Publication 304, North Central Project 117, Monograph 16. Madison: Research Division, College of Agricultural and Life Sciences, University of Wisconsin.

Duvick, Susan. 1996. "GEM Quarterly Report for the Corn-Sorghum Basic Research Committee of ASTA." Agricultural Research Service, Ames, Iowa. Unpublished memo.

Frey, Kenneth. 1996. *National Plant Breeding Study I: Human and Financial Resources Devoted to Plant Breeding Research and Development in the United States in 1994.* Special Report 98. Ames: Iowa Agriculture and Home Economics Experiment Station, Iowa State University.

———. 1997. *National Plant Breeding Study II. National Plan for Promoting Breeding Programs for Minor Crops in the U.S.* Special Report 100. Ames: Iowa Agriculture and Home Economics Experiment Station, Iowa State University.

———. 1998. *National Plant Breeding Study III. National Plan for Genepool Enrichment of U.S. Crops.* Special Report 101. Ames: Iowa Agriculture and Home Economics Experiment Station, Iowa State University.

Fuglie, Keith O., Nicole Ballenger, Kelly Day, Cassandra Klotz, Michael Ollinger,

John Reilly, Utpal Vasavada, and Jet Yee. 1996. *Agricultural Research and Development: Public and Private Investments Under Alternative Markets and Institutions,* Agricultural Economics Report 735. Washington, DC: Economic Research Service, U.S. Department of Agriculture.

(GEM) Germplasm Enhancement of Maize. 1994. *Annual Report.*

———. 1995. *Annual Report.*

———. 1996. *Annual Report.*

Huffman, Wallace. 1984. "The Role of the Public Sector in Corn Inbred Line Development." Staff Paper 146. Ames: Department of Agricultural Economics, Iowa State University.

Klotz, Cassandra, Keith O. Fuglie, and Carl E. Pray. 1995. "Private-Sector Agricultural Research Expenditures in the United States." Staff Paper SP9525. Washington, DC: Economic Research Service, U.S. Department of Agriculture.

Knudson, Mary K. 1988. "The Invention and Diffusion of Two Competing Technologies: Semi-dwarf and Hybrid Wheat." Ph.D. dissertation, University of Minnesota, St. Paul.

———. 1990. "The Role of the Public Sector in Applied Breeding R&D: The Case of Wheat in the U.S." *Food Policy* 15(3): 209–17.

———. 1994. "Agricultural Diversity, Do We Have Resources to Meet Future Needs?" Paper presented at the conference on Global Environmental Change and Agriculture—Assessing the Impacts. Sponsored by the Economic Research Service and Farm Foundation, Rosslyn, VA.

Knudson, Mary K., and Carl E. Pray. 1991. "Plant Variety Protection, Private Funding, and Public Sector Research Priorities." *American Journal of Agricultural Economics* 73:882–86.

Perrin, Richard K., K.A. Kunnings, and L.A. Ihnen. 1983. *Some Effects of the U.S. Plant Variety Protection Act of 1970.* Economics Research Report 46. Raleigh: Department of Economics and Business, North Carolina State University.

Pray, Carl E., Mary K. Knudson, and Leonard Masse. 1993. "Impact of Changing Intellectual Property Rights on U.S. Plant Breeding." Department of Agricultural Economics, Rutgers University, New Brunswick, New Jersey. Mimeo.

Ruttan, Vernon W. 1982. *Agricultural Research Policy.* Minneapolis: University of Minnesota Press.

Salhuana, Wilfredo. 1988. "Seed Increase and Germplasm Evaluation." In *Recent Advances in the Conservation and Utilization of Genetic Resources: Proceedings of the Global Maize Germplasm Workshop.* Mexico, DF: International Maize and Wheat Improvement Center, pp.29–38.

Salhuana, Wilfredo. 1995. "Latin America Maize Project Leaves Untapped Legacy of Agricultural Riches," *Diversity* 2(4): 6.

Salhuana, Wilfredo, Quentin Jones, and Ricardo Sevilla. 1991. "The Latin American Maize Project: Model for Rescue and Use of Irreplaceable Germplasm." *Diversity* 7(1–2): 40–41.

Schmid, Andrew A. 1985. "Biotechnology, Plant Variety Protection and Changing Property Institutions in Agriculture." *North Central Journal of Agricultural Economics* 7 (July): 129–38.

10

STATE AGRICULTURAL EXPERIMENT STATIONS AND INTELLECTUAL PROPERTY RIGHTS

MARY K. KNUDSON, RICHARD L. LOWER, AND RICHARD JONES[1]

PLANT BREEDING research and development have undergone significant changes in the last few decades. Because of the emergence of the various protection mechanisms and intellectual property rights (IPRs) for plant varieties and plant germplasm,[2] the era of the free exchange of crop germplasm may be drawing to a close. While IPRs are designed primarily to encourage private-sector research and innovation, within the last decade they have also had a significant effect on research at public institutions. Many federal agencies and state research institutions now regularly seek IPRs for new inventions developed by their scientists as a means of promoting more rapid commercialization of new technology, raising revenues for research, and rewarding scientists.

The use of IPRs by public institutions has raised questions about the potential effect of these protections on the goals and mission of public research. The role of public agricultural research institutions in plant breeding has historically been to conduct research that increases public welfare. Such research does not necessarily result in profit. For this reason, public institutions have long been the primary source not only of important basic research but also of applied research that private firms are unwilling to undertake because they are unable to capture the economic returns on their investment.

The use of IPRs raises institutional issues for public-sector research. To acquire, license, and monitor the use of IPRs, public institutions must establish intellectual property offices and staff them with experts knowledgeable about IPR laws and procedures. They must develop policies regulating the use of IPRs and the distribution of revenues the IPRs bring in. These issues can be especially daunting for state agricultural experiment stations (SAESs), many of which have only recently initiated IPR policies. And, given the rapid advances taking place in plant biotechnology and private-sector capacity in plant breeding research, the legal and institutional structure for protecting products of crop improvement programs is continually changing and evolving.

This chapter provides preliminary results of a survey of the use of IPRs for plant varieties and germplasm at SAESs. A similar though less extensive survey conducted in 1988 by Brooks (1989) serves as a baseline for documenting changes that have occurred at public agricultural research institutions over the past decade. The present survey, which examines the ways SAESs use, manage, and profit from IPRs for new plant germplasm, covers specific crops; types of plant breeding programs for each crop (that is, whether the program is done in collaboration with other public institutions or with private companies); and the organization, benefits, and costs of the IPR process for all SAESs. We conducted the survey in order to better understand how reliant SAESs have become on IPR revenue, how much they are spending to administer IPRs and distribute earned revenue, and whether the new protections are affecting the traditional research mission at these public institutions—that is, whether they continue to do research that focuses on the public good rather than on private welfare.

The chapter first discusses the institutional history of the SAES system and describes how new IPR laws have affected plant breeding research at the institutions. It then reviews the results of Brooks's 1988 survey. The third section presents the methodology of the authors' survey, and the fourth section offers some preliminary results. The final section draws some conclusions and makes suggestions for further research.

STATE AGRICULTURAL EXPERIMENT STATIONS AND INTELLECTUAL PROPERTY RIGHTS FOR PLANTS

State agricultural experiment stations were developed in the late 1800s to enhance the public research environment. They were developed through a series of federal laws—in particular the 1862 Morrill Act, the 1887 Hatch Act, and the Second Morrill Act of 1890 (Ruttan 1982). The culmination of these acts was the creation of land-grant colleges, with which SAESs are affiliated, and a parallel system of agricultural colleges free of racial discrimination, commonly known as the 1890 Schools. Other laws, notably the McIntire-Stennis Act of 1962 and Section 1433 of the Food and Agriculture Act of 1977, provided federal funds for agricultural research at forestry and veterinary schools. These two acts, along with the Hatch Act, provided formula funds, or unrestricted block grants, to SAESs, forestry schools, and veterinary schools associated with land-grant colleges of agriculture for research (Ruttan 1982; Fuglie et al. 1996).

Despite this public mandate, SAESs have experienced funding constraints in recent years, largely because of the need to reduce federal deficits and fund other public priorities (Fuglie et al. 1996). As chapter 5 points out, between

1990 and 1994 public plant breeding programs steadily lost scientists at an annual rate of about 2.5 scientist-years, or 0.5 percent per year. With the slowdown in public funding, SAESs have had to look to other sources, including the private sector. In 1994, nongovernment sources provided SAESs with $418 million of the $2,132 million in funds available for agricultural research (Fuglie et al. 1996). This money came in the form of direct industry grants ($153 million), product sales including technology license fees ($117 million), and other sources ($148 million).

Funding constraints are one reason behind the increased use of IPRs at state research institutions. Intellectual property rights may offer a potential and significant source of income by allowing SAESs to license new plant varieties and germplasm to private firms or other users for a fixed annual charge or a share of market sales.

The Evolution of Intellectual Property Rights for Plant Germplasm

Historically, living things have been excluded from protection by patent laws because they are considered "products of nature." But several categories of plant protection mechanisms are now available for plants (table 10.1). The earliest form of protection for plants was a special category of patents established by an act of Congress in 1930. These plant patents covered new varieties of asexually reproduced plants and were used primarily for fruit trees and ornamentals. Tuber crops were specifically excluded from the act.

In 1970 Congress enacted a special category of protections for sexually reproduced crops (crops produced from botanical seed) with the Plant Variety Protection Act. The act did not extend to hybrid seed, since the progeny of hybrid seed do not exhibit the same characteristics as the parent plant.[3] In 1994, the act was amended to include tuber crops and hybrids.[4] Under the act, breeders can receive plant variety protection certificates (PVPCs), which require only that a new variety be distinct, uniform, and stable. The PVPCs contain special exemptions for farmers, giving them the right to keep seed for their own use, and for researchers, allowing them to use protected seed as parent material to breed new varieties.

Another type of protection now available for plants is the utility patent. The Supreme Court reversed the historical interpretation of patent law in 1980 with the landmark ruling *Diamond v. Chakrabaty,* which allows the patenting of genetically modified microorganisms. In 1985, the U.S. Patent Office concluded in *Ex Parte Hibberd* that the 1980 Supreme Court decision also covered plants, plant parts, plant genes, and tissue cultures (Fuglie et al. 1996).

Of the different types of protection available for plants, PVPCs and plant patents complement one another: PVPCs cover primarily plants grown from seed, and plant patents cover plants reproduced vegetatively. Utility patents, however, offer a fundamentally new and stronger form of intellectual property

Table 10.1. Types of intellectual property protection for new crop varieties and germplasm

Type of intellectual property protection	Enabling legislation	Criterion	Coverage
Plant patent	Plant Patent Act of 1930	New variety must be novel and have utility	Plants reproduced asexually (vegetative clone), except tuber crops. Used mainly for —fruit trees —sugar cane —ornamentals and floriculture
Plant Variety Protection Certificate	Plant Variety Protection Act of 1970	New variety must be distinct, uniform, and stable	Plants reproduced from botanical seed and tubers. Used mainly for —self-pollinating field crops —self-pollinating vegetable crops
Utility patent	Patent Act of 1790; interpreted by Supreme Court in 1980 to include genetically engineered organisms	New invention must be nonobvious and have utility	Since 1985, utility patents may be obtained for seeds, plants, plant parts, plant genes, and tissue cultures. Used mainly for —biotechnology process inventions —plant genes and genetically modified crops —inbred parents of hybrid varieties
Trade secret	State law	New invention must be kept out of the public domain	Hybrid seed, by restricting access to inbred parent lines. Used mainly for —hybrid seed (e.g., corn, sorghum, tomato) —some process inventions

protection, because they do not offer either of the exemptions contained in the PVPCs. Utility patents tend to be used in plant breeding for genetically modified plants, biotechnology processes, and inbred parent lines of hybrid varieties (Fuglie et al. 1996), although PVPCs may also be pursued for genetically modified plants.

While utility patents seem to be a better alternative than PVPCs, they are more difficult to acquire. As we have seen, PVPCs require breeders to show only that new varieties are distinct, uniform, and stable. About 90 percent of applications for PVPCs have been granted (Pray, Knudson, and Masse 1993). To obtain a utility patent, a breeder must also demonstrate that the invention is *nonobvious*—that is, entails a significant new "inventive step" not immediately evident to an expert in the field. This requirement is meant to exclude from patent protection minor improvements to prior inventions. Applying for a utility patent can also be more time-consuming and costly than applying for a PVPC. In practice, only about 20 percent of patent applications that include claims on plants have been granted (Pray, Knudson, and Masse 1993).

The Effect of IPRs on State Agricultural Experiment Stations

Intellectual property rights provide more than a potential source of new funds for public research institutions. They provide an institute's scientists with personal rewards, encouraging economically important research and preventing a "brain drain" of able scientists to the private sector. In addition, they can be important in promoting the rapid commercial application of new science and technology developed at public institutions. The offer of an exclusive license to a new technology may entice a private company to make greater investments in adaptive research, product scale-up, and market development (see chapter 8 of this volume). Finally, with cross-licensing agreements, IPRs can become important bargaining chips for a public institute to gain access to technology developed elsewhere. Such bargaining is especially important for key "enabling" technologies such as genetic engineering tools and processes, which are essential to a successful crop improvement program.

The new types of protection for plant germplasm significantly affected public-sector plant breeding programs. The public sector received 13 percent of the PVPCs issued between 1970 and 1994 and 18 percent of the utility patents issued for new plants or plant parts between 1985 and 1994 (Fuglie, Klotz, and Gill 1996). Future trends may see increasing levels of public-sector involvement. In previous surveys of SAESs, many reported that they planned to increase their use of utility patents, particularly in wheat, corn, and soybeans (Brooks 1989; Pray and Knudson 1994). The impact of IPRs on public research has not been limited to quantitative changes. A study by Knudson and Pray (1991) showed that PVPCs had influenced the direction of public research in soybeans, wheat, hybrid corn, hybrid sorghum, and cotton by encouraging more research on crops with larger potential revenues from technology licensing arrangements.

A RECAP OF THE 1988 BROOKS SURVEY

In 1988 Howard Brooks of the Agricultural Research Service (ARS) of the United States Department of Agriculture (USDA) surveyed SAESs on their use of IPRs for plant germplasm. Brooks asked the SAESs about their current use and their intent to use various IPRs. He included questions concerning the administration of IPRs and plant germplasm releases. Forty-nine SAESs responded to the survey. The results showed that SAESs relied largely on public releases for their crops, with the exception of turf grass (table 10.2). Among the protection mechanisms used, PVPCs were the most popular, followed by exclusive licenses, research or tag fees, patents, and branding.

Wheat and triticale, soybeans, and forage and range grasses were the crops for which PVPCs were most frequently issued. All three crops produce seeds that reproduce plants that are essentially identical to the parent plant. This trait makes it difficult to appropriate returns on a variety, since once it is available in the public domain, it can easily be multiplied by other users. For this reason, public institutions invest in these crops more heavily than the private sector (Frey 1996). Crops such as hybrid corn come with built-in trade secrets, as they do not reproduce identically from generation to generation. The private sector began investing relatively heavily in hybrid corn even before IPRs became available for new plant varieties (Huffman and Evenson 1993).

Most of the utility and plant patents were issued for small fruits and grapes, tree fruits and nuts, and ornamental crops. Many breeders applying for these patents were taking advantage of the 1930 Plant Protection Act, which covered vegetatively propagated crops. Utility patents were not available for

Table 10.2. Types of intellectual property protection used by SAESs for crop germplasm releases, 1988

Crop type	Public releases only[a]	Plant variety protection[b]	Utility or plant patent	Exclusive license[c]	Tag fee[d]	Branding[e]
Cereal grains	92	30	3	13	24	2
Oil seed crops	25	16	3	5	10	0
Fiber crops	7	4	1	4	2	0
Forages and grasses	31	29	1	28	5	0
Turf grass	0	4	3	5	1	1
Tobacco	5	0	0	1	0	0
Potato (Irish)	10	0	2	1	0	0
Vegetables	61	16	4	19	23	0
Small fruits and grapes	14	0	7	1	0	0
Tree fruits and nuts	16	0	13	3	3	0
Ornamentals	10	1	7	3	6	0
Other crops	10	1	0	2	1	0
Number of programs	281	101	44	85	76	3
(% of total programs)	(68%)	(24%)	(11%)	(20%)	(18%)	(<1%)

Figures give number of crop improvement programs (out of a total of 415) that used intellectual property right.

[a]Public release—germplasm released without any fee or restriction on the use of the germplasm.
[b]Plant variety protection—applied for Plant Variety Protection Certification.
[c]Exclusive license—germplasm released on an exclusive basis to one or more companies to commercialize the variety.
[d]Tag fee—additional fees charged on seed sales specifically to support research programs.
[e]Branding—exclusive rights sold to a company for an unnamed variety to commercialize the variety.

Source: Brooks (1989).

plants until the 1985 *Ex Parte Hibberd* ruling. Utility patents mainly affected corn and turf grass. Utility patents for corn covered inbred parent lines of hybrid varieties, which were not eligible for protection under PVPCs. The private sector has also been active in seeking utility patents for corn. Patent office statistics reveal that between 1985 and 1994, 83 of 286 utility patents awarded to plants and plant parts had claims covering corn, more than for any other crop (Fuglie, Klotz, and Gill 1996).

At the time Brooks conducted his survey, IPRs policies at many public institutions were changing. About half the SAESs said that they intended to seek PVPCs for germplasm in the future, especially for cereal grains, oil seed crops, forages and grasses, and vegetables (table 10.3). Only a small proportion planned to seek utility or plant patents for germplasm. Most of the patents were intended for hybrid field crops like corn and sorghum or for vegetatively propagated plants like small fruits and grapes, tree fruits and nuts, ornamental crops, and some types of vegetables.

The Brooks survey also asked the SAESs how they managed IPRs. Just over half (26 out of 49 respondents) had written policies on managing IPRs and other revenue-generating mechanisms. Twenty-eight reported having policies for distributing the revenues from IPRs. Many of these policies had

Table 10.3. Intended future use of intellectual property protection for crop germplasm releases by SAESs, 1988

	Plant variety protection		Utility or plant patents	
Crop category	Intend to use	Undecided or will not use	Intend to use	Undecided or will not use
Cereal grains	50	60	12	99
Oil seed crops	24	13	6	31
Fiber crops	6	5	5	6
Forages and grasses	36	20	7	49
Turf grasses	3	4	2	5
Tobacco	3	4	0	7
Potato (Irish)	0	0	3	10
Vegetables	25	52	11	72
Small fruits and grapes	0	0	11	11
Tree fruits and nuts	0	0	19	11
Ornamentals	1	6	14	9
Other crops	4	8	2	14
Total programs	152	172	92	324
(% of total)	(47%)	(53%)	(22%)	(78%)

Figures indicate the number of crop improvement programs at State Agricultural Experiment Stations intending or not intending to use intellectual property right.

Source: Brooks (1989).

formulas for sharing revenues among the inventor, the inventor's research program, and the institution. Five of the 49 respondents reported restricting the free exchange and use of crop germplasm, breeding lines, or numbered selections. Eleven had separate policies specifically for handling germplasm involving genetically engineered material.

Finally, Brooks asked about the types of research programs active at the SAESs. Forty-five SAESs reported having cooperative crop improvement programs with other states, and 46 reported having such programs with ARS.

METHODOLOGY OF THE AUTHORS' SURVEY

In January 1998, we mailed questionnaires to each of the 60 SAESs operating in the United States and its territories. The questionnaire consisted of two parts. The first focused on types of plant breeding programs and the use of IPRs for plant breeding products at the SAESs. The second dealt with the management of the IPR process.

The survey differs from the Brooks study in several ways. First, because our survey was conducted a decade later, it captures changes in SAES policies toward IPRs during a time when these policies were undergoing significant internal review. Brooks's survey occurred just after many SAESs had begun not just to realize the potential biotechnology would have for plant science research and the significance of the changes in IPR laws but also to recognize that public institutions would have to look for new funding sources. The survey examines changes in SAESs since the late 1980s, particularly in the kinds of plant breeding research they conduct and in their use and management of IPRs.

Second, our survey examines the relationships between SAESs and the private sector. We ask, for example, whether the SAESs shared any research and development programs with the private sector. Third, our survey contains greater detail on the ways SAESs manage the IPR process. These questions ask how SAESs set up their intellectual property offices, who manages them, how much they cost, the actual revenues IPRs raise, and how these revenues are distributed. This information may help SAES directors and other policy makers understand and manage the IPR process more effectively.

Of the 60 SAESs surveyed, 43 had responded by August 1998, for a response rate of 72 percent (table 10.4). Among the respondents, the 34 that have active breeding programs became the focus of the survey results presented below. We divided the SAESs into four regions to examine regional differences—the western region (20 SAESs), the southern region (15 SAESs), the north central region (12 SAESs), and the northeast region (13 SAESs). The response rate is high for all regions except the west, where only half of the SAESs responded to the survey. The results of the survey should be con-

Table 10.4. Number of SAESs and responses to authors' survey by region

Region	Total number of SAESs	Responses to survey	Responses with plant breeding programs
West	20	10	6
South	15	13	13
North central	12	10	9
Northeast	13	10	6
All United States	60	43	34

sidered preliminary, as we hope to increase the number of responses in the coming months, especially from the western region.

RESULTS OF THE SURVEY

While SAESs have taken advantage of new opportunities to use IPRs, the great majority of plant germplasm released by SAESs continues to be made without any form of intellectual property protection. Most of SAES breeding programs continue to be conducted solely by the stations, suggesting that SAESs have maintained considerable autonomy in program management and direction. Nevertheless, SAESs are collaborating more frequently with private breeders. For most SAESs, the cost of maintaining intellectual property offices and distributing revenues obtained from IPRs on plant germplasm are modest. Most of these revenues were used to support the research program of the inventing scientist.

Plant Breeding Programs and IPRs

Of the types of plant breeding programs at SAESs, 51 percent were operated by the SAES alone (table 10.5). Programs operated jointly with other institutions include cooperative programs with other states (18 percent), cooperative programs with the ARS (16 percent) and cooperative programs with the private sector (15 percent). Note that SAESs reported almost as many cooperative programs with the private sector as with other states or ARS. The high share of programs operated with the private sector supports other evidence that the private sector plays a strong role in plant breeding research and development at many SAESs.

Overall, wheat had the most programs (58). Soybeans came in second (46), followed by corn (39), oats (38.25), large fruited vegetables (33), turf grass (29), forage and range grasses (26), potatoes (25), barley (23.75), and beans and peas (23.5). With the exception of corn, these are crops where private incentives for plant breeding are relatively weak. They include major crops such

Table 10.5. Number and types of crop improvement programs at SAESs, 1994–96

Crop category	Own in-house program	Cooperative programs with other SAESs	Cooperative programs with USDA/ARS	Cooperative programs with private sector	*Total*
Cereal grains	93	40	44	22	189
Oil seed crops	37	17	17	15	78
Fiber crops	9	5	5	5	19
Forages and grasses	52	13	18	17	84
Vegetables	72.5	29.5	11.5	27	123.5
Small fruits	19	3	4	6	30
Tree fruits and nuts	22	7	2	4	25
Woody ornamentals	17	1	1	1	20
Floriculture	7	2	2	3	14
Other crops	19.2	7.2	3	4.1	27.5
Total programs	347.7	124.7	107.5	104.1	610

Numbers in the *Total* column may be less than the sum of its row if there are overlapping programs.

Source: Authors' survey.

as wheat for which it is difficult to appropriate the returns to research, and crops with limited markets, such as turf grass. For corn, SAESs work to create inbred parents or to answer basic questions on disease or resistance to environmental stresses. Much biotechnology research has also been done on corn and may be included in some of these programs.

Most new plant germplasm developed at SAESs is released publicly without IPR protection (table 10.6). Seventy-eight percent of all SAES germplasm releases between 1994 and 1996 had no IPR protection. Of the rest, 10 percent were protected by PVPCs, 6 percent by plant patents, 3 percent by utility patents, and 1 percent by trade secrets. Compared with the Brooks survey, these trends indicate a decline in the use of IPRs for plant germplasm, despite the stated intentions of the SAES programs surveyed in 1988.

Nonetheless, many SAESs continue to employ IPRs for plant germplasm. Among the types of IPRs obtained, PVPCs were favored over utility patents by a margin of 134 to 40. The PVPCs are the most widely used form of protection. The crops for which they are generally sought, such as wheat and soybeans, are among those to which the public sector devotes the most research resources (Frey 1996). Brooks also found in his survey that wheat and soybeans were among the crops for which PVPCs were most often sought.

Plant patents are used primarily for vegetables, forages and grasses, fruits, and ornamentals. Brooks also found this result for crop categories. The relatively low use of utility patents is also consistent with Brooks's findings, which indicated that many SAESs were reluctant to use this form of IPR protection. This reluctance may stem from the fact that utility patents, with their

Table 10.6. Types of intellectual property protection used by SAESs for crop germplasm releases, 1994–96

Crop category	Public release without protection	Plant variety protection	Plant patent	Utility patent	Trade secret	*Total releases*[a]
Cereal grains	708	50	6	3	5	789
Oil seed crops	38	35	1	3	0	77
Fiber crops	101	1	1	0	0	106
Forages and grasses	52	17	12	1	0	88
Vegetables	120	20	23	4	6	173
Small fruits	4	0	11	2	0	17
Tree fruits and nuts	4	1	9	0	2	16
Woody ornamentals	1	2	11	0	0	14
Floriculture	1	1	2	2	0	6
Other crops	13	7	2	25	0	47
Total releases	1,042	134	78	40	13	1,333
(% of total releases)	(78.2)	(10.0)	(5.9)	(3.0)	(1.0)	(100)

[a]Total releases reported for a category may be more than the sum of the row as in some cases respondents did not indicate what type of protection was sought for a release.

Source: Authors' survey.

requirement that an innovation contain a significant inventive step, are relatively difficult and costly to obtain.

Managing the Intellectual Property Process

The availability of IPRs for plant germplasm has led more and more SAESs to establish specialized offices for managing intellectual property. The implementation of IPR policies raises many questions: Who should be responsible for managing the IPRs? Should all types of intellectual property rights be overseen by the same body? What rules should govern access to protected technology? How should licensing fees and regulations be determined? How should revenues from intellectual property rights be distributed among research programs, the administration, and participating scientists?

Administering an Intellectual Property Office

All but one of the SAESs with plant breeding programs reported having an intellectual property office. Most of these offices are located at their affiliated land-grant university: 26 are located in the central university while 5 are located both within the SAES and university. One SAES reported having an intellectual property office only within the SAES, and one reported having an office located both at the SAES and the college of agriculture.

Table 10.7 shows the history of intellectual property offices by region. Offices established before 1970 dealt mainly with plant patents for asexually re-

Table 10.7. Establishment of intellectual property offices for crop germplasm by region

When established	Western region	Southern region	North central region	Northeastern region	*United States*
Before 1970	1	5	2	0	8
1970–79	1	2	0	0	3
1980–84	1	1	1	1	4
1985–89	0	4	1	3	8
1990–94	0	3	5	1	9
Since 1995	1	1	0	0	2
Total	4	16	9	5	34

Source: Authors' survey.

produced plants under the Plant Patent Act of 1930, including forages and grasses, vegetables, small fruits, tree fruits and nuts, woody ornamentals, and floriculture. These crops are of particular importance in the South. About half of the offices have been established since 1985, when utility patents for plants first became available.

The growth in intellectual property offices reflects the importance not only of new IPR laws but also of licensing and other collaborative arrangements, including the memorandum of understanding (MOU) and the material transfer agreement (MTA) with cross-licensing (table 10.8). An MOU is a traditional way of transferring material and is in keeping with the public mission of the SAESs. Under these agreements, SAESs transfer germplasm at no charge, often without restricting use of the material. The MTAs, however, often limit the way the receiving party can use the germplasm, prohibiting the receiver from seeking IPRs on shared germplasm or stipulating a revenue-sharing arrangement for germplasm that is successfully commercialized. The MTAs are often used to support the commercialization of technologies or to help an SAES gain access to important technologies at other institutions for use in its own research program.

Our survey showed that a majority of the SAESs use the MOU or an MTA with cross-licensing when working with another U.S. public institution or private company. The MTAs are more often used for collaborations with the private sector, suggesting these agreements are more appropriate for marketing a commercial product.

SAESs varied in the mechanisms they use to collaborate with international institutions. Some use the MOU; far fewer use MTAs with cross-licensing. The fact that MTAs are used less frequently in dealing with international institutes suggests that these institutes are not viewed as future marketing partners.

Table 10.8. Use of technology transfer mechanisms for crop germplasm by SAESs

Crop germplasm transfer agreement	Type of institution to which germplasm was transferred	Number of state agricultural experiment stations that have used this agreement with this type of institution				
		West	South	North central	Northeast	All U.S.
Memorandum of understanding	Other U.S. public institution	4	9	8	1	22
	U.S. private company	3	6	6	3	18
	International institution	3	5	6	1	15
Material transfer agreement with cross-licensing	Other U.S. public institution	0	4	6	2	12
	U.S. private company	2	7	6	3	18
	International institution	1	2	4	2	9
Assigned material without cross-licensing	Other U.S. public institution	0	2	1	2	5
	U.S. private company	1	2	2	3	8
	International institution	0	0	1	2	3

Source: Authors' survey.

The Cost of Administering Intellectual Property Agreements

An intellectual property office at an SAES or affiliated university typically employs between two and three full-time equivalents (FTEs) of administrative staff (table 10.9). However, some offices can be much smaller, employing as few as 0.05 FTEs, or much larger, requiring as many as 12 FTEs. Assuming $100,000 per FTE (including salary, office expenses, and legal fees), maintaining a typical intellectual property office costs between $200,000 and $300,000 per year.

The amount of revenue SAESs earn from the licensing of plant germplasm varies widely. Two SAESs earned more than $1 million from plant germplasm in 1996, but the median amount was $150,000 per year (table 10.10). We can break down the amounts of revenue earned by source—for example, PVPCs, plant patents, utility patents, and other sources. Of the 28 responses to this question, 13 indicated that the germplasm revenue came solely from PVPCs and 4 that it came solely from plant patents. Seven SAESs received at least two-thirds of their germplasm income from PVPCs and one-third from plant

Table 10.9. Cost of operating intellectual property (IP) office for crop germplasm

Region	Average staff size of IP office (full-time equivalents)	Estimated cost of operating IP office ($ thousand/year)[a]
West	0.8	80
South	3.4	340
North central	1.8	180
Northeast	2.5	250
All United States	2.3	230

[a] Assumes $100,000/year per full-time intellectual property officer to operate an intellectual property office.

Source: Authors' survey and estimates.

patents. Three SAESs receive most of their earned germplasm revenue from either utility patents or other sources, with less than 30 percent coming from PVPCs and plant patents. One SAES reported that most of its germplasm income came from utility patents (in this case, two-thirds of its income came from utility patents and one-third from PVPCs).

SAESs' earnings from IPRs on plant germplasm are distributed among the inventor, the inventor's research program, and other research programs at the SAES or university. They may also be allocated for administrative expenses. Many of the survey respondents indicated that they have a two-tier formula for revenue sharing, depending on whether the amount of revenue from an IPR is above or below a certain threshold. On average, the largest shares of earnings from PVPCs are allocated to the breeders' research programs, to the breeders as personal income, and to the SAESs or university administrations. Similar trends were found for plant patent and utility patent revenues, al-

Table 10.10. Revenues earned by SAESs from crop germplasm, 1996–97

Revenues ($ thousand)	Number of SAESs earning revenues in this range (responses=33)[a]
0	7
1 to 100	7
101 to 250	9
251 to 500	5
501 to 750	2
751 to 1,000	1
1,000 to 2,000	1
greater than 2,000	1
Median revenue	$150,000

[a] Five SAESs reported 1997 earnings instead of 1996 earnings.

Source: Authors' survey.

though less of these monies are allocated to the inventor and more to SAESs or university administrations.

Seventeen of the 33 respondents with plant breeding programs indicated that they handled IPRs for genetically engineered constructs in the same manner as other plant products. Ten SAESs indicated that they adopted special procedures for these IPRs—about the same number (11) identified by Brooks (1989). Seven SAESs did not respond to this question.

CONCLUSIONS AND IMPLICATIONS

The results of this survey show that while SAESs have taken advantage of the opportunity to use strengthened IPRs, the overwhelming majority of plant breeding releases by SAESs continue to be made without any form of intellectual property protection. The SAESs are using PVPCs, plant patents, and utility patents less than the 1989 Brooks survey anticipated. When SAESs do seek protection for intellectual property, they favor PVPCs, which allow protected products to be used freely within other plant breeding programs. In this sense, SAESs are still working for the public good, since information released in this way can be used freely in the breeding programs of other institutions.

The survey results also suggest that SAESs are collaborating more frequently with private breeders. Breeding programs conducted cooperatively with other institutions are about evenly divided among other SAESs, the ARS, and the private sector. But about half of the SAES breeding programs continue to be conducted solely by the stations, suggesting that SAESs have maintained considerable autonomy in program management and direction.

The SAESs have similar policies for exchanging intellectual property with other public institutions and private firms in the United States. Memoranda of understanding as well as material transfer agreements with cross-licensing are used with both types of organizations. Memoranda of understanding generally support free exchange of germplasm. Material transfer agreements with cross-licensing can be used to support commercialization of technologies or help an SAES gain access to important technologies at other institutions for use in its own research program. Material transfer agreements may be more commonly used when research products have greater commercial potential.

For most SAESs, the cost of maintaining intellectual property offices and distributing revenues obtained from IPRs on plant germplasm are modest. Most of these revenues were used to support the research program of the inventing scientist. Revenue shares are also used to provide scientists with incentives and to support other research and administrative costs at the SAES or university.

The use of IPRs at public agricultural research institutions is still evolving. We could continue to see changes in how SAESs use and manage intellectual property rights in the near future. Future surveys (in 5 or 10 years) will show whether SAESs are still making changes in the way they manage IPRs and distribute earned revenues and whether plant breeding programs are still following the traditional mission of public research institutions. So far, it seems that SAESs have kept that mission in mind.

NOTES

1. The authors would like to thank Keith Fuglie, Eldon Ortman, Carl Pray, Kenneth Frey, and Douglas Dion for their comments and insights on this project.

2. *Plant germplasm* refers to any plant material, including finished varieties, advanced breeding material, and unimproved accessions.

3. Because hybrid seed does not reproduce exactly, intellectual property in hybrid seed can easily be protected as a trade secret by keeping parent inbred lines of hybrid seed out of the public domain. In this way a plant breeder can protect his or her investment in research.

4. Either parent inbred lines or F1 hybrid varieties can now receive plant variety protection under the PVPA. To satisfy the stability requirement of the PVPA, a deposit of parent seed is required in order to protect an F1 hybrid.

REFERENCES

Brooks, H.J. 1989. "Questionnaire on Agricultural Experiment Station Policies on Release of Plant Materials." Agricultural Research Service, U.S. Department of Agriculture, Beltsville, MD. Unpublished mimeo.

Frey, K. 1996. *National Plant Breeding Study-I: Human and Financial Resources Devoted to Plant Breeding Research and Development in the United States in 1994.* Special Report 98. Ames: Iowa Agriculture and Home Economics Experiment Stations, Iowa State University.

Fuglie, K., C. Klotz, and M. Gill. 1996. "Intellectual Property Rights Encourage Private Investment in Plant Breeding." *Choices* (First Quarter): 22–23.

Fuglie, K., N. Ballenger, K. Day, C. Klotz, M. Ollinger, J. Reilly, U. Vasavada, and J. Yee. 1996. *Agricultural Research and Development: Public and Private Investments under Alternative Markets and Institutions.* Agricultural Economics Report 735. Washington, DC: Economic Research Service, U.S. Department of Agriculture.

Huffman, W.E, and R.E. Evenson. 1993. *Science for Agriculture.* Ames: Iowa State University Press.

Knudson, M.K., and C.E. Pray. 1991. "Plant Variety Protection, Private Funding, and Public Sector Research Priorities." *American Journal of Agricultural Economics* 73:882–886.

Pray, C.E., and M.K. Knudson. 1994. "Impact of Intellectual Property Rights on Genetic Diversity: The Case of U.S. Wheat." *Contemporary Economic Policy* 12 (January): 102–11.

Pray, C.E., M.K. Knudson, and L. Masse. 1993. "Impact of Changing Intellectual Property Rights on U.S. Plant Breeding R&D." Department of Agricultural Economics and Marketing, Rutgers University, New Brunswick, NJ. Unpublished mimeo.

Ruttan, V.W. 1982. *Agricultural Research Policy.* Minneapolis: University of Minnesota Press.

IV. International Spillovers in Agricultural Research

11

AGRICULTURAL TECHNOLOGY SPILLOVERS

ROBERT E. EVENSON

THE INCREASING role of multinational corporations in the development of agricultural technologies has led to a rapid diffusion of the fruits of research around the world. These developments are discussed in some detail in chapter 2 and to a lesser extent in the overview and the conclusion of this volume. This spillover of new technologies is occurring through both the public and private sectors, operating in both the donor and recipient countries. Typically, plant breeding—one of the key forms of agricultural invention—is done at agricultural experiment stations receiving public and perhaps private funds (Huffman and Evenson 1993). In many developed and some developing countries, systems of experiment stations have evolved with a strong regional or geographic focus. National agricultural research systems (NARSs) engage in plant breeding and agronomic and related research. Research units similar to the state agricultural experiment stations (SAESs) in the United States often operate at the regional and provincial levels. And for many individual commodities, international agricultural research centers (IARCs) have been built. The statistical significance of spillovers between the public research systems of the United States and the countries of the European Community are directly estimated in the next chapter by Schimmelpfennig and Thirtle.

This chapter provides some context for the results of the following chapters by defining and then reviewing different measures of various types of spillovers. First, it classifies spillovers according to their scope, using three criteria—whether they are direct or indirect, whether they are generated within the sector in which they are used, and whether they cross geographic boundaries. The next two sections present measures of spillovers that are based on crop varietal and patent data, respectively. The third section provides an introduction to a new database of private-sector R&D activity in the agricultural inputs industry (Evenson and Johnson 1997). The next section examines studies of agricultural productivity that incorporate spillovers into the analysis. The final section offers suggestions for further research. An appendix reviews statistical sources and methods for measuring spillovers.

TYPES OF AGRICULTURAL SPILLOVERS

There are two types of technology spillovers, *direct* and *indirect*. Direct technology spillovers occur when improved technology developed in one region or industry is used in another. Indirect, or germplasmic, technology spillovers occur when research that produces a technology (or the technology itself) spurs research programs in another region or industry to develop improved technology. The distinction between direct and indirect spillovers is important because agricultural research programs affect each other in both ways. In some cases, the distinction is incorporated into the design of research programs: Researchers may specialize in "preinvention science" to support other researchers who specialize in inventions.[1]

Both direct and indirect spillovers fall into four categories:

- Type 1: intraindustry/intraregional
- Type 2: intraindustry/interregional (international)
- Type 3: interindustry (sectoral)/intraregional
- Type 4: interindustry (sectoral)/interregional (international)

Type 1 spillovers occur when an individual, a firm, or a research station develops an invention that is used (either through licensing or simply by copying) by farms or firms in the same geographic region. Most intraindustry spillovers in the agricultural sector begin at an experiment station designed to produce inventions for farms in the region.[2] Occasionally an invention may begin with a farmer, but such cases are rare.

Type 2 spillovers occur when an intraindustry invention is used outside the region or country for which it was designed or targeted. Because experiment stations tend to target a regional ecosystem or regional geoclimatic conditions, direct type 2 spillovers may occur less frequently than type 1 spillovers. Differences in soil, climatic, economic, and institutional conditions among regions may also inhibit direct spillovers. If the differences between regions are sufficiently great, they can prevent direct interregional spillovers altogether. Indirect spillovers are generally less affected by regional differences, so that type 2 indirect spillovers are more common.

Type 3 and 4 spillovers occur when an invention generated by one industry is adapted by another industry or economic sector. Generally, the invention is embodied in a product that is manufactured in the inventing industry and then sold to another. This type of spillover is one of the most important in industrial economies and is becoming more important to many developing economies. (See table 11.1 for a list of some industrial patents used in Canadian agriculture in the last 20 years. See also the appendix to this chapter for some notes on the patent data discussed in this chapter.) Direct type 3 and 4 spillovers can be inhibited in much the same way as type 2 spillovers. When interregional barriers are high, only direct type 3 spillovers occur. Again, in-

Table 11.1. Patents used in agriculture by major industries of manufacture in Canada, 1978–93

IOM (Industry of manufacture)	Agric. Livestock	Agric. Crops	Agric. F. and Vegs.	Agric. Hort.	Agric. Serv. live	Agric. Serv. crops	Agric. Other	Agric. Total	Agric. share of IOM
10—Food industry	179	2	2	0	14	0	0	197	0.11
1053—Feed industry	174	1	1	0	14	0	0	190	0.70
12—Tobacco	0	8	0	0	0	0	0	8	0.02
15—Rubber products	3	0	1	1	1	0	0	6	0.00
16—Plastic products	38	11	15	54	20	13	4	155	0.02
1610—Foamed plastic	24	2	5	0	11	0	0	42	0.03
1699—Other plastic	11	7	7	44	9	5	3	86	0.02
17—Leather products	13	0	2	0	11	0	0	26	0.08
1719—Other leather	12	0	1	0	5	0	0	18	0.30
18—Primary textiles	0	1	0	1	2	1	0	5	0.00
19—Textile products	4	3	0	4	0	0	1	12	0.01
24—Clothing industry	2	1	0	0	1	0	0	4	0.02
25—Wood industry	15	1	0	2	2	0	0	20	0.02
26—Furniture industry	4	1	0	0	3	0	0	8	0.00
27—Paper industry	8	2	10	6	4	1	1	32	0.01
2731—Folding cartons, boxes	1	0	8	5	3	0	0	17	0.02
29—Primary metals	0	1	0	0	0	0	0	1	0.00
30—Fabricated metals	44	17	11	14	34	11	3	134	0.01
3032—Prefab metal buildings	4	1	0	2	4	0	0	11	0.03
3063—Hand tools	4	0	2	2	1	6	0	15	0.04
3092—Metal valves	4	0	3	0	3	0	0	10	0.00
3099—Other fabricated metal	9	1	1	2	2	1	1	17	0.01
31—Machinery (nonelectric)	400	1,477	106	130	162	700	29	3,004	0.04
3111—Agricultural implements	251	1,266	81	66	99	606	18	2,387	0.79
3190—Other machinery	49	79	5	0	31	31	1	196	0.01
3192—Mach. for handling mat.	12	32	3	8	1	8	1	65	0.01
3199—Other machinery nec.	84	72	16	56	30	50	8	316	0.01
32—Transportation equipment	1	15	0	0	0	2	1	19	0.00

Table 11.1. *(continued)*

IOM (Industry of manufacture)	Agric. Livestock	Agric. Crops	Agric. F. and Vegs.	Agric. Hort.	Agric. Serv. live	Agric. Serv. crops	Agric. Other	Agric. Total	Agric. share of IOM
33—Electrical products	40	22	4	9	10	8	7	95	0.00
3359—Other electronic equip.	11	3	0	5	1	2	0	22	0.00
3379—Other electrical equip.	13	9	2	0	2	1	1	28	0.00
35—Nonmetallic minerals	3	1	0	2	1	0	0	7	0.00
37—Chemical products	191	830	59	33	291	311	34	1,749	0.03
3712—Industrial organic chem.	2	8	0	0	0	15	12	37	0.00
3722—Mixed fertilizer	11	105	4	13	0	18	3	154	0.84
3729—Other agricultural chem.	34	494	24	5	12	209	6	784	0.74
3741—Pharm. and medical chem.	99	29	4	0	257	23	7	419	0.03
3799—Other chemicals	43	184	27	10	18	36	6	324	0.03
39—Other manufacturing	63	31	6	39	106	10	14	269	0.01
3910—Scient. and profess. equip.	2	0	0	0	16	0	0	18	0.01
3911—Controlling instruments	7	9	1	1	20	2	2	42	0.00
3912—Other instruments	5	3	0	2	46	3	4	63	0.00
3999—Other manuf., nec.	41	19	5	34	21	3	8	131	0.03
40—Building and contracting	9	2	0	17	11	4	0	43	0.21
4020—Nonresidential	3	0	0	0	7	1	0	11	0.50
4021—Light industrial	4	0	0	4	4	0	0	12	0.44
4022—Commercial	0	2	0	12	0	1	0	15	0.21
41—Industrial construction	0	1	0	0	0	2	0	3	0.01
42—Trade contracting	1	0	0	2	0	0	0	3	0.02
49—Other utilities	5	1	0	0	0	1	2	9	0.13
4999—Other utilities, nec.	5	1	0	0	0	1	2	9	0.26
59—Other wholesale products	1	0	0	0	0	0	0	1	0.20

This table is based on actual Canadian patent office assignments.

direct spillovers are less affected by interindustry barriers and thus are often type 4.

USING CROP VARIETAL DATA TO MEASURE INTRASECTORAL SPILLOVERS

Evenson and Gollin (1997) studied releases of indica and japonica rice varieties between 1962 and 1991, classifying a total of 1,709 according to the releasing country and release date (table 11.2 summarizes these varieties by releasing country). Analyzing the genealogies (parentage) of each release allowed for further characterization of breeding strategies and landrace complexities.[3] The International Rice Research Institute (IRRI), an IARC, made a number of the crosses from which these varieties were selected but officially released only a few varieties. India, with 26 rice breeding programs, led all countries in number of releases (643). In all, varieties from more than 100 breeding programs were released. Releases occurred at a rate of approximately 20 per year in the early "green revolution" period (1966–70), but rose to nearly 80 per year in 1976–80. They have remained steady at around 75 per year since.

From these data we can see something of the scope of direct intrasectoral exchange of varieties. Table 11.3 provides a comparison of the locations of the

Table 11.2. Number of varieties of rice by country of release and time period of release

Country/region	Pre-1965	1966–70	1971–75	1976–80	1981–82	1986–91	*Total*
Africa	3	7	8	17	26	42	101
Bangladesh	1	7	8	11	4	33	34
Burma	0	4	6	21	37	8	76
China	0	1	8	30	31	12	82
India	10	67	136	139	125	166	643
Indonesia	1	2	5	21	10	9	48
Korea	0	5	11	35	40	15	106
Latin America	7	9	48	32	43	100	239
Nepal	0	0	1	10	4	2	17
Oceania	0	1	4	1	0	0	6
Pakistan	0	4	2	3	3	0	12
Philippines	3	4	13	23	8	2	53
Sri Lanka	3	14	4	8	21	3	53
Taiwan	0	3	0	3	0	0	6
Thailand	1	2	4	8	5	3	23
USA	2	5	18	17	3	6	51
Vietnam	0	16	6	16	16	5	59
Other Southeast Asia	2	1	8	7	6	5	29
Other	0	7	15	15	15	19	71
Total	32	150	284	394	377	378	1,709

breeding programs where the crosses were made with the locations of the programs that released varieties based on these crosses. Table 11.4 summarizes the data, which show that IRRI was an important producer of the crosses from which subsequent releases were made (panel I). In the early green revolution period, IRRI made 25 percent of all crosses leading to varieties. This percentage declined to 12 percent in 1986–91, but IRRI's plant breeding program remains a potent contributor to varietal development.

Table 11.5 presents comparable data for varietal parents that provide a measure of indirect intrasectoral spillover (this information is summarized in table 11.4, panel II). Here we see that IRRI produced the crosses from which 24 percent of varietal parents were selected. Other NARSs produced the crosses from which an additional 18 percent of varietal parents were selected.

Further data on breeding strategies in panel III of table 11.4 show that the most widely used and most successful breeding strategy over this period used one parent from IRRI and one from a NARS. The importance of international exchanges in rice breeding is shown by the relatively low percentage of varietal releases for which all parental material came from national sources (most of these releases were made in India).

The landrace content of released varieties has increased. The average number of landraces in a given release has risen from under 3 to around 8, although some recently released varieties contain more than 25 landraces in their genealogies (panel IV of table 11.4). More than 70 percent of these landraces were brought into the genealogies through an IRRI ancestor.

Panel V of table 11.4, which shows another dimension of IRRI's role in plant breeding, gives the number of new landraces (by period and originating source) introduced into the landrace pool. These data reveal two facts. First, genetic resources consisting of an impressive number of new landraces (and one or two wild species) have been introduced into the pool of successful varieties. The fact that the 1,709 releases included 838 landraces that were not contained in the landrace pool before 1965 shows that genetic resource collections have been valuable to breeding programs.

Second, IRRI has actually introduced very few landraces into the pool. Only 80 of the 838 new landraces were introduced via IRRI crosses. By contrast, all of the landraces in released varieties—roughly 70 percent—were introduced via an IRRI cross. Two factors are responsible for this situation. One is that IRRI's powerful breeding lines incorporate many landraces that were first brought in through a NARS cross. The other is that the widespread use of IRRI crosses as breeding lines multiplies the use of the landraces they contain.

Similar data exist for wheat and other crops. These data show relatively low rates of direct international intrasectoral spillover and relatively high rates of indirect (germplasmic) international intrasectoral spillover.

Table 11.4. Flows of international genetic resources, by time period

	Pre-1965	1966–70	1971–75	1976–80	1981–85	1986–91	*Total*
I. *Released varieties, percentage based on*							
IRRI cross (through INGER)	3	25	19	22	18	12	17
	(0)	(0)	(2)	(13)	(14)	(11)	(8)
Other NARS cross (through INGER)	16	7	6	6	6	5	6
	(0)	(0)	(0)	(2)	(4)	(3)	(3)
Own NARS cross	81	68	75	72	76	83	77
II. *Parents of released varieties, percentage with one or more parents*							
IRRI cross (through INGER)	0	24	29	33	23	19	24
	(0)	(0)	(0)	(9)	(20)	(15)	(10)
Other NARS cross (through INGER)	27	25	21	15	18	20	18
	(7)	(2)	(5)	(9)	(15)	(15)	(10)
Own NARS cross	73	51	50	52	59	6	58
III. *Frequency of parental/cross, percentage with no foreign genetic resource*							
All NARS parents	24	11	8	6	7	10	8
IV. *Landrace content of released varieties, parent greater than*							
4	10	31	47	67	62	56	55
9	0	3	13	39	34	32	27
15	8	0	3	21	18	18	14
Average number of landraces	2.55	4.01	5.29	8.15	7.49	7.23	6.36
Percentage from IRRI	3	3	59	79	74	71	68
V. *Landrace introduction*							
Number from IRRI	0	16	14	21	11	13	80
Number from NARSs	21	87	126	146	171	180	758

IRRI = International Rice Research Institute; INGER = International Network for the Genetic Evaluation of Rice; NARS = National Agricultural Research System.

USING PATENT DATA TO MEASURE INTERSECTORAL SPILLOVERS TO AGRICULTURE

A large and growing source of intersectoral spillovers is the activities of multinational corporations. In addition to companies who produce agricultural inputs and the emerging players in a possible future life-sciences industry, both industrial and service firms undertake R&D and produce inventions that are used in agriculture, creating intersectoral spillovers with potential interregional and international effects. For example, agricultural equipment invented in the machinery sector may have spillover benefits for the agricultural sector. These inventions, and some preinvention R&D, may also create indirect spillovers. For instance, a mechanical or chemical invention may lead public-sector researchers to develop a new hybridization technique and thus a new variety of rice.

The Yale Technology Concordance (YTC) provides a probability-based assignment of inventions that have an International Patent Classification (IPC) designation, assigning them to industries of manufacture (IOM) and sectors of

Table 11.3. Matrix of varietal exchange

Releasing country	Source of varietal cross: Other	Latin America	Oceania	Bangladesh	Africa	Burma	USA	China	India	Indonesia	IRRI	Korea	Southeast Asia	Nepal	Pakistan	Philippines	Sri Lanka	Taiwan	Thailand	Vietnam	*Total*
Other	9	7	0	0	0	1	0	5 (2)	16	3 (2)	18 (9)	6	5	0	0	0	0	1	0	0	71 (13)
Latin America	3 (1)	185 (2)	0	0	1 (1)	0	5	0	2 (2)	0	39 (15)	1	0	0	0	2	1 (1)	0	0	0	239 (22)
Oceania	0	0	1	0	0	0	0	0	0	0	5	0	0	0	0	0	0	0	0		6 (0)
Bangladesh	0	0	0	17	0	0	0	1	4 (1)	0	11 (3)	0	1	0	0	0	0	0	0	0	34 (4)
Africa	1	1 (1)	0	0	69 (1)	0	0	0	1 (1)	0	26 (12)	0	0	0	0	1	2 (2)	0	0	0	101 (17)
Burma	2	0	0	1 (1)	0	33	0	1 (1)	1	1 (1)	18 (5)	0	2 (1)	0	0	1	2 (2)	0	12 (6)	2 (2)	76 (19)
USA	0 (2)	1 (2)	0	0	0	0	48	0	0	0	2	0	0	0	0	0	0	0	0	0	51
China	0	0	0	0	0	0	0	66	1	0	13 (8)	1	0	0	0	0	1 (1)	0	0	0	82 (9)
India	5 (33)	0 (3)	0 (1)	0 (37)	1	0	0	0	573	0	53	0	1	0	1	1	4	2	1	1	643
Indonesia	0	0	0	0	0	0	0	0	0	29	18 (13)	0	0	0	0	1	0	0	0	0	48 (13)
Southeast Asia	0	0	0	0	0	0	0	0	1	0	7 (2)	0	21	0	0	0	0	0	0	0	29 (2)
Korea	0	0	0	0	0	0	0	0	0	0	1	105	0	0	0	0	0	0	0	0	106 (0)
Nepal	0	0	0	0	0	0	0	0	0	0	5 (2)	0	1	8	0	0	1 (1)	0	0	0	15 (3)
Pakistan	0	0	0	0	0	0	0	0	0	0	7	0	0	0	5	0	0	0	0	0	12 (0)
Philippines	0	0	0	0	0	0	0	0	0	0	25 (15)	0	0	0	0	26	0	0	0	1	52 (15)
Sri Lanka	0	0	0	0	0	0	0	0	0	0	2	0	0	0	0	0	51	0	0	0	53 (0)
Taiwan	0	0	0	0	0	0	0	0	0	0	0	0	0	0	0	0	0	6	0	0	6 (0)
Thailand	0	0	0	0	0	0	0	0	0	0	0	0	0	0	0	0	0	0	23	0	23 (0)
Vietnam	0	0	0	0	0	0	0	0	0	0	44 (27)	0	0	0	0	0	0	0	0	15	59 (27)
Total	20 (1)	194 (3)	1 (0)	18 (1)	71 (2)	34 (0)	53 (0)	73 (3)	599 (4)	33 (3)	294 (146)	113 (0)	31 (1)	8 (0)	6 (0)	32 (0)	62 (10)	9 (0)	36 (7)	19 (2)	1,706 (183)

Numbers in parentheses represent borrowings through the International Network for the Genetic Evaluation of Rice (INGER); IRRI = International Rice Research Institute.

Table 11.5. Matrix of parental borrowing of crop germplasm used in plant breeding

Releasing country	Other	Latin America	Oceania	Bangladesh	Africa	Burma	USA	China	India	Indonesia	IRRI	Korea	Southeast Asia	Nepal	Pakistan	Philippines	Sri Lanka	Taiwan	Thailand	Vietnam	*Total*
	Source of varietal cross																				
Other	28 (2)	2	0	0	0	1	1	0	42 (23)	4	48 (13)	1	7	0	0	4 (3)	0	2 (1)	2	0	142 (42)
Latin America	116 (26)	73 (20)	0	0	4 (3)	0	22 (3)	0	62 (58)	3 (5)	161 (68)	4	0	0	0	8	6 (3)	5 (2)	10 (3)	4 (3)	478 (194)
Oceania	0	0	0	0	2	0	2	0	5	0	0	0	0	0	0	0	0	1	2	0	12 (0)
Bangladesh	4 (1)	0	0	1 (1)	0	0	2	1	18 (11)	2	34 (9)	0	2	0	0	2	0	1 (1)	0 (1)	1	68 (24)
Africa	33 (20)	3 (2)	0	0	34 (13)	0	0	2 (2)	57 (51)	6 (4)	42 (24)	0	6 (3)	0	0	3	3 (3)	5 (1)	5	3 (3)	202 (126)
Burma	26 (15)	0	0	0	0	1	0	1	68 (68)	5 (1)	38 (29)	0	1	0	0	7 (5)	0	2	0 (4)	3	152 (122)
USA	28	2	0	0	0	0	47 (1)	0	15 (15)	0	6 (4)	0	0	0	0	0	1	1	2 (4)	0	102 (24)
China	42 (13)	0	0	0	0	0	0	20 (7)	27 (29)	1 (2)	72 (52)	0	0	0	0	0	1	1	0	0	164 (103)
India	150 (32)	1	0	3 (2)	4 (4)	0	2 (1)	10 (3)	637 (101)	14 (7)	351 (174)	0	23 (13)	0	0	15 (6)	10 (4)	57 (14)	7 (4)	0	1,284 (365)
Indonesia	11 (3)	0	0	0	0	0	0	1	22 (12)	22 (1)	35 (14)	0	1	0	0	2	0	0	2	0	96 (30)
Southeast Asia	0	0	0	0	0	0	0	1	9 (5)	3	17 (7)	0	9 (1)	0	0	3 (2)	0	2	2	0	46 (15)
Korea	0	0	0	0	0	0	0	0	3 (3)	0	49 (15)	74 (8)	0	0	0	0	0	0	0	0	126 (26)
Nepal	9	0	0	1	1	0	0	0	7 (4)	1	10 (4)	0	2	0	0	1	2 (2)	0	0	0	34 (10)
Pakistan	1	0	0	0	0	0	0	2	6 (3)	0	12 (4)	0	0	0	0	1	0	0	2	0	24 (7)
Philippines	20	0	0	0	0	0	4	0	11 (2)	5 (1)	45 (13)	0	2	0	0	18 (2)	0	0	0	0	105 (15)
Sri Lanka	24 (3)	0	0	0	0	0	1 (1)	1 (1)	4 (3)	3 (1)	22 (4)		0	0	0	0 (5)	50	1	0	0	106 (18)
Taiwan	2	0	0	0	0	0	0	0	3 (2)	0	5	0	0	0	0	0	0	2	0	0	12 (2)
Thailand	12 (5)	0	0	0	0	0	0	0	12 (11)	2	7 (5)	0	0	0	0	1	0	1	11 (1)	0	46 (22)
Vietnam	11 (1)	0	0	0	0	0	0	3	34 (22)	3	63 (23)	0	1	0	0	1	0	0	0	2	118 (46)
Total	517 (121)	81 (22)	0	5 (3)	45 (20)	2	81 (6)	42 (13)	1,042 (423)	74 (22)	1,017 (462)	79 (8)	54 (17)	0	0	66 (18)	73 (17)	81 (19)	45 (12)	13 (11)	3,317

Numbers in parentheses represent borrowings through the International Network for the Genetic Evaluation of Rice (INGER); IRRI = International Rice Research Institute.

use (SOU).[4] The probabilities are based on assignments made by the Canadian Patent Office over a period of 25 years, during which time IPC, IOM, and SOU codes were attached to over 300,000 patents. Evenson and Johnson (1999) have developed a new version of the YTC that not only updates the probabilities by including more data but considers simultaneous IOM and SOU probabilities. Previous studies were limited to calculations of total patents by IOM or total patents by SOU, but this work makes possible full IOM-SOU tables, or input-output tables of inventions based on patent data (see appendix).

This advance enables a measure of direct invention spill-ins (or spillovers) to agricultural sectors by source (for instance, the number of inventions used in agricultural subsectors that originated in the machinery industry). Using data on private-sector R&D expenditures by sector, we can also construct financial measures of direct and indirect R&D spill-ins, or the dollar amount of private-sector R&D spilling over to agriculture through spending in other sectors and other nations.[5]

Table 11.1 shows the most important elements of the IOM-SOU matrix for Canada. Not surprisingly, the most important industrial sources of patents for agricultural use are the food industry (especially the animal feed subsector), machinery (in particular agricultural implements), and chemicals (notably fertilizer, other agricultural chemicals, and pharmaceuticals). However, many other industries that may not be obvious at first glance also provide inventions for agricultural use, including plastics, fabricated metals, and electrical products.

Despite the fact that the YTC is based on Canadian data, the industrial structure of Canadian inventions is not imposed when the probabilities are applied to patent data for other countries. The probabilities of the YTC indicate a relationship between an IPC (a product/process definition) and an IOM or SOU (an industry definition). But they also allow enormous flexibility in using the data—for instance, to display the industrial composition of patenting in a specific economy. As expected, in every country, chemicals and machinery (including agricultural implements) are the primary providers of patents for agricultural use.

Measuring Direct Intersectoral National Spillovers

Not surprisingly, patenting activity within developed countries spills over between sectors in that country. More surprising is the level of these spillovers in some developing countries. The numbers of domestic inventions used in food and agriculture vary among countries, largely because of nuances in the definition of a patent (table 11.6 gives a sample of developing countries). Some countries permit larger claims within one patent and so have fewer patents for the same amount of protected knowledge. However, we can make some comparisons. For example, Brazil, Korea, and China engage in far more

Table 11.6. Domestic-origin patents used in agriculture and food subsectors (all with IPCs)

Nation	Year	Agric. Livestock	Agric. Crops	Agric. F. & Vegs.	Agric. Hort.	Agric. Serv. live	Agric. Serv. crops	Agric. Other	Food Meat	Food F. & Vegs.	Food Dairy	Food Cereals	Food Beverages	Food Tobacco	Food Other	Agric. Total	Food Total	A+F Total	*All Patents*
Argentina	1974	1	3	0	0	1	2	0	1	1	0	0	1	0	2	6	4	11	118
Argentina	1980	5	23	2	5	3	12	0	7	5	4	2	13	6	16	50	51	102	1,002
Argentina	1985	1	2	0	0	1	1	0	1	0	0	0	1	0	2	6	5	11	218
Argentina	1990	1	4	1	1	1	2	0	1	0	1	0	2	3	1	10	8	18	180
Brazil	1974	18	70	7	5	11	37	1	34	17	17	8	40	22	59	150	197	346	4,335
Brazil	1980	14	66	6	6	11	34	1	15	9	7	6	28	10	34	138	108	246	3,183
Brazil	1985	14	46	4	6	9	21	1	13	7	7	4	22	5	33	102	92	194	2,885
Brazil	1990	14	42	4	6	9	22	1	14	8	9	6	16	6	31	99	90	189	2,996
Brazil	1995	12	41	4	6	8	20	2	10	8	7	4	18	9	32	94	88	182	2,457
Canada	1978-1993	87	237	21	52	52	121	7	81	20	30	21	62	36	115	576	365	941	14,818
China	1990	69	136	15	30	39	45	13	65	34	26	38	76	43	228	347	510	857	10,426
China	1995	129	248	27	48	124	72	27	160	75	81	121	282	68	825	675	1,612	2,287	15,280
Egypt	1977	1	5	0	1	1	2	0	0	0	0	1	0	0	1	10	2	12	143
Egypt	1980	0	1	0	0	0	0	0	0	0	0	0	0	0	1	2	1	3	83
Egypt	1985	0	2	0	0	0	1	0	0	0	0	0	0	0	3	4	5	9	77
Egypt	1990	0	2	0	0	0	1	0	1	0	0	0	0	0	0	4	2	5	67
Egypt	1995	1	3	0	0	1	1	0	2	0	0	0	1	0	3	7	6	13	157
Hungary	1974	6	15	1	4	3	6	3	3	2	1	3	4	2	10	39	24	62	917
Hungary	1980	11	19	2	4	6	7	1	3	3	2	3	5	2	8	50	25	75	878
Hungary	1985	24	42	4	8	12	15	3	9	3	6	6	9	3	22	107	58	165	1,546
Hungary	1990	28	57	6	17	16	22	14	13	7	5	9	11	8	40	161	93	254	1,933
Hungary	1995	5	13	1	10	5	6	20	5	2	2	2	3	10	12	61	35	96	756
India	1976	2	8	1	1	2	4	1	3	2	4	3	9	2	18	18	41	59	1,239
India	1980	3	16	1	1	1	5	0	1	0	1	1	1	0	7	27	11	39	568
India	1985	4	15	1	1	2	5	0	6	2	4	2	3	4	17	27	38	66	1,021
India	1992	3	9	1	1	2	3	0	3	2	2	2	3	2	11	18	25	44	690
Japan	1990	713	1,847	197	354	469	815	119	930	403	470	447	776	249	2,921	4,514	6,197	10,710	432,599
Kenya	1975–1989	0	0	0	0	0	0	0	0	0	0	0	0	0	0	0	0	0	1
Korea	1979	2	7	1	0	2	2	0	1	1	2	1	2	1	6	14	15	29	683
Korea	1985	4	8	1	3	2	3	1	7	4	3	4	8	1	35	22	62	84	1,275
Korea	1990	14	12	2	3	9	5	1	22	11	8	10	15	5	60	48	130	178	3,546
Korea	1995	23	33	4	11	12	15	4	29	17	13	17	27	19	108	103	230	333	8,988
Malaysia	1974	0	0	0	0	0	0	0	0	0	0	0	0	0	0	0	0	0	1
Malaysia	1980	0	0	0	0	0	0	0	0	0	0	0	0	0	0	0	0	0	0
Malaysia	1987	0	0	0	0	0	0	0	0	0	0	0	1	0	0	0	1	1	16
Mexico	1981	1	8	0	1	1	4	0	1	0	2	1	2	1	5	15	12	28	479
Mexico	1985	4	7	1	1	2	3	0	5	2	5	3	4	2	19	18	42	59	603
Mexico	1992	2	2	0	0	1	1	0	2	0	1	1	2	0	7	6	14	20	394
Philippines	1976	3	12	1	1	3	5	1	3	3	0	1	2	0	5	25	14	39	303

Table 11.6. *(continued)*

Nation	Year	Agric. Livestock	Agric. Crops	Agric. F. & Vegs.	Agric. Hort.	Agric. Serv. live	Agric. Serv. crops	Agric. Other	Food Meat	Food F. & Vegs.	Food Dairy	Food Cereals	Food Beverages	Food Tobacco	Food Other	Agric. Total	Food Total	A+F Total	*All Patents*
Philippines	1980	2	8	1	1	1	3	0	1	1	1	1	2	0	7	16	15	31	258
Philippines	1985	1	5	0	1	1	1	0	1	0	0	1	2	1	5	9	11	20	211
Philippines	1990	1	3	0	0	1	1	0	2	0	0	1	2	3	4	6	12	18	157
Poland	1974	18	31	3	3	10	13	1	24	10	5	8	9	3	34	80	94	175	5,047
Poland	1980	69	119	10	13	27	50	4	81	24	24	35	45	16	104	293	329	622	15,765
Poland	1985	43	104	8	15	19	46	4	66	15	19	24	25	14	87	240	249	489	11,391
Poland	1990	29	66	5	12	12	30	3	26	9	15	13	14	6	57	156	139	296	7,162
Poland	1995	30	67	6	8	13	28	2	21	11	24	10	12	4	53	155	135	290	5,212
Turkey	1979	0	0	0	0	0	0	0	0	0	0	0	0	0	0	0	0	0	1
Turkey	1985	1	4	0	1	0	1	1	0	0	0	0	0	2	2	7	4	11	116
Turkey	1990	0	2	0	0	0	1	0	0	0	0	0	1	3	2	4	6	9	118
Turkey	1995	1	1	0	0	0	1	0	0	1	0	0	0	1	1	3	3	6	132
USA	1990	323	599	67	70	267	260	28	305	91	203	93	345	140	694	1,614	1,871	3,485	74,056
Zambia	1968-1994	5	23	2	1	6	7	0	2	1	2	2	5	4	17	44	33	77	797

Values may not sum due to rounding.

invention for agriculture and the food industries than do the other developing countries in the sample. Hungary and Poland also engage in significant invention for food and agriculture.

Measuring Direct Intersectoral International Spillovers

On average, 5 percent of all domestic and foreign patenting activity in developing countries involves agricultural uses, and 3 percent involves food. The shares of patents of foreign origin in these countries illustrates the importance of direct intersectoral international spillovers. Patents originating in the United States account for an average of one-quarter of all patents, including 23 percent of all those in agriculture (table 11.7). Patents of Japanese origin account for only 4 percent of all patenting but represent 6 percent of agriculture. (Their importance is greatest in crops and crop services, where 9 percent of all patents are of Japanese origin.) The United States and Japan take out the most foreign patents, with all other countries together taking out 45 percent of all patents and 37 percent of agricultural patents.

Domestic patenting averages 33 percent of all patenting, accounting for 34 percent of all patenting for agricultural uses and 38 percent of patenting for food uses. Horticulture is the subsector with the most domestic patents (an average of 44 percent), while tobacco and general agricultural crops have the fewest (29 percent). On average, domestic invention is less focused on livestock and tobacco, with more emphasis on horticulture and beverages. American-origin patents are weighted much more heavily toward tobacco, while Japanese-origin patents are heavily centered on crops and cereals.

Within the agricultural sector, half of all agricultural inventions, from domestic or foreign sources, go directly to crops and 17 percent to crop services. Livestock averages another 14 percent of all agricultural-use patents and services to livestock earn a further 10 percent. The remaining 10 percent is split evenly between fruits and vegetables and horticulture.

Table 11.8 shows international flows in more detail for nine countries of the Organization for Economic Cooperation and Development (OECD) with the agricultural sector as the SOU in 1987. This table shows an asymmetry among countries, with the United States as the dominant supplier. Domestic patenting in Japan is high because of claim limitations in Japanese patent law.

Four nations in the YTC sample (Brazil, China, Hungary, and Korea) have, at some point, used utility model protection that restricts technological spillovers. Utility models give intellectual property rights to minor inventions or adaptations of existing technology and require a smaller "inventive step" than full patents.[6] Table 11.9 presents both the number of utility models by SOU and the ratio of utility models to full domestic patents in each SOU. In these four countries, an average of 7 percent of all utility models were in agriculture and food, a proportion slightly larger than for full patents. More of the utility models were in the food sector than in agriculture, a trend not seen in

Table 11.7. Domestic share of patents for use in agriculture and food subsectors

Nation	Year	Agric. Livestock	Agric. Crops	Agric. F. & Vegs.	Agric. Hort.	Agric. Serv. live	Agric. Serv. crops	Agric. Other	Food Meat	Food F. & Vegs.	Food Dairy	Food Cereals	Food Beverages	Food Tobacco	Food Other	Domestic Share of Ag+Food	US Share of Ag+Food	Japan Share of Ag+Food	Other Share of Ag+Food	Domestic Share of All patents
Argentina	1974	0.24	0.36	0.38	0.25	0.22	0.41	0.19	0.16	0.36	0.05	0.11	0.20	0.06	0.17	0.23	0.32	0.00	0.45	0.14
Argentina	1980	0.10	0.12	0.12	0.36	0.14	0.18	0.10	0.14	0.31	0.17	0.10	0.33	0.25	0.17	0.16	0.29	0.04	0.50	0.17
Argentina	1985	0.29	0.18	0.22	0.22	0.22	0.24	0.22	0.16	0.22	0.28	0.21	0.45	0.08	0.37	0.24	0.30	0.01	0.45	0.35
Argentina	1990	0.17	0.23	0.28	0.59	0.17	0.30	0.37	0.16	0.18	0.53	0.07	0.51	0.30	0.14	0.26	0.23	0.05	0.46	0.26
Brazil	1974	0.31	0.27	0.30	0.39	0.33	0.38	0.20	0.42	0.52	0.43	0.33	0.46	0.32	0.37	0.35	0.23	0.03	0.38	0.30
Brazil	1980	0.30	0.25	0.24	0.41	0.42	0.35	0.24	0.35	0.50	0.32	0.36	0.47	0.33	0.38	0.32	0.18	0.06	0.43	0.28
Brazil	1985	0.32	0.21	0.21	0.40	0.38	0.28	0.14	0.32	0.50	0.36	0.37	0.44	0.10	0.39	0.28	0.25	0.06	0.41	0.30
Brazil	1990	0.37	0.26	0.28	0.54	0.44	0.38	0.18	0.39	0.54	0.41	0.41	0.40	0.17	0.43	0.35	0.24	0.05	0.36	0.34
Brazil	1995	0.29	0.27	0.28	0.49	0.26	0.37	0.21	0.24	0.46	0.23	0.23	0.33	0.28	0.27	0.29	0.29	0.06	0.36	0.28
China	1990	0.71	0.61	0.62	0.76	0.62	0.61	0.66	0.63	0.72	0.52	0.68	0.67	0.47	0.64	0.63	0.16	0.08	0.13	0.48
China	1995	0.72	0.65	0.65	0.79	0.71	0.61	0.80	0.79	0.81	0.72	0.83	0.86	0.57	0.83	0.77	0.05	0.07	0.12	0.51
Egypt	1977	0.11	0.10	0.09	0.26	0.11	0.11	0.12	0.14	0.06	0.20	0.34	0.11	0.03	0.11	0.11	0.24	0.06	0.59	0.22
Egypt	1980	0.05	0.02	0.02	0.23	0.10	0.02	0.06	0.26	0.18	0.23	0.23	0.27	0.13	0.22	0.05	0.20	0.10	0.64	0.19
Egypt	1985	0.12	0.08	0.09	0.25	0.09	0.09	0.08	0.35	0.36	0.13	0.32	0.15	0.01	0.44	0.15	0.34	0.00	0.51	0.15
Egypt	1990	0.11	0.11	0.13	0.21	0.12	0.11	0.11	0.60	0.12	0.07	0.05	0.27	0.02	0.06	0.12	0.21	0.08	0.59	0.19
Egypt	1995	0.23	0.13	0.13	0.18	0.27	0.18	0.22	0.63	0.31	0.23	0.23	0.25	0.02	0.52	0.22	0.25	0.11	0.42	0.30
Hungary	1974	0.37	0.18	0.18	0.66	0.43	0.24	0.57	0.50	0.68	0.50	0.51	0.56	0.25	0.46	0.31	0.08	0.02	0.58	0.37
Hungary	1980	0.49	0.41	0.43	0.54	0.62	0.40	0.31	0.41	0.62	0.34	0.45	0.67	0.40	0.48	0.46	0.08	0.01	0.44	0.47
Hungary	1985	0.63	0.37	0.39	0.80	0.65	0.38	0.68	0.52	0.59	0.69	0.61	0.56	0.52	0.59	0.50	0.11	0.04	0.35	0.56
Hungary	1990	0.55	0.33	0.35	0.80	0.38	0.38	0.82	0.67	0.84	0.68	0.65	0.79	0.80	0.64	0.49	0.09	0.04	0.37	0.44
Hungary	1995	0.50	0.34	0.37	0.92	0.35	0.38	0.95	0.59	0.58	0.46	0.41	0.34	0.91	0.48	0.54	0.08	0.02	0.36	0.37
India	1976	0.26	0.30	0.31	0.36	0.38	0.38	0.50	0.40	0.57	0.43	0.35	0.54	0.25	0.42	0.39	0.22	0.05	0.34	0.40
India	1980	0.53	0.76	0.81	0.58	0.41	0.80	0.39	0.50	0.41	0.29	0.43	0.36	0.18	0.54	0.60	0.16	0.02	0.22	0.48
India	1985	0.46	0.55	0.47	0.48	0.32	0.54	0.37	0.50	0.43	0.61	0.44	0.39	0.28	0.47	0.47	0.20	0.03	0.30	0.43
India	1992	0.42	0.52	0.51	0.66	0.32	0.49	0.48	0.37	0.44	0.34	0.41	0.33	0.36	0.32	0.39	0.29	0.03	0.29	0.30
Kenya	1975–1989	0.00	0.00	0.00	0.00	0.00	0.00	0.00	0.00	0.00	0.00	0.00	0.00	0.00	0.00	0.00	0.27	0.03	0.70	0.00
Korea	1979	1.00	1.00	1.00	1.00	1.00	0.99	0.98	0.98	0.96	1.00	0.99	0.99	1.00	0.99	0.99	0.01	0.00	0.00	0.97
Korea	1985	0.75	0.52	0.59	0.76	0.66	0.54	0.73	0.83	0.89	0.79	0.85	0.79	0.48	0.88	0.77	0.11	0.10	0.03	0.55
Korea	1990	0.48	0.33	0.45	0.53	0.45	0.34	0.44	0.55	0.64	0.54	0.53	0.54	0.45	0.51	0.49	0.12	0.32	0.07	0.34
Korea	1995	0.47	0.37	0.43	0.68	0.37	0.42	0.59	0.66	0.71	0.51	0.63	0.56	0.71	0.60	0.54	0.11	0.22	0.13	0.50
Malaysia	1974	0.00	0.00	0.00	0.00	0.00	0.00	0.00	0.00	0.00	0.00	0.00	0.00	0.00	0.00	0.00	0.36	0.22	0.42	0.00
Malaysia	1980	0.00	0.00	0.00	0.00	0.00	0.00	0.00	0.00	0.00	0.00	0.00	0.00	0.00	0.00	0.00	0.30	0.10	0.59	0.00
Malaysia	1987	0.00	0.00	0.00	0.00	0.00	0.00	0.00	0.00	0.00	0.00	0.00	0.15	0.00	0.00	0.01	0.32	0.21	0.46	0.01
Mexico	1981	0.20	0.36	0.22	0.38	0.14	0.32	0.21	0.08	0.12	0.27	0.22	0.18	0.09	0.29	0.23	0.49	0.01	0.27	0.18
Mexico	1985	0.39	0.31	0.33	0.43	0.39	0.31	0.35	0.42	0.60	0.52	0.45	0.33	0.31	0.49	0.41	0.43	0.05	0.12	0.30
Mexico	1992	0.15	0.08	0.08	0.05	0.07	0.08	0.04	0.09	0.07	0.10	0.17	0.08	0.03	0.14	0.10	0.79	0.02	0.09	0.10
Philippines	1976	0.29	0.27	0.29	0.47	0.41	0.34	0.45	0.50	0.65	0.25	0.26	0.41	0.12	0.25	0.32	0.36	0.07	0.24	0.30

Table 11.7. (continued)

Nation	Year	Agric. Livestock	Agric. Crops	Agric. F. & Vegs.	Agric. Hort.	Agric. Serv. live	Agric. Serv. crops	Agric. Other	Food Meat	Food F. & Vegs.	Food Dairy	Food Cereals	Food Beverages	Food Tobacco	Food Other	Domestic Share of Ag+Food	US Share of Ag+Food	Japan Share of Ag+Food	Other Share of Ag+Food	Domestic Share of All patents
Philippines	1980	0.22	0.16	0.17	0.30	0.13	0.22	0.21	0.27	0.58	0.25	0.27	0.30	0.03	0.35	0.22	0.38	0.08	0.32	0.22
Philippines	1985	0.07	0.08	0.08	0.20	0.06	0.08	0.09	0.12	0.15	0.08	0.15	0.21	0.07	0.16	0.10	0.45	0.11	0.34	0.13
Philippines	1990	0.07	0.07	0.06	0.04	0.07	0.08	0.03	0.20	0.09	0.05	0.12	0.29	0.28	0.11	0.11	0.47	0.07	0.35	0.09
Poland	1974	0.91	0.81	0.82	0.92	0.95	0.86	0.80	0.92	0.95	0.85	0.89	0.85	0.60	0.89	0.87	0.02	0.02	0.09	0.87
Poland	1980	0.69	0.38	0.37	0.70	0.68	0.46	0.50	0.85	0.80	0.72	0.80	0.79	0.64	0.77	0.60	0.08	0.02	0.30	0.78
Poland	1985	0.69	0.53	0.49	0.84	0.72	0.64	0.62	0.87	0.90	0.88	0.81	0.89	0.73	0.86	0.71	0.08	0.02	0.19	0.86
Poland	1990	0.82	0.64	0.68	0.89	0.83	0.68	0.77	0.85	0.95	0.96	0.89	0.94	0.83	0.91	0.79	0.05	0.04	0.12	0.92
Poland	1995	0.56	0.53	0.47	0.59	0.32	0.57	0.42	0.49	0.57	0.56	0.46	0.32	0.16	0.45	0.48	0.12	0.01	0.39	0.52
Turkey	1979	0.00	0.00	0.00	0.00	0.00	0.00	0.00	0.00	0.00	0.00	0.00	0.00	0.00	0.00	0.00	0.34	0.02	0.64	0.00
Turkey	1985	0.10	0.08	0.04	0.33	0.07	0.10	0.49	0.09	0.10	0.10	0.04	0.11	0.28	0.26	0.12	0.18	0.08	0.63	0.02
Turkey	1990	0.09	0.11	0.11	0.18	0.14	0.14	0.16	0.19	0.11	0.25	0.15	0.31	0.32	0.33	0.19	0.13	0.02	0.66	0.02
Turkey	1995	0.11	0.06	0.07	0.05	0.16	0.08	0.05	0.05	0.30	0.02	0.12	0.02	0.09	0.06	0.07	0.27	0.03	0.62	0.12
Zambia	1968–1994	0.30	0.31	0.30	0.31	0.24	0.31	0.28	0.46	0.46	0.42	0.35	0.28	0.17	0.57	0.32	0.19	0.01	0.48	0.27
LDC, NIC Average		0.34	0.29	0.30	0.44	0.33	0.32	0.34	0.39	0.44	0.37	0.37	0.40	0.29	0.40	0.34	0.23	0.06	0.37	0.33

Table 11.8. International flows of patents with agricultural sectors-of-use (SOU), 1987

	Granting								
Applying	US	Canada	Japan	UK	Germany	France	Switzer-land	Nether-lands	Belgium
US	1,001	780	112	34	23	16	4	2	0
Canada	20	225	2	1	0	0	0	20	0
Japan	105	137	3,719	24	16	16	3	20	0
UK	51	105	38	210	9	3	2	21	1
Germany	111	109	65	20	597	35	21	10	3
France	50	51	26	6	13	401	5	25	2
Switzerland	13	38	13	2	7	3	42	20	0
Netherlands	23	32	11	10	13	5	2	116	1
Belgium	2	32	0	1	4	2	0	21	0
Other	106	174	45	32	47	23	20	10	1
Total	1,482	1,683	4,031	340	729	504	99	145	8

full agricultural-use patenting, but similar to domestic patenting for agricultural uses, which is a little less agricultural and a little more food-oriented than all domestic and foreign patenting activity in developing countries. Refer to the discussion of table 11.7 at the beginning of this subsection.

The ratio of utility models to full domestic patents differs among the four countries. Hungary had only one utility model for every three domestic patents, while China had twice as many utility models as domestic patents. While relatively few utility models were used in agriculture in Hungary, roughly equally as many utility models as domestic patents were put to agricultural uses in Brazil, China, and Korea. Both Brazil and Korea relied much more on utility models for livestock use than for crop use (relative to full patents), but China relied most heavily on utility models in the crop services.

Measuring Indirect International Intersectoral Spillovers

An examination of spending on R&D in the United States and Japan is the first step in understanding the relevance of this type of spillover to other (developing) countries (table 11.10). The vast majority of indirect international spillovers are from U.S. sources and are highest to Brazil, China, and Mexico; they are also significant to the Korean and Philippine food industries. Japanese spillovers are more geographically focused, affecting primarily China, Korea, Malaysia, and the Philippines (and also Brazil) in the food sector. A measure of the significance of this R&D spending can then be calculated using the proportion of American patents that are also patented in developing countries. In Brazil, for example, 70 percent of American patents are relevant to the local market, suggesting that 70 percent of American R&D is relevant to the Brazilian economy. Spillovers from R&D performed in the United

Table 11.9. Utility models

Nation	Year	Agric. Live-stock	Agric. Crops	Agric. Fruit & Vegs.	Agric. Hort.	Agric. Serv. live	Agric. Serv. crops	Agric. Other	Food Meat	Food Fruit & Vegs.	Food Dairy	Food Cereals	Food Bever-ages	Food Tobacco	Food Other	Agric. Total	Food Total	A+F Total	All Patents
Number of Utility Models																			
Brazil	1995	19.04	22.83	3.98	5.83	13.26	11.94	0.82	12.47	5.68	13.20	2.51	26.16	5.90	37.38	77.70	103.29	180.99	2,695
China	1995	135.88	257.32	36.11	63.96	103.14	140.50	9.10	101.59	62.35	58.73	46.05	170.78	95.27	355.79	746.02	890.56	1,636.58	34,106
Hungary	1995	1.00	1.78	0.25	0.83	0.59	0.91	0.06	0.95	0.36	0.94	0.66	1.79	0.22	3.44	5.42	8.36	13.78	257
Korea	1979	4.95	3.47	0.58	0.66	3.24	1.96	0.17	2.67	1.11	1.14	1.57	2.63	1.94	6.77	15.02	17.85	32.87	603
Proportion of Utility Models to Domestic Patents																			
Brazil	1995	1.57	0.55	0.89	0.95	1.61	0.60	0.51	1.28	0.68	1.84	0.56	1.47	0.64	1.18	0.83	1.17	0.99	1.10
China	1995	1.05	1.04	1.36	1.33	0.83	1.95	0.34	0.63	0.83	0.73	0.38	0.61	1.39	0.43	1.11	0.55	0.72	2.23
Hungary	1995	0.19	0.13	0.19	0.08	0.12	0.15	0.00	0.19	0.18	0.55	0.32	0.67	0.02	0.29	0.09	0.24	0.14	0.34
Korea	1979	2.93	0.48	0.86	1.36	1.66	0.93	0.62	1.80	2.01	0.56	1.40	1.33	1.33	1.07	1.05	1.19	1.12	0.88
Average		1.43	0.55	0.83	0.93	1.06	0.91	0.37	0.98	0.93	0.92	0.67	1.02	0.85	0.74	0.77	0.79	0.74	1.14

States and Japan to Brazil in each manufacturing sector can therefore be estimated as

Spillovers from U.S. chemical industry to Brazilian agriculture =
(U.S. chemical industry R&D) ×
(share of U.S. chemical inventions used in agriculture) ×
(share of U.S. chemical patents also protected in Brazil).

These spillovers can be summed across all industrial sectors for a measure of all Japanese or U.S. spillovers to Brazilian agriculture. This procedure has been applied to all fields of invention (Johnson and Evenson 1997). Estimates of total R&D flows to developing countries were in the 30 to 40 billion dollar range in 1995. This is at least four times the level of technology payments (licensing and other payments) made by developing countries in total.

Table 11.10. International intersector spillovers to agriculture and food from industrial R&D (millions of 1990 U.S. dollars)

	To agriculture		To food	
	From US	From Japan	From US	From Japan
Argentina, 1990	5.42	0.22	1.99	0.02
Brazil, 1990	32.12	2.57	19.78	0.43
China, 1990	32.36	4.76	28.11	3.60
Egypt, 1990	4.02	0.39	1.02	0.01
Hungary, 1990	19.45	1.87	3.23	0.27
India, 1992	6.95	0.28	6.24	0.11
Korea, 1990	11.08	4.63	10.02	4.63
Malaysia, 1987	7.76	1.37	3.39	0.47
Mexico, 1992	22.64	0.25	27.13	0.17
Philippines, 1990	24.12	1.20	8.16	0.31
Poland, 1990	8.20	1.76	1.16	0.07
Turkey, 1990	2.03	0.08	0.48	0.01

SPILLOVERS IN STUDIES OF AGRICULTURAL PRODUCTIVITY

A recent review of studies of returns to agricultural research and extension shows that few incorporate intersectoral spillovers (Evenson, 1999). Further, few incorporate indirect (germplasmic) spillovers in a structured way, although a number do incorporate interregional effects.

Intraregional Intrasectoral Studies

Most agricultural studies of rates of return (ROR) have utilized cross-sectional data. The analyst in this case defines an appropriate research variable to assign to each observation. In contrast, research ROR studies often have time series units of observation on an aggregate of farms, usually at the district, state, or national level. Almost all national aggregate time series studies have simply used a research "stock" variable constructed by summing past real investments in the public sector NARS program. Few such studies have attempted to "deflate" the research stock (see Latimer 1964).

However, a considerable number of research ROR studies have utilized a cross-sectional time series panel. Studies of U.S. agricultural productivity have been based on observations utilizing a state aggregate for a number of years. The cross-sectional dimension of the panel requires that a research variable be constructed to match this unit of observation. The early practice was to presume that there were few or no research spillovers from one state to another, or that spill-ins canceled out the spill-outs. This procedure was clearly not very satisfactory (see Griliches 1964).

Interregional Intrasectoral Studies

Two methods of measurement have been developed for interregional intrasectoral spillovers. The first uses geoclimatic region and subregion classifications; the second uses measures of barriers to spillovers. Evenson (1989) and Huffman and Evenson (1993) have used the first method for studies of agricultural productivity in the United States. Because geoclimatic regions and subregions cross state boundaries, in these studies the research variable for state i is defined as

$$R_i = R_i^s + \alpha R_i^{ss} + \beta R_i^{sR}$$

where

$$R_i^s$$

is the research stock, based on research conducted in state i,

$$R_i^{ss}$$

is the research stock, based on research conducted in other states with subregions similar to those in state i, and

$$R_i^{sR}$$

is the research stock, based on research conducted in other states with regions similar to those in state i (excluding the subregions).

Using this method, we can construct variables for each state. Nonlinear least-squares iterative methods can then be used to estimate α and β, and these results can be interpreted as proportionate spill-in parameters. This method has been applied in a number of studies (Evenson 1999).

The second method requires the construction of "spillover barrier indexes" between locations. An index for two locations, i and j is

$$SB_{ij} = 1 - C_{ii}/C_{ij}$$

where

C_{ii} is the minimum cost of producing a unit of product in location i using the optimal cost-minimizing technology, and

C_{ij} is the minimum cost of producing a unit of product in location i when producers are constrained by the optimal cost-minimizing technology used in location j.

If the same technology is cost minimizing in both locations, i.e., $SB_{ij} = 0$, it would be reasonable to assume that new technology targeted to location j would freely spill in to location i.

Spillover barrier (SB) measures can then be used in productivity studies to define the research variable for state i as

$$R_i = \sum_j SB_{ij}^{\omega} R_j$$

where S is a parameter scaling the SB measure into productivity units.

Spillover barrier (SB) measures can be computed from yield time-series data. Evenson (1994) used rice trial-yield data for different locations in India to compute SB indexes. In these yield trials a common set of rice varieties (or advanced lines) is planted in different trial locations. We can therefore compare the yield of the varieties that performed best in location j with the best varietal performers in location i. Evenson (1982) and Avila and Evenson (1995) used a roughly similar procedure in a study focusing on Brazil.

Intrasectoral International Studies

International within-sector spill-in studies could, in principle, estimate free parameters. For example, productivity in country or state i could be explained by research stocks in that country or state and, say, 10 other countries or states. In this way spill-in from each source (country or state) could be identified. This unstructured or free-parameter approach is subject to two difficulties, however. The first is that multiple spill-in routes may be involved. Country i may have spill-ins from countries j and k. If there is an indirect spillover from j to k there is an additional indirect spill-in from country j to country i through k. Two coefficients, one for research in j and one for research in k, are unlikely to capture these complex flows correctly.

The second difficulty is that the research variables in the several countries are likely to be highly collinear, making it very difficult to identify parameters. The remedy to this problem (which also exists with time lags) is to impose structure on the spill-in weights. The geoclimatic and SB methods discussed above impose such structure and provide a rationale for the weights. (See the following section for related intersectoral structures.) There is also a further specification that has been used in international studies to address the first problem. This specification involves creating an interaction variable between domestic and foreign research that will take indirect spill-ins more fully into account. The basic idea here is that spill-in R&D can complement (or be a substitute for) domestic R&D.[7]

Intersectoral Intraregional Studies

Early intrasectoral studies used personal experience and a judgment approach to identify potential sources of spill-ins to the agricultural sector from private-sector R&D. As we saw in the discussion of table 11.3, the agricultural implement, fertilizer, and agricultural chemical industries are the source of only slightly more than one-half of all the inventions relevant to agriculture. The YTC procedure provides a much more reliable measure.

Several national studies have utilized private-sector R&D variables. Imposing a spillover structure within a national economy on these variables is difficult, however. Most studies have used observation input weights to impose spillover structure (that is, the machinery share of costs is used to weigh machine R&D or invention). See Scherer 1984, Schmookler 1996, Griliches 1992, Mohnen 1990, and Keller 1995 for other treatments of interindustry spillover.

For international studies, a natural weighting is that used to construct table 11.10. Here the foreign R&D spilling in to country i from country j is based on the proportion of country j's inventions protected in country i. This weighting is thus based on decisions by invention holders about the likely market for an invention in the foreign country. One useful approach to empirical work is to use this weighting as a starting point and then to test for departures from these weights.

SUGGESTIONS FOR FURTHER WORK

The importance of spillovers in agricultural technology development is clear. But we need better methods of measuring them and of specifying them in empirical studies. The studies reviewed here suggest some avenues for further research. Two especially promising areas are the measurement of spillover barriers and the analysis of patent and research publication citations.

There is considerable scope for measuring spillover barriers using crop yield data. National and international experimental (and farmers' field) trials can be developed for a number of crops. These in turn can be associated with soil and climate data. New econometric techniques can be developed for estimation of barrier effects from this data.

Citation analysis is particularly important in identifying indirect spillovers, particularly when applied research publications cite both applied and basic research publications in other regions. Patent documents also cite inventions that have already been patented and related scientific work that may have been carried out in another region or country. Most data on citations can be interpreted using methods for interpreting information on indirect spillovers.

Technology spillovers also raise complex issues for public-private relationships in agricultural research. Will the private sector fund agricultural research that it is unable to appropriate the gains from? Is the future role of the public sector to produce public goods research with a high spillover potential within its jurisdiction. What will happen if those public goods typically also flow outside of its borders? As these pathways become better understood, they may become a source of a type of financial assistance for some countries in the future.

APPENDIX: NOTES ON STATISTICAL SOURCES AND METHODS

The details of the techniques used to create these data sets are explained in *Economic Systems Research* 9(2) (June 1997), which is devoted to the topic. Interested readers are referred to the introductory article of that volume (Evenson and Johnson 1997) and to the more technical piece by Kortum and Putnam (1997).

As discussed in the section of this chapter on "Using Patent Data to Measure Intersectoral Spillovers to Agriculture," the new Yale Technology Concordance (YTC) provides a probability-based assignment of inventions, given an International Patent Classification (IPC) designation, assigning them to industries of manufacture (IOM) and sectors of use (SOU). For counts of patents in each IPC for any nation, the YTC distributes them across IOM-SOU space, using the probability that any given product (or IPC) originates from a particular industry or (IOM) and is destined for a particular sector (or SOU). Thus, the technical nature of the relationship between product types and industries is embodied in the concordance, but does not impose the industrial structure of Canadian inventions on the patterns of other nations. For evidence, see Kortum and Putnam (1997).

This advance enables a measure of direct invention spill-in to agricultural sectors by source (i.e., the number of inventions used in agricultural subsectors which originate in the machinery industry). It also permits construction of financial measures for direct and indirect R&D spill-ins (i.e., the dollar amount of private sector R&D spilling in to agriculture via spending in other sectors and/or other nations).

Patent data for 80 nations across time are available from the International Patent Document (INPADOC) database via STN online services, with copyrights in North America to the original data held by the American Chemical Society. Since nations report patent applications and granted protection differently, publishing upon application or upon approval or both, selection of which intellectual property rights documents to study was delicate. Wherever possible, we consider only granted patents.

The INPADOC database registers and counts patents by International Patent Classification (IPC), information which is valuable in distinguishing the product or process involved in a patent, but which may lead to measurement error since some patents may either have no IPC assigned or multiple IPCs assigned. Counts of total patents (with or without IPC) by nation of origin and granting nation are not subject to this "multiple assignment" problem and were used to partially correct the double-counting problem. For example, if Argentina issued the following two patents to Americans in 1974,

Patent 1 with an IPC of AO1B
Patent 2 with two IPCs of AO1B and BO2C

then a count by IPC would list three patents (two to IPC AO1B and 1 to BO2C). A supplementary search to count all Argentinean patents to Americans in 1974 would count only two patents in total. The result is that American-origin and Japanese-origin totals may be overstated, while the "Other applicant" category will be understated since it is calculated as a residual. In practice the double-counting problem is rare, so in most cases the "Other applicant" category will be very close to the European applicant total, since domestic, American, Japanese, and European applications account for the vast majority of the patents in any nation.

NOTES

1. Most inventions are adaptations of other inventions and hence dependent to some degree on prior inventions. This type of spillover may literally be in the form of germplasm for crop varietal technology. Preinvention discoveries are sometimes referred to as "recharge science."
2. For example, in the United States each state supports a state agricultural experiment station (SAES) system with regional or branch stations within the state.
3. Landraces are the farmer-selected type of cultivars that constitute the bulk of the genetic resources used in conventional plant breeding.
4. The YTC is available to users from the Economic Growth Center at Yale University.
5. DeBresson and Murray (1984) developed a similar data set for innovations.

6. Utility models are sometimes described as "petty" patents because of low "inventive step" requirements.

7. Evenson and Kislev (1973, 1975) report an early study of international research complementarily.

REFERENCES

Avila, A. Flavio and R. Evenson. 1995. "Productivity Change and Technology Transfer in the Brazilian Grain Sector." *Rivista de Economia Rural* 34(2): 93–109.

DeBresson, C., and B. Murray. 1984. "Innovation in Canada." In *Innovative Capability of Canadian Industry Strengths and Weaknesses Analysis.* A Report to the Science Council of Canada. Vols 1 and 2. Ottawa, Canada.

Evenson, R.E. 1982. "Observations on Brazilian Agricultural Research and Productivity." *Revista de Economia Rural* 20:368–401.

———. 1989. "Spillover Benefits of Agricultural Research." *American Journal of Agricultural Economics* 71(2): 447–52.

———. 1994. "Analyzing the Transfer of Agricultural Technology." In J.R. Anderson, ed. *Agricultural Technology: Policy Issues for the International Community.* Wallingford, Oxon OX10 8DE, UK: CAB International.

———. 1999. "Economic Impact Studies of Agricultural Research and Extension." In Bruce Gardner and Gordon Rausser, eds. *Handbook of Agricultural Economics.* Elsevier Science B.V. Amsterdam: North Holland Publishing Co.

Evenson, R.E., and D. Gollin. 1997. "Genetic Resources, International Organizations, and Rice Varietal Improvement." *Economic Development and Cultural Change* 45(3): 471–500.

Evenson, R.E., and D. Johnson. 1997. "Introduction: Invention Input-Output Analysis." *Economic Systems Research* 9(2): 149–60.

———. 1999. "R&D Spillovers to Agriculture: Measurement and Application." *Contemporary Economic Policy* 17(4):432–456.

Evenson, R.E., and Y. Kislev. 1973. "Research and Productivity in Wheat and Maize." *Journal of Political Economy* 81(6):1309–29.

———. 1975. *Agricultural Research and Productivity.* New Haven and London: Yale University Press.

Griliches, Z. 1964. "Research Expenditures, Education and the Aggregate Agriculture Production Function." *American Economic Review* 54:961–74.

———. 1992. "The Search for R&D Spillovers." *The Scandinavian Journal of Economics* 94:29–47.

Huffman, W.E., and R.E. Evenson. 1993. *Science for Agriculture.* Ames, Iowa: Iowa State University Press.

Keller, W. 1995. "Technological Change: Identification, Theory, and Evidence." Ph.D. Dissertation, Yale University.

Kortum, S., and J. Putnam. 1997. "Assigning Patents to Industries: Tests of the Yale Technology Concordance." *Economic Systems Research* 9(2): 161–75.

Latimer, R. 1964. "Some Economic Aspects of Agricultural Research and Extension in the U.S." Ph.D. Dissertation, Purdue University.

Mohnen, P. 1990. "New Technologies and Intraindustry Spillovers." *STI Review* 7:131–47.

Scherer, F.M. 1984. "Using Linked Patent and R&D Data to Measure Interindustry Technology Flows." In Z. Griliches, ed. *R&D Patents and Productivity*. Chicago: University of Chicago Press.

Schmookler, J. 1996. *Invention and Economic Growth.* Cambridge, MA: Harvard University Press.

12

SIGNIFICANCE OF INTERNATIONAL SPILLOVERS FROM PUBLIC AGRICULTURAL RESEARCH

DAVID E. SCHIMMELPFENNIG
AND COLIN G. THIRTLE[1]

THE PIONEERING work of Jorgenson and Nishimizu (1978), on intercountry comparisons of total factor productivity (TFP) has led to a literature on multilateral total factor productivity indices that allow competitiveness to be measured both intertemporally and interspatially. The theoretical issues are discussed in Caves, Christensen and Diewert (1982) and applied to agriculture by Capalbo, Ball, and Denny (1990) and Capalbo et al. (1991). Empirical work on agriculture in the European Community (EC) can be found in Terluin (1990), Bureau, Butault, and Barkaoui (1992), and Ball et al. (1996), whose indices are used in this study.

Evenson, Landau, and Ballou (1987) show that changes in TFP can be explained by means of "determining" variables such as R&D, extension, and farmer education. They call this approach to explaining technical change the "two-stage decomposition," as opposed to the "integrated" approach, in which the "determining" variables are incorporated directly in the estimation of the production, cost, or profit function. Both approaches are common in the considerable literature on the returns to agricultural R&D that has been surveyed by Echeverria (1990). However, previous estimates of the returns to R&D for European agriculture (see Rutten 1992; Thirtle and Bottomley 1989; and Khatri and Thirtle 1996) fail to allow for the spillovers between research jurisdictions, due to lack of data. Evenson and his colleagues (Evenson and Pray 1991; Huffman and Evenson 1993; and chapter 11 in this volume) have shown that spillovers can be important.

Indeed, endogenous growth theory rests on the notion of positive spillovers, resulting from the nonrival, partially excludable nature of new technology, combined with the inability of firms to fully appropriate the returns to their research investments, which results in increasing returns at the aggregate level. Thus, technological spillovers form the basis of the increasing returns in Romer's (1986, 1990) models, just as they provide the rationale for public investment in agricultural research. Lucas (1988) introduces increasing returns in the aggregate production function by allowing individual human capital investments to have positive spillovers on the productivity of the human capital investments of others. In both the Romer (1986) and Lucas (1988) models, policies that impede investment in human and physical capital reduce growth, and appropriate public policies accelerate growth.

The particular policy focus of this paper is the USDA's concern with effective coordination of public and private research in the United States (Fuglie et al. 1996). For this purpose private as well as public R&D must be considered in the global environment of the multinationals, which have played a major role in internationalizing agricultural R&D. Agricultural productivity in one country might depend on its own research efforts and the level of the pool of international knowledge that it has access to, so modeling international spillovers is essential.

We find that there is convergence in the levels of TFP between countries and that international spillovers are partly responsible. The United States and the European Community (EC) countries with more-advanced research systems (Netherlands, Denmark, France, and Belgium) form a high-growth club, probably due to technological and geographical proximity, in which spillovers from the United States are a powerful force. Germany, Luxembourg, Greece, Italy, the United Kingdom and Ireland converge in a slow-growth group, which does not benefit directly from intercontinental spillovers, except for the United Kingdom. Thus, the UK is the exception, in that it belongs to the high-technology set, but is now joining the low-growth convergence club, and its poor performance is partly attributable to the reductions in public R&D in the 1980s.

The next section of this chapter compares the multilateral TFP indices for the 10 EC countries and the United States. Then, section three explains the data and the model, which allows for externalities between the public national agricultural research systems (NARSs) in the European Community, for intercontinental spill-ins from the U.S. system and also for private-sector research activities. In section four the data is treated as a panel, which gives aggregate results for the "United States of Europe". In section five, seemingly unrelated regressions (SURE) are used to identify spillovers at the country level and a research hierarchy is determined, which correlates closely with the convergence results. The rate of return calculations, reported in section six, suggests that the return to EC R&D is around 35 percent, but that return calculations are seriously biased if spillovers are ignored.

MULTILATERAL PRODUCTIVITY COMPARISONS

Bureau, Butault, and Barkaoui (1992) constructed Fisher TFP indices for 10 EC countries and the United States, for 1973–89.[2] Then, to allow international comparisons, agricultural sector purchasing power parity (PPP) exchange rates had to be calculated to make the outputs and inputs of the 11 countries comparable. The spatial index, for 1985, was used to calibrate the time series for each country, giving multilateral indices. This work has been updated by Ball et al. (1996), whose results are summarized in table 12.1.

Table 12.1. Agricultural total factor productivity (TFP) comparisons for 10 EC countries and the United States, 1973–93

Belg.	Denm.	Irel.	Fran.	Germ.	Gre.	Italy	Lux.	Neth.	UK	US	EC10
Starting Level, 1973–75 Average (Base is EC10=100 in 1990)											
117.3	91.1	54.9	78.6	63.0	59.0	63.4	30.0	101.7	80.3	93.7	72.3
Final Level, 1991–93 Average (Base is EC10=100 in 1990)											
147.4	138.1	77.3	128.7	87.3	77.7	93.6	37.2	139.7	105.3	156.9	106.0
Growth Rate, 1973–93											
1.3	2.6	1.9	3.0	1.9	1.5	2.2	1.4	1.8	1.6	3.0	1.92

Ball et al. (1996) show that in 1973 the United States was more efficient than all the European countries, except the Netherlands and Belgium/Luxembourg. However, there are considerable annual variations, so it is better to base the comparison on an average. Thus, the first row of table 12.1 shows the efficiencies of the EC countries and the United States, relative to the aggregate of the EC countries, averaged over 1973–75. On this basis, the Netherlands and Belgium performed better than the United States, but Italy and Greece were at less than 70 percent of the U.S. efficiency level and Luxembourg was at less than one-third of the U.S. productivity level. When an unweighted average was used, the EC countries achieved 77 percent of the U.S. efficiency level.[3]

The growth rates shown in the last row indicate that only France has grown as fast as the United States, with Denmark not far behind, while all the rest of the EC countries have fared considerably worse. Thus, as the spatial index in the penultimate row shows, by 1991–93, the United States had overtaken the Netherlands and Belgium, which both grew slowly, the French remain just as far behind the United States, and all the other countries have fallen still further behind. Since the growth rate of the 10 EC countries in aggregate was only 1.9 percent, they are at only 67 percent of the U.S. efficiency level at the end of the period.

These data allow a test of simple versions of the convergence hypothesis. Dowrick and Nguyen (1989) found that for the Organisation for Economic Co-operation and Development (OECD) countries, since 1950, "TFP catch-up stands out as a dominant and stable trend." For their sample, initial levels of income were negatively related to growth rates and accounted for over 50 percent of the variance. Thus, the richer countries grew more slowly, and there is a tendency for income levels to converge. This does not seem to be suggested by the figures in table 12.1. Indeed, regressing the TFP growth rates on the starting values shows that there is no clear relationship, since the estimated coefficient is not significantly different from zero.

However, two qualifications should be added. First, when we turn to actually explaining these interspatial and intertemporal efficiency differences, adding the starting points to the other explanatory variables shows that they are highly significant in explaining deviations in the variance in TFP, indicating that conditional convergence is in operation. Second, later versions of the convergence hypothesis are far richer in empirically testable content. For instance Quah (1997) considers the possibility that countries of close technological proximity will form convergence clubs, whereas those who are more technologically backward may fail to keep up. This more interesting case is supported by Figures 12.1 and 12.2.

Figure 12.1 shows the TFP indices for the high technology club: the United States, France, the Netherlands, Belgium, Denmark, and the United Kingdom. The first five follow convergent paths with high levels of TFP. The United Kingdom appears to be a member of the group until 1984 and then is left far behind as its growth rate falls considerably. This drop in TFP may be associated with the decline in public agricultural R&D investment in the United Kingdom between its peak in 1983 and 1989 when it was 12.5 percent lower (Thirtle, Piesse, and Smith, 1997).

Figure 12.1. Convergence of the high-level productivity group

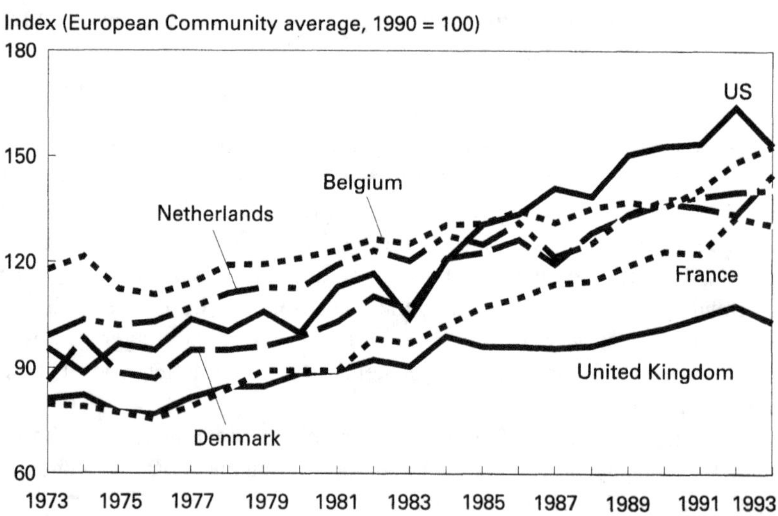

Figure 12.2. Convergence of the low-level productivity group

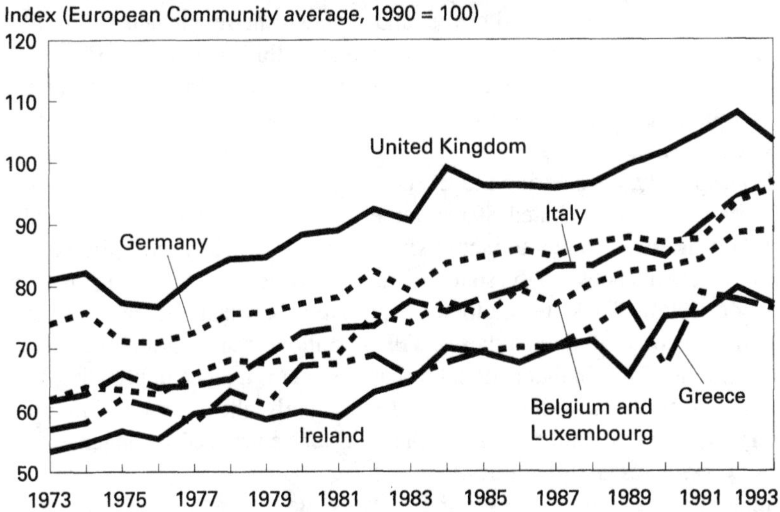

The United Kingdom is the link between figure 12.1 and figure 12.2, in which it is the leading country in the low TFP level convergence club. These countries have the less technologically advanced research systems and the technological distance from the leaders may be proving too great for them to stay in touch with the leading group. Their average growth rate is 1.5 percent per annum as compared with 1.75 percent for the leading group, so if they converge it will be to a lower TFP level.

We now continue by explaining TFP growth, using public and private R&D, international spillovers, extension, education, and the weather. The pattern of the spillovers suggests that there is an R&D hierarchy, with a leading group that are in touch with the United States and a trailing group for whom the technological distance is too great. These countries benefit only from secondary spillovers. Comparisons between the high technology group and a bicycle racing "peloton" may be unavoidable. Thus, the pattern of spillovers explains why there are two TFP growth clubs, supporting the convergence results.

EXPLAINING TFP GROWTH: THE MODEL AND THE DATA

Changes in the TFP index may be explained by the "conditioning" factors that shift the static production function over time. In the basic closed economy model these are public sector R&D expenditures that generate new technology, extension expenditures that transmit the results to the farmers, thereby diffusing the technology, and the education level of the farmers, which affects both their own creative and managerial abilities and their skill in appraising and adapting exogenous technologies. The weather is normally included, as it explains some of the residual errors, and several other explanatory variables have also been used in the extensive literature in this area. This study incorporates technological spillovers between the EC countries and from the United States as well as spillovers from the private sector, especially the input industries.

Conceptually, considering the EC countries as a group is similar to working with data for the United States, rather than handling the states individually. This has been shown to matter since there are considerable spillovers of research benefits between the state jurisdictions (Evenson 1989; Huffman and Evenson 1993). Thus, technological spillovers between the EC countries should be incorporated in studies of TFP. If spillovers are important there are policy implications, in that national research systems may be duplicating each others' efforts and wasting resources. Alternatively, there may be important synergies. Either situation supports the case for centralized management, or even EC-wide research facilities.[4]

Including the United States in the analysis allows for the possibility of intercontinental spillovers. Technical change in the input industries should be captured in the input series, but these are unlikely to fully account for quality changes (Cooper, Rayner, and Greenaway 1993), so private-sector activity is measured first by international patent data. Even this is a poor variable for capturing all the efforts of the private sector; the lack of better data alternatives leaves us with only one other way of accounting for private activities. This is to use foreign TFP indices as the proxy for the foreign knowledge stocks, instead of knowledge stocks constructed from the foreign R&D expenditures. This was done by Bouchet, Orden, and Norton (1989) to avoid the difficulties of determining the lags for U.S. spillovers into French agriculture. Such a simplification is far more necessary here, as we are dealing with spillovers between 11 countries.

However, using the TFPs rather than the knowledge stocks is crucially different in that the TFPs pick up all new technologies, from all sources, whereas the knowledge stocks are for public R&D only. This distinction is illustrated in figure 12.3. The figure is a modification of the familiar declining schedule of rates of return (ROR) for increasing levels of expenditure that is used to discuss market failures and the case for public intervention in R&D. In this case, the actual ROR to public research (R5) is shown as being negative as no public R&D organizations even come close to fully recovering costs, let alone showing a profit.[5] The actual ROR to public and private R&D, shown as R4, is what is normally claimed to be measured in closed economy versions of the model being developed here, with the private sector bringing basic public research advances to market. However, despite arguments that estimated RORs (R1) are not biased upward, figure 12.3 shows some reasons why this is to be expected. First, only if there is a variable to capture the effects of foreign public research will the estimated ROR be R2. Similarly, only if foreign private research is also captured will the estimated ROR be R3. To the extent that it is easier to count the productivity gains from research than the costs that come from many different sources (in fact all sources benefiting the home country—including for example, the training of foreign graduate students as we will see in the penultimate chapter), estimates will still exceed the actual ROR by R3-R4.

Figure 12.3. Estimated and actual rate of return (ROR) to research with and without international spillovers

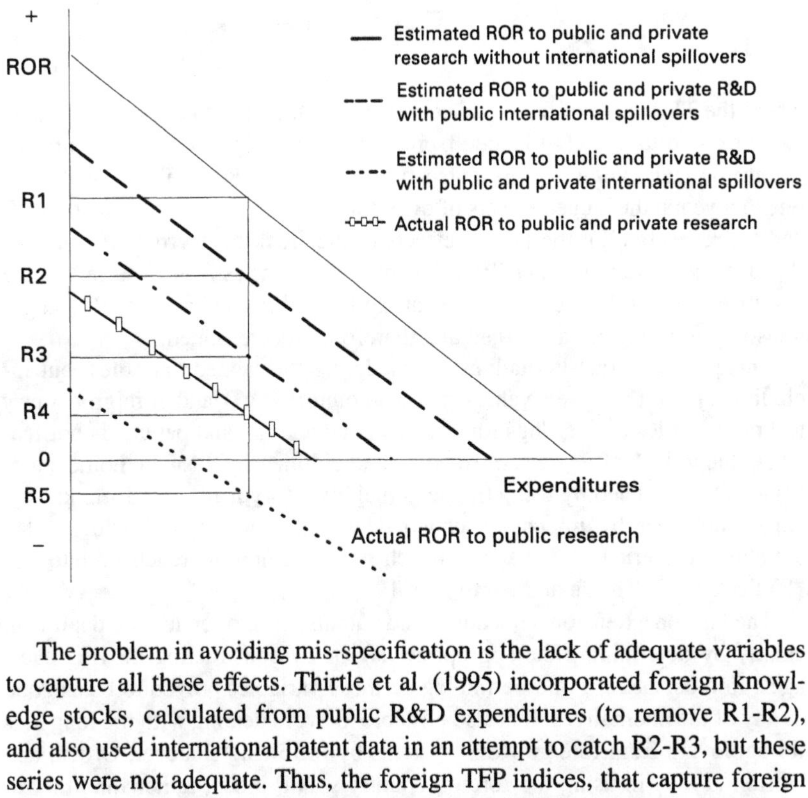

The problem in avoiding mis-specification is the lack of adequate variables to capture all these effects. Thirtle et al. (1995) incorporated foreign knowledge stocks, calculated from public R&D expenditures (to remove R1-R2), and also used international patent data in an attempt to catch R2-R3, but these series were not adequate. Thus, the foreign TFP indices, that capture foreign private activities as well as public research, proved to be the best variables for measuring the spillovers.[6] This approach is taken here as it adequately catches all the spillovers, but it does not lead to accurate estimates of the ROR to public R&D, since private domestic R&D is still not directly accounted for.

The R&D expenditures for each country are measured in constant 1980 PPP U.S. dollars and are updated from Pardey and Roseboom (1989); the extension data are largely from Evenson and Pray (1991) and are in constant 1980 U.S. dollars; education is an index of years of secondary education, constructed from various issues of the World Bank World Tables; the weather variable for each country is an index of cereal yield deviations around a time trend; and private-sector chemical and mechanical patents pertaining to agriculture are patent counts for all U.S. and foreign patents registered in the United States.

All the variables except the weather have lagged effects on TFP, so the model is

$$TFP^h_t = \sum_{i=1}^{m} \alpha_i RD^h_{t-i} + \sum_{j=1}^{n}\sum_{f=1}^{o} \beta_{jf} T^f_{t-j}$$
$$+ \sum_{k=1}^{p} \delta_k X^h_{t-k} + \sum_{e=1}^{r} \gamma_e E^h_{t-e} + \sum_{g=1}^{s} \phi_g P_{t-g} + \theta W^h_t + u_t$$

where the TFP index of the home country (h) at time t is a function of its own R&D expenditures (RD^h) lagged from one to m periods; foreign-country (or countries) technologies, represented by their TFP indices (T^f), lagged from one to n years; the lagged effects of own-country extension and education (X^h and E^h, respectively); the lagged effects of the stock of international patents (P); and the current weather (W^e_t). Lastly, u_t is a stochastic error term. All the variables, except the weather index, are in logarithms and are weakly exogenous because they are all lagged and therefore predetermined.

There are several alternatives for modeling the lagged variables, but including up to 15 lagged values of own-country R&D and foreign-country technology, plus shorter lags for extension, education, and patents is not feasible, due to lack of degrees of freedom and collinearity. Instead, home country R&D is modeled by using the perpetual inventory method to form knowledge stocks, or by imposing second degree polynomially distributed lag structures (inverted "U"-shapes), which is a common approach (White and Havlicek 1982; Thirtle and Bottomley 1989).

The lags on extension, education, and patents are shorter and are dealt with simply by determining the appropriate lag or by constructing moving averages. In most cases, a one-period lag on these variables proved to be as good as any of the alternatives. It is also possible to avoid dealing with the long lags for spillovers from foreign R&D systems by following Bouchet, Orden, and Norton (1989) in using the foreign country's TFP index, rather than the research expenditures, as the indicator of potential technology spillovers. Indeed, foreign TFPs, lagged one period, gave better results than distributed lags of foreign R&D in determining the spillovers. One reason the TFPs are stronger is that they incorporate the effects of both private- and public-sector research, which tends to weaken the effect of the private-sector patent series.

The choice of model for each country is still complicated by the fact that the length of the polynomial lag and the lags on the other variables need to be jointly determined. The model selection criteria used were the Akaike Information Criteria (AIC) and the Schwartz Criteria (SC), which were in agreement in almost all cases.[7] These were employed in combination with variable deletion tests which indicated the variables that could be jointly deleted from the equations.

Some amount of prior knowledge was used in applying these standard model selection criteria. Schimmelpfennig and Thirtle (1994) found strong evidence of cointegrating relationships for all the countries in this sample, using Johansen tests, the cointegrating regression Durbin Watson, and the Dickey Fuller and augmented Dickey Fuller tests. Their results, on orders of integration of the series and the lags at which R&D is Granger prior to TFP, are used in constructing the models. However, repeating the same exercise with the spillovers included was not helpful, due to the proliferation of variables and, particularly, because the cointegration techniques do not accommodate lag structures. Since the lag length and shape is essential to estimating the returns to R&D, which is the main objective of this study, traditional distributed lag formulations are used. However, there is now justification for this approach since the Monte Carlo results presented by Pesaran and Shin (1995) "provide strong evidence in favour of a rehabilitation of the traditional ARDL [autoregressive distributed lag] approach to time series econometric modeling" (p.32).

PANEL DATA ESTIMATES FOR THE UNITED STATES AND EUROPE

Panel data estimation gives ample degrees of freedom and both the cross-sectional and time-series variances help to determine the parameter estimates. The cost is in terms of imposing restrictions, as the alternative models show. Thus, pooled ordinary least squares (OLS),

$$Y_{it} = \alpha + \beta X_{it} + u_{it} \quad \text{where} \quad \alpha,\beta = \alpha_{it}, \beta_{it}$$

means assuming that both the intercepts and slope coefficients are the same for all the countries. The fixed effects model (FEM),

$$Y_{it} = \alpha_i + \beta X_{it} + u_{it} \quad \text{where} \quad \alpha_i,\beta = \alpha_{it}, \beta_{it}$$

assumes that the intercepts vary, but the slope coefficients are the same. [This is equivalent to least squares with dummy variables (LSDV)]. The random effects model (REM) treats the fixed effect as a random variable, allowing the country-specific effect to enter the equation through the error term, rather than being deterministic, as it is in the FEM.

$$u_{it} = \mu_i + \epsilon_{it}, \text{ where } Var(\mu_i) = \sigma_\mu^2, \; Var(\epsilon_{it}) = \sigma_\epsilon^2$$

The pooled OLS model was clearly rejected, relative to the FEM/LSDV and the REM, and Hausman tests led to rejection of the REM, relative to the LSDV/FEM. Tests using the Schwartz criterion led to acceptance of the FEM restriction that α_i, $\beta_i = \alpha_i$, β. However, tests showed both serial correlation and heteroscedastic errors, so weighted, generalized, least squares with first-order autoregressive corrections within the country groups were applied to deal with these problems.

The FEM/LSDV model was tested against the random coefficients model and the two-factor error components model, which are less restrictive alternatives, but these models gave poor results. The time-series properties of the data were also checked and nonstationary variables that were only stationary in first differences were generally not used. The FEM/LSDV results are reported in table 12.2.

Table 12.2. Pooled regression results for agricultural knowledge spillovers among 10 EC countries

	National effects					Spillovers of foreign knowledge stocks					
Variable	R&D	Ext	Ed	Wea	START	Fran.	Germ.	Italy	Neth.	UK	US
Elasticity	0.19	.0028	0.05	0.17	−0.023	.042	.065	.068	.023	.047	.061
t-Stat	12.3′	11.9	1.03	11.4	−25.4	9.70	10.01	14.5	0.62	10.4	10.2

Pooling may be viewed as a means of aggregation, so the results in table 12.2 indicate the effects of the variables for the 10 EC countries taken as a group. The parameter estimates are thus the same for all the countries; for any parameter, β_{ij}, where i is the country and j is the parameter identifier, the restriction is that $\beta_{ij} = \beta_j$. Specification of the pooled regression draws on Schimmelpfennig and Thirtle (1994, 1999), who used these data to find that R&D expenditures were causally prior to TFP with lags of 2, 9, 10 and 11 years. This suggests that, for the pooled sample, imposing the inverted "U" shape of the second degree polynomial lag structure is not appropriate. This does not contradict the results of the last section. Rather, it suggests that for the pooled sample, the short lags for the countries that do adaptive research show clearly and so do the long lags for those involved in basic research. The two cases are better handled individually, rather than imposing a lag structure that is the same for all countries.

So, we assume that there are two distinct outputs from the NARSs. There is adaptive, near-market research, such as field trials, which affect TFP after a very short lag of only two years, and more basic activity, such as plant breeding programs, that have an average gestation period of 10 years. In treating the two separately, there is the usual problem that the lagged R&D terms are collinear and are also collinear with extension. The best results were obtained by aggregating R&D lagged two periods with extension, to produce a "near-market" research variable, and using a moving average of R&D lagged 9, 10, and 11 periods, to pick up "more basic" research activities.[8] The spillovers were allowed for by using foreign TFPs lagged one period, and tests showed that the results do not seem to be particularly sensitive to the lag length.

Thus, as table 12.2 shows, for the "United States of Europe," own-country, long-term R&D expenditures have an elasticity of 0.19, which is a little lower than the unweighted average of 0.256 for the individual countries. Extension and short-term R&D expenditures have a small elasticity of .0028, but are highly significant.[9] Education is not significant, which is almost certainly due to lack of data specific to the rural population, but the weather index is positive and significant. The last national effect, labeled START, is the estimated coefficient on the starting value of the national TFP index, from table 12.1. While the simple convergence test in the second section of this chapter found no relationship, once the other explanatory variables are taken into account, TFP growth is negatively related to the starting level. Thus, the countries with high levels of TFP have overcome the disadvantage of a high starting point by working harder on technology generation and on exploiting foreign information, as we will show in the next section, and are undergoing conditional convergence with each other.

Spillover effects are shown on the right of table 12.2. The variables used for the spillovers are the TFPs rather than R&D expenditures, and the results are aggregate effects for the group of countries. French, German, and American technology spillovers, which affected several other countries, remain important, but Danish R&D (not shown) was found to have an insignificant effect, and the same is true for the Netherlands. However, U.K. research does have an effect on the group that could not be established by looking at the individual countries. The total elasticity of the significant intra-EC spillovers is 0.22 (not shown), which is already greater than the direct effects of the NARSs. Including the transatlantic spillovers from the United States gives a total spillover elasticity of 0.28 (not shown), so the spillover effects taken together easily outweigh the own-country R&D effects. Indeed, the combined elasticities for the spillovers are almost 50 percent greater than the direct elasticity. Chemical and mechanical patents are not significant, but this does not mean that private research is unimportant, since the foreign TFP indices reflect the availability of new technologies, whether produced by the public or private sectors.

RESULTS AT THE NATIONAL LEVEL: WITH AND WITHOUT SPILLOVERS

Closed Economy Model

The well-established method for explaining the growth of TFP is followed here, with some refinements later. The top half of table 12.3 reports the results, explaining TFP using extension, education, the weather, and a second degree polynomial lag to model R&D expenditures. These are individual country equations with no allowance for spillovers and are included here only to show the extent of the errors that result. The equations are individually estimated, which takes no account of the interrelationships that we think exist. Thus, the next step is to exploit the panel nature of the data, but impose less severe restrictions than in the FEM, by using seemingly unrelated regressions (SURE).

All the variables except the weather index are in logarithms, so the coefficients may all be interpreted as elasticities, with the sum of the lagged values reported for R&D (the total elasticity of the distributed lag). Row one of table 12.3 shows that the lags vary from 4 to 18 years, and the next row reports the elasticities of R&D, which are all positive, (except for Ireland), ranging from 0.191 to 0.591, and significantly different from zero.[10] For Ireland, it appears to be extension, the weather, and private R&D that explain TFP growth. Extension is positive and significant for five countries, not significant for three, and excluded for the remaining two countries. It was omitted on the basis of the tests for the order of integration and of the fact that it was not significant. Education is positive and significant for only two countries, insignificant for two, and excluded for the others, in three cases (Ireland, Greece, and Italy) because of the order of integration and in the others (Belgium, Germany, and the UK) because of collinearity with R&D. The weather variable performed well in eight cases, but was not significant for Greece and Italy.

The adjusted R^2s show that the majority of the variance is explained in all cases, with an average value of 0.85 (not shown). Although Germany, Greece, and the United States fit less well than the rest, over 90 percent of the variances are explained in six of nine cases. The Durbin-Watson statistics show that the errors are serially correlated for Belgium, Germany, Greece, and the Netherlands. This deficiency is not removed by using autoregressive correction techniques; instead the misspecified equations are improved by allowing for research spillovers.

Open Economy Model with Spillovers and Panel Data

The lower part of table 12.3 reports the results for a considerably better model, where spillovers are included and the panel is again used to produce the best possible estimates. Seemingly unrelated regressions (SURE) allow national effects to differ and are thus less restrictive than the pooled panel

Table 12.3. Explaining agricultural TFP growth with and without knowledge spillovers among 10 EC countries and the United States

Variables	Belg.	Denm.[a]	Irel.	Fran.[b]	Germ.[c]	Gre.[d]	Italy	Neth.	UK[e]	US
					OLS without Spillovers					
Lag Length	13	7	7	11	13	4	12	14	9	18
R&D Coeff.	0.312	0.585	−0.667	0.591	0.384	0.523	0.191	0.261	0.365	0.243
t-Stat.	3.98	6.43	−8.13	17.7	6.08	5.81	3.56	2.24	5.94	5.94
Ext Coeff.	N.S.	0.639	0.322	0.179	N.S.		0.351		0.433	N.S.
t-Stat.		2.42	9.28	3.24			3.83		6.33	
Ed Coeff.		0.313		N.S.				0.556		N.S.
t-Stat.		1.78						1.88		
Weather	0.153	0.306	0.209	0.227	0.463	N.S.	0.108	0.169	0.233	0.211
t-Stat.	2.52	3.03	5.85	2.61	1.99		1.17	2.09	3.92	3.71
Patents		Chem.	0.151		Mech.	0.128	Chem.	0.158		
t-Stat.		pats	5.85		pats	1.46	pats	2.05		
Adj. R^2	0.87	0.95	0.97	0.96	0.71	0.78	0.97	0.93	0.95	0.78
D-W Stat.	0.89	2.22	2.24	2.12	0.46	0.78	1.88	1.26	1.99	1.62
					SURE Model with Spillovers					
Lag length	5	10	10	10	7	4	8	9	15	17
R&D Coeff.	0.006	0.252	−0.41	0.442	0.121	0.556	0.127	0.838	0.499	0.188
t-Stat.	0.92	3.00	−3.13	2.74	1.56	3.11	3.21	31.55	2.35	22.13
Ext Coeff.	N.S.	N.S.	N.S.	N.S.	N.S.	N.S.	N.S.	N.S.	N.S.	N.S
Ed Coeff.	N.S.	N.S.	N.S.	N.S.	N.S.	N.S.	N.S.	N.S.	N.S.	N.S
Weather	0.144	0.179	0.110	0.121	0.151	0.263	N.S	0.063	0.213	0.920
t-Stat.	3.85	2.37	1.96	3.42	5.49	8.31		2.44	9.87	14.12
Denmark TFP								0.107		
t-Stat.								2.67		
France TFP	0.856				0.364		0.931		0.643	
t-Stat.	22.90				5.67		27.82		13.33	
Germany TFP			0.836	0.574				0.920		
t-Stat.			13.75	9.42				14.40		
Italy TFP						0.891				
t-Stat.						7.64				
Neth. TFP					0.560					
t-Stat.					7.85					
US TFP		0.821		0.306				0.125	0.144	
t-Stat.		10.86		6.24					3.60	3.65
Chem. Patents	0.406					0.80				
t-Stat.	2.13					1.970				
Mech. Patents										
Adj. R^2	0.88	0.91	0.89	0.94	0.89	0.88	0.97	0.96	0.95	0.78
D-W Stat.	1.74	1.80	1.82	1.70	1.54	1.17	2.02	2.34	2.28	2.09

Except where otherwise noted, extension, education, and private patents are lagged one period. Luxembourg is omitted due to lack of data. All variables are in logarithms, except the weather index. Critical *t* values with 8 degrees of freedom (one-tailed test), are 90 percent = 1.4, 95 percent = 1.86.

[a] Danish education is lagged two years.
[b] The French extension is a four-year moving average.
[c] For Germany, there is a three-year lead time before the lagged R&D effects begin.
[d] Mechanical patents for Greece are lagged four years.
[e] The United Kingdom extension is lagged three years.

models used in a previous section, but handle all 11 countries at once and allow for interactions through the error mechanism.

Some of the changes are quite startling, but entirely plausible. For instance, once the French TFP is included in the Belgian equation, it is significant, with a large elasticity of 0.856, and so are private-sector chemical patents. The lag on Belgian R&D shortens to five years and it becomes insignificant. Perhaps the small Belgian NARS uses the output of the very much larger French system and the private chemical industry and does adaptive research, to tailor the French technology to suit Belgian conditions. Danish TFP is explained by Danish R&D, the weather, and a considerable spill-in from the U.S. system, rather than from European neighbors. It is certainly possible that the advanced Danish animal sector draws on the technology generated by the world's largest and generally most advanced research system, discussed in some detail in chapter 7 of this volume. Ireland's productivity growth appears to be a function of the weather and German technology. It should be no surprise that the small Irish NARS adapts foreign technology, but the Bayesian prior would be that the main provider would be the UK; however, the U.K. R&D series tends to be I(2) (because of a reduction to near-zero growth from 1982 onward) and does not cointegrate with the Irish TFP, whereas the German TFP does.

The other countries have spill-ins that can be explained by obvious factors such as size, efficiency, geographical proximity, and common interests in terms of crops and other enterprises.[11] Thus, for France, it is Germany (the large next-door neighbor) and the United States (more basic research) that matter. For Germany, the contributors are France (next door) and the Netherlands (advanced animal production and a leader of this sample in terms of productivity). Greek productivity depends on Italian R&D far more than on the NARS, which is likely since Italy produces similar products, tends to lead technologically, and has a far larger NARS. Private-sector chemical patents, lagged four years, are also significant for Greece. For the Italians, French technology seems to dominate the NARS, while the Netherlands benefits from Germany (the large neighbor), Denmark (the specialized animal producer), and the United States. Lastly, the UK draws on French and U.S. technology, and perhaps surprisingly, U.K. technology does not appear to affect any continental TFPs.

Figure 12.4 shows that there is a hierarchical pattern to the spillovers, with U.S. technology spillovers affecting the TFPs of the more obviously innovative countries, which perhaps have more effective NARSs that are capable of exploiting the U.S. output of technology; these are Denmark, France, the Netherlands, and the United Kingdom. These countries plus Belgium form the high productivity convergence club (TFP levels of 130–150), but the U.K. growth rate collapsed after 1984 due to reductions in public R&D, and the United Kingdom slipped into the low productivity group. Belgium shows that direct spillovers from the United States are not a prerequisite for belonging to the high technology growth club. Instead, Belgium gains from close collaboration with the French research system, which is much larger and is geographically and culturally close.

The remaining countries in the figure are not in the high technology club. In terms of convergence in the levels, they are (excepting Luxembourg) at a TFP level of 80–100, which really is a long way behind. The United Kingdom

Figure 12.4. Technical spillover and research hierarchies

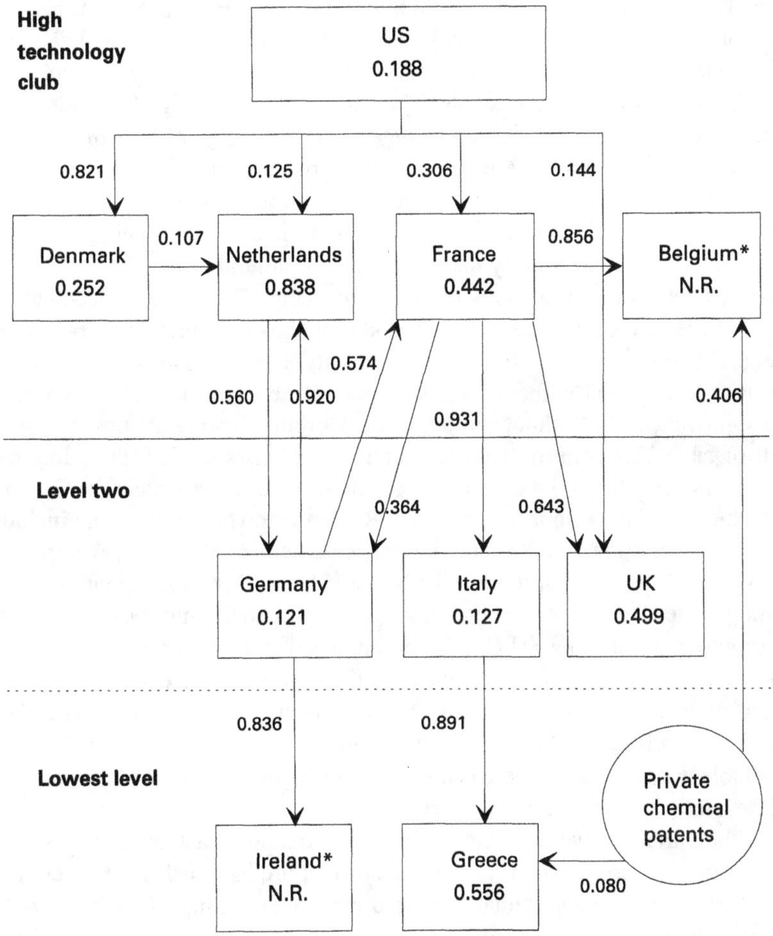

All values are elasticities. Values by arrows are spillovers and values in boxes are direct effects of national R & D. *N.R. = not reported.

would normally be regarded as having a large and advanced research system, but its failure to keep pace is explained above. It remains the leader of the group of laggards, followed by Germany and Italy, which are clearly in the second rank of the research hierarchy, drawing on European neighbors, rather than the United States. Thus, the German system does not draw on the United States, but it does exchange technologies reciprocally with both France and the Netherlands, and it is this link that keeps Ireland in the international system, if only at the lowest level of the hierarchy and with no obvious returns to its NARS.

Italy is next in the low productivity group, gaining from spillovers from the French system and serving as the link with Greece, which is also at the bottom of the international hierarchy. The technology gap between Italy and Greece is not too large, and the similarities in climate and crop mix are substantial, so this is highly possible. The short lag and the significance of private-sector chemical patents both suggest that the Greek NARS may be inclined toward adaptive research (as may be true of Belgium). Finally, Luxembourg, which is not even converging with the low productivity group, could not be included in the regressions due to lack of data. This is perhaps not surprising, as the country barely has an agricultural sector.

The size of the spillovers is perhaps surprising. Their total effects, calculated as the sum of the elasticities is 8.08 (not shown), which appears to outweigh the direct impacts of the NARSs of only 2.48 (not shown). Comparing the results with and without spillovers makes the consequences of ignoring the spillovers clear. Without spillovers, the Durbin-Watson statistics in the top half of table 12.3 confirm that the equations are mis-specified. Including the spillovers cures the serial correlation problems in all cases except for Greece, for which the data is poor. The adjusted R^2s in the top half of the table include three relatively poor results and on average 90 percent of the variance in TFP is explained. With the spillovers, only the U.S. result remains poor, at 0.78. Simply stated, the regression equations are considerably improved by including spillovers and using SURE. The problem is that the poor quality of the extension and education data becomes clear once alternative explanatory variables are included. The public extension data is incomplete, and by the end of the period private extension is playing an increasing role and no data are available. For education, there is no rural data at the EC level, and the national aggregates prove to be a poor proxy.

If the variables that are omitted from the equations that are mis-specified are positively correlated with the variables that are included, the expectation is that there will be a systematic upward bias to the estimated coefficients of the variables that are included. This would seem to be the case here; in 7 out of 10 cases the R&D elasticity for the NARS expenditures are lower when the spillovers are allowed for, and the average of the elasticities (excluding Ireland) is 0.40 without spillovers (not shown) and 0.30 when spillovers are included (counting Belgium as zero). This suggests that, all else being equal, the elasticity of R&D is biased upward by 33 percent if spillovers are ignored. Thus, not including spillovers may be a significant cause of the generally high rates of return to R&D noted by Echeverria (1990).

The remaining aspect of figure 12.4 that has not been mentioned is the total lack of any feedback from the European systems to the United States. This is in contrast with other research findings, such as those by Pardey et al. (1996), who show substantial returns to U.S. involvement in the CGIAR program.

RATES OF RETURN TO R&D

The coefficients of the R&D variables, however they are estimated, are output elasticities relating R&D expenditures to the TFP index, but they can be converted to marginal value products to allow calculation of the marginal internal rate of return (MIRR) to R&D. Thus, the α_is in the equation

$$\alpha_i = \frac{\partial LnTFP_t}{\partial LnRD_{t-i}} = \left[\frac{\partial TFP_t}{\partial RD_{t-i}}\right]\left[\frac{\overline{RD}_{t-i}}{\overline{TFP}_t}\right] \quad where \quad \overline{RD}, \overline{TFP}$$

are the (geometric) mean values, so that the marginal product of R&D in year i is

$$MP_{RD_{t-i}} = \alpha_i \left[\frac{\overline{TFP}_t}{\overline{RD}_{t-i}}\right]$$

However, this equation is still in terms of the effect of R&D on TFP, and for a rate of return to be calculated, the change in productivity must be converted into a value. Since the TFP index is the logarithmic ratio of an output index to an input index, it is reasonable to use the value of the surplus of output value minus the value of the inputs. Thus, both sides of the equation are multiplied by the change in the value of the surplus from the beginning to the end of the period (ΔS), divided by the change in the value of TFP (ΔTFP) over the period.[12] This gives a value marginal product, with both R&D and the surplus being measured in 1980 U.S. dollars.

$$VMP_{RD_{t-i}} = \alpha_i \left[\frac{\overline{TFP}_t}{\overline{RD}_{t-i}}\right]\left[\frac{\Delta S}{\Delta TFP}\right]$$

Note that although the last two terms are averages, α_i varies over the lag period, giving a series of marginal returns resulting from a unit change in

R&D expenditure. The marginal internal rate of return (MIRR) can be calculated from the equation,

$$\sum_{i=1}^{n} \left[\frac{VMP_{t-i}}{(1+r)^{i}} \right] - 1 = 0$$

in which, i is the length of the lag, for each expenditure term, and the MIRR for a one dollar change in R&D expenditure is calculated by solving for r.

The polynomial lags reported in table 12.3 give poor results for these short time series, so we confine ourselves to the pooled panel results for the 10 EC countries together. The panel data tests in Schimmelpfennig and Thirtle (1994) support specific lags of 2 and 9–11 years, rather than the polynomial form. Thus, calculation of the MIRR for the 10 EC countries together, from the panel elasticity estimates, reported in table 12.2, may be the best estimate of the ROR for the EC countries. The MIRR of R&D alone, calculated from the previous equations and assuming an average lag of 10 years, is 35 percent; and the MIRR to extension, with a 2-year lag, is minus 17.5 percent. The combined MIRR of R&D and extension together is 28.5 percent.[13] This would suggest that the EC has some way to go before facing any danger of overinvesting, but the returns are far lower than the conventional polynomial estimates, even though the intra-EC spillovers are internalized. The importance of the spillovers may indicate that more collaboration between the EC research systems would reduce duplication and improve efficiency.

SUMMARY AND CONCLUSIONS

This chapter compares multilateral TFP indices for 10 EC countries and the United States. At the start of the period the EC was at 77 percent of the U.S. productivity level. But over the period, the U.S. growth rate was 3 percent while the 10 EC countries managed an annual average rate of only 1.92 percent. Thus by the end of the period, the productivity gap had widened substantially so that the European countries were only at two-thirds of the U.S. efficiency level.

These aggregate figures hide an unexpected division of the EC countries into two groups. The high technology countries, which are the Netherlands, Denmark, France, and Belgium, actually converge with the United States to productivity levels between 130 and 150. These countries appear to have formed a club based on technological proximity and are growing faster than the remainder of the European Community. The second group that are converging to low productivity levels between 80 and 100 are Germany, Italy, Greece, Ireland, and the United Kingdom. The United Kingdom was a member of the technology club until the mid 1980s when budget cuts, or other factors, drastically reduced its growth rate. Thus the only other country that had benefited directly from U.S. spillovers slipped into the second rank of the hierarchy. At the lowest level are Greece and Ireland, who are linked to the international system only by spillovers from second rank countries (Italy and Germany).

The changes in TFP are explained by public R&D, extension, education, private-sector research, and the weather, allowing for spillovers between research jurisdictions. Ignoring spillovers between countries leads to relatively poor cointegrating regressions, some of which are serially correlated. Including the spillovers between countries improves the regressions, curing the serial correlation and giving greater explanatory power. Failing to include the spillovers between the members of "the United States of Europe" and from the United States biases the national elasticities of R&D upward, so ignoring spillovers could be a major cause of possibly inflated estimates of the returns to investments in NARSs that are often reported.

Pooling the sample of countries shows that the total effect of the R&D spillovers is 47 percent greater than the direct impact of the NARSs on national TFPs. For the 10 EC countries together the MIRR on public R&D is 35 percent, and the return to research and extension together is estimated at 28.5 percent. The spillover variables appear to have greater explanatory power than commonly used explanatory variables like extension and education, which contribute relatively little to explaining TFP growth for these European countries. The results suggest that calculations of the returns to R&D at the national level may be misleading unless international flows of agricultural technology are taken into account. The growth of collaborative EC research and the increasing role of multinationals will exacerbate this tendency.

One surprising result is that there appears to be absolutely no feedbacks from the European system to the United States. From a policy point of view this makes the USDA's attempts to coordinate public and private research in the United States somewhat simpler, in that foreign spill-ins may not be that significant. On the other hand, there is a policy dilemma for the United States in that collaboration in R&D, which is normally not a zero sum game, in this case seems to be benefiting a powerful competitor. Indeed, the United States may need to justify the gift of agricultural technology as a form of aid to rich allies or seriously consider collecting fees for its contribution. For the EC, there is also a policy problem in that collaborative R&D funding can either be granted to the top performers to maximize the efficiency of its use or be given to the poorer nations in an attempt to reduce inequality.

APPENDIX: MULTILATERAL TFPs: THEORY AND DATA

The Fisher index, which is the geometric average of the well-known Laspeyres and Paasche indices, is used because it satisfies more desirable properties than the Tornqvist, when used for intercountry comparisons.

Diewert (1992) shows that both indices can be viewed as ideal approximations, in the sense that they are consistent with economic theory. The sectoral PPPs are spatial price indices that compare several countries in the same year, with a base of 100 for a particular country. The method is similar to GDP-based PPPs, but the basket of goods is different, being just the inputs and outputs of the agricultural sectors.

Since a matrix of bilateral Fisher indices can give inconsistent comparisons, multilateral index numbers have been developed to construct sets of consistent indicators. The multilateral version of the Fisher index is the Elteto, Koves, and Szulc (EKS) index, which is widely used in international comparisons, especially by the OECD (Elteto and Koves, 1964; Szulc, 1964). The EKS index is the geometric mean of the ratios of bilateral indices. It satisfies the form symmetry test, is transitive, and although not directly derived from a flexible functional form, it is close to a superlative index (Diewert, 1976).

The source of the data for the 10 EC countries is Eurostat's Economic Accounts for Agriculture, Forestry, and Fisheries. If TFPs are to be compared internationally, it is also crucial that the data be comparable. Eurostat has tried to impose this uniformity on the EC10, and the U.S. series were reconstructed to make them compatible with the European data. The most obvious differences occur in calculating capital stocks of buildings and machinery, so these were all specially constructed by Ball et al. (1993).

NOTES

1. We thank Eldon Ball of the USDA for the TFP data; Han Roseboom of the International Service for National Agricultural Research (ISNAR) for providing research expenditure data and John Cantwell of University of Reading for the agricultural patent series; participants at the 72nd Annual Western Economic Association International Conference, in Seattle, on July 12, 1997; and the Economic Research Service and Farm Foundation sponsored workshop on Public-Private Collaboration in Agricultural Research held in Rosslyn, Virginia, on August 5, 1997, for many helpful discussions. David Hadley of University of Reading and Rob Townsend of the World bank and University of Pretoria, and Ellen Bielema and Charlie Hallahan at USDA provided excellent econometric assistance. The views expressed are not necessarily those of the USDA.

2. It is not appropriate to describe the methods and data in any detail, since these are available in the published sources, but for the convenience of the reader, the main technical issues are summarized in the Appendix.

3. Note that an average with weights to reflect output shares would be considerably higher. As it is, Luxembourg, which has a small agricultural sector, carries the same weight as France and Germany.

4. The EC allocated 333 million Euros, for 1990–94, to a broad-ranging Agriculture and Agro-Industrial Research Programme.

5. This situation is changing, and some elements of public systems have been profitable. This was true of the Plant Breeding Institute in the U.K. system, which was thus one of the first public institutions to be sold to the private sector.

6. Note that spillovers may well be reciprocal and multicountry. Any simultaneity in the TFP variables is eliminated by using lagged values as the spillover variables.

7. The $AIC_i = (MLL_i - k_i)$ and $SC_i = (MLL_i - 0.5k_i(\ln N))$, where MLL_i is the maximized log-likelihood for model i, k_i is the number of parameters, and N is the sample size.

8. This split is no more arbitrary than the division between research and extension. In the United Kingdom the near-market category would be in-house research of The Ministry of Agriculture, Fisheries, and Foods, plus all the work of the Agricultural Development Advisory Service, which is very difficult to split between research and extension. More basic research would be primarily the work of the Agricultural and Food Research Council.

9. Attempts to keep extension entirely separate from the R&D variable led to negative elasticities for extension as a result of collinearity with R&D.

10. The short series causes two problems here. Firstly, the lag length should be determined by including unrestricted lag terms in the regressions, but this leaves no degrees of freedom, so the polynomial had to be used, with the length decided on the basis of the Akaike and Schwartz criteria. Secondly, it is likely that the short time series tends to truncate the lags.

11. The approach taken here is to let the data determine the spillovers, but the results correspond well to the explanations of spillovers, based on technological closeness, that are listed by Huffman and Evenson (1993).

12. For both the surplus and the change in TFP, the average for 1973–75 is subtracted from the average for 1987–89 in order to minimize the effects of atypical years. The ratio of the change in the surplus to the change in TFP is simply a conversion factor, from TFP changes to real value changes.

13. Adding one-quarter of the R&D expenditures (a guess at the near-market portion) to the extension expenditures, thus cutting its lag from 10 to 2 years, increases the combined MIRR to 30 percent. The effect is small because the shorter discounting period is offset by the much smaller elasticity.

REFERENCES

Ball, V.E., A. Barkaoui, J.C. Bureau, and J.P. Butault. 1996. "Agricultural Productivity in Developed Countries: A Comparison Between the United States and the European Community." In *Conference Proceedings, Invited Papers.* Melbourne, Australia: The Conference Secretariat of the Global Agricultural Science Policy for the Twenty-first Century, pp.21–50.

Ball, V.E., J.C. Bureau, J.P. Butault, and H.P. Witzke. 1993. "The Stock of Capital in European Community Agriculture." *European Review of Agricultural Economics* 20:437–50.

Bouchet, F.C, D. Orden, and D.W. Norton. 1989. "Sources of Growth in French Agriculture." *American Journal of Agricultural Economics* 71:280–93.

Bureau, J.C., J.P. Butault, and A. Barkaoui. 1992. "Productivity Gaps Between European and United States Agriculture." In S. Narayayanam and J. King, eds. *Measuring Agricultural Productivity and Related Data for Regional, National and International Comparisons.* Ottawa: Agriculture Canada.

Capalbo, S., V. Ball, and M. Denny. 1990. "International Comparisons of Agricultural Productivity: Development and Usefulness." *American Journal of Agricultural Economics* 72:1292–97.

Capalbo, S., M. Denny, A. Hoque, and C. Overton. 1991. *Methodologies for Comparisons of Agricultural Output, Input and Productivity.* Washington, DC: Economic Research Service, U.S. Department of Agriculture.

Caves, D.W., L.R. Christensen, and W.E. Diewert. 1982. "Multilateral Comparisons of Output, Input and Productivity Using Superlative Index Numbers." *The Economic Journal* 92(365): 73–86.

Cooper, A., A.J. Rayner, and D. Greenaway. 1993. "Constant-Quality Price Indices for Agricultural Inputs: Tractors and Fertilisers Revisited." *Journal of Agricultural Economics* 44:67–81.

Diewert, W. 1976. "Exact and Superlative Index Numbers." *Journal of Econometrics* 4:115–46.

Diewert, W. 1992. "Fisher's Ideal Output, Input and Productivity Indexes Revisited," *The Journal of Productivity Analysis* 3:211–48.

Dowrick, S., and D.T. Nguyen. 1989. "OECD Comparative Economic Growth 1950-85: Catch-Up and Convergence." *American Economic Review* 79: 1010–30.

Echeverria, R. ed. 1990. *Methods for Diagnosing Research System Constraints and Assessing the Impact of Agricultural Research, Vol.2.* The Hague, Netherlands: International Service for National Agricultural Research.

Elteto, O., and P. Koves. 1964. "On a Problem of Index Number Computation Relating to International Comparisons." *Statsztikia Szemle* 42:507–18.

Evenson, R.E. 1989. "Spillover Benefits of Agricultural Research: Evidence from U.S. Experience." *American Journal of Agricultural Economics* 71:447–52.

Evenson, R.E., and C.E. Pray, eds. 1991. *Research and Productivity in Asian Agriculture.* Ithaca, NY: Cornell University Press.

Evenson, R.E., D. Landau, and D. Ballou. 1987. "Agricultural Productivity Measurement for U.S. States." Miscellaneous Publication 52-1987. St. Paul: Minnesota Agricultural Experiment Station, University of Minnesota.

Fuglie, K., N. Ballenger, K. Day, C. Klotz, M. Ollinger, J. Reilly, U. Vasavada and J. Yee. 1996. *Agricultural Research and Development: Public and Private Investments Under Alternative Markets and Institutions.* Agricultural Economics Report 735. Washington, DC: Economic Research Service, U.S. Department of Agriculture, May.

Huffman, W.E., and R.E. Evenson. 1993. *Science for Agriculture: A Long Term Perspective.* Ames: Iowa State University Press.

Jorgenson, D.W., and M. Nishimizu. 1978. "U.S. and Japanese Economic Growth, 1952–1974: An International Comparison." *Economic Journal* 88:707–26.

Khatri, Y., and C. Thirtle. 1996. "Supply and Demand Functions for UK Agriculture: Biases of Technical Change and the Returns to Public R&D." *Journal of Agricultural Economics* 47 (September): 338–54.

Lucas, R. 1988. "On The Mechanics of Economic Development." *Journal of Monetary Economics* 22(1): 3–42.

Pardey, P.G,. and J. Roseboom. 1989. *ISNAR Agricultural Research Indicator Series: A Global Data Base on National Agricultural Research Systems.* Cambridge, UK: Cambridge University Press.

Pardey, P.G., J.M. Alston, J.E. Christian, and S. Fan. 1996. "Summary of a Productive Partnership: The Benefits from US Participation in the CGIAR." EPTD Discussion Paper No 18. Washington, DC: International Food Policy Research Institute.

Pesaran, M.H., and Y. Shin. 1995. "An Autoregressive Distributed Lag Modeling Approach to Cointegration Analysis." Paper presented at The Symposium at the Centennial of Ragnar Frisch, the Norwegian Academy of Science and Letters, Olso, March.

Quah, D.T. 1997. "Empirics for Growth and Distribution: Stratification, Polarization and Convergence Clubs." Discussion Paper No. 1586, Centre for Economic Policy Research, London, March.

Romer, P. 1986. "Increasing Returns and Long-Run Growth." *Journal of Political Economy* 94:1002–37.

Romer, P. 1990. "Endogenous Technical Change." *Journal of Political Economy* 98(5): S71–S102.

Rutten, H. 1992. "Productivity Growth of Dutch Agriculture, 1949–1989." Working Paper 470. The Hague, Netherlands: Agricultural Economics Research Institute.

Schimmelpfennig, D., and C. Thirtle. 1994. "Cointegration and Causality: Exploring the Relationship between Agricultural R&D and Productivity." *Journal of Agricultural Economics* 45:220–31.

———. 1999. "The Internationalization of Agricultural Technology: Patents, R&D Spillovers and Their Effects on Productivity in the European Union and United States," *Contemporary Economic Policy* 17(4):457–468.

Szulc, B., 1964. "Indices for Multiregional Comparisons." *Przeglad Statystyczny,* 3: 239–254.

Terluin, I.J. 1990. "Comparison of Real Output, Productivity and Price levels in Agriculture in the EC: A Reconnaissance." In *Onderzoekverslag 69.* The Hague, Netherlands: Agricultural Economics Research Institute LEI.

Thirtle, C., and Bottomley, P. 1989. "The Rate of Return to Public Sector Agricultural R&D in the UK, 1965-80." *Applied Economics* 21:381–400.

Thirtle, C.G., J. Piesse and V. Smith. 1997. *An Economic Approach to the Structure, Historical Development and Reform of Agricultural R&D in the United Kingdom.* Reading, UK: Centre for Agricultural Strategy, University of Reading.

Thirtle, C., Ball, E., Bureau, J-C. and R. Townsend. 1995. "Accounting for Efficiency Differences in European Agriculture: Cointegration, Multilateral Productivity Indices and R&D Spillovers." In G.H. Peters and D.D. Hedley, eds. *Agricultural Competitiveness: Market Forces and Policy Choice.* Proceedings of the Twenty-Second International Conference of Agricultural Economists, Harare, Zimbabwe, 1994. Aldershot, UK: Dartmouth Publishing Company, pp. 652–69.

White, F.C., and Havlicek, J. 1982. "Optimal Expenditures for Agricultural Research and Extension: Implications of Underfunding." *American Journal of Agricultural Economics* 64:47–55.

13

THE PRIVATE SECTOR AND INTERNATIONAL TECHNOLOGY TRANSFER IN AGRICULTURE

CARL E. PRAY AND KEITH O. FUGLIE

THE RATE at which agricultural technology is diffused around the world is growing rapidly. Improvements in scientific infrastructure, advances in communication technology, and a reduction in barriers to trade and foreign direct investment have made transferring technology across national borders easier. The private sector in particular is playing an increasingly important role in the development and international diffusion of new agricultural technology, largely through the activities of multinational companies that develop improved seed, livestock, agricultural chemicals, farm machinery, and food processing technology to serve global markets.

The purpose of this chapter is to examine the role of the private sector in the international transmission of agricultural technology, especially between industrial and developing countries. We define the private sector to include for-profit companies and other institutions that conduct research using funds from private commercial sources in order to develop proprietary technology for the food and agricultural sector.[1] Thus, we do not include research at public universities or national research institutes funded by commodity groups or private donations. Nor do we consider research funded by private philanthropic organizations. The incentives that influence private agribusiness firms to investment in agricultural research and technology transfer include a number of economic, political, and technological factors. Moreover, private-sector international technology transfers can take place through several means, including trade in agricultural inputs, technology licensing, foreign direct investment, and international networks of research stations.

Measures of international technology transfer indicate that the United States is not only a large exporter of agricultural technology but also a large importer. And as an ever-larger share of global agricultural research investment is conducted outside U.S. borders, the country is increasingly relying on foreign sources of new technology to improve productivity and meet consumer demand. This trend has important implications for U.S. farmers, agribusinesses, and consumers, as well as for government policy makers. International technology transfer not only promotes more rapid economic growth globally but also improves productivity and expands markets of U.S. farms and agribusiness firms. Efforts by the United States to reduce international barriers to trade and foreign investment and to strengthen the protection of intellectual property rights (IPRs) worldwide should increase opportunities for private companies to invest in agricultural research and technology transfer.

International technology transfer has important implications for research "spillovers"—benefits from research that accrue to those who don't share in its costs. The emergence of global markets for improved agricultural technology helps reduce spillovers from private research. But at the same time they increase international spillovers from public research. The rapid pace at which new technologies are flowing across national borders raises concerns that governments may underinvest in public agricultural research due to the free rider problem—national governments may become increasingly reluctant to fund agricultural research, instead relying on technology transfers from other countries. Meeting the challenges presented by the growth in technology transfer will require greater international collaboration in funding and conducting agricultural research. Collaboration is particularly needed in the basic agricultural sciences, where international spillovers are likely to be especially large (Ruttan 1986).

The first section of the chapter describes the economic, technological and political determinants of private agricultural research and technology transfer. The second section presents evidence from the United States and developing countries on the flow of agricultural technology across national borders, with special emphasis on the growing role of multinational companies that supply agricultural inputs. The third section discusses future trends in international technology transfer based on recent changes in the underlying determinants of private agricultural research. The final section discusses some conclusions and implications for U.S. trade, development, and research policies.

DETERMINANTS OF PRIVATE RESEARCH AND TECHNOLOGY TRANSFER

Private-sector investments in agricultural research and international technology transfer are influenced by several economic, technological, and political factors. Profit-maximizing firms will undertake research when they determine that it will increase their product sales at prices sufficient to allow them to recoup their investments in research and market development. Three key determinants of the private profitability of research are (1) the potential market size of products that embody the new technology, (2) technology opportunity, or the likelihood that an investment in research will result in a successful innovation, and (3) appropriability, or the ability to capture the economic benefits from research (Pakes and Schankerman 1984).

Economic Determinants: Market Size and the Demand for New Technology

Economic and market factors determine the size of expected sales of new technology. Firms will not conduct research into a new product or reduce production costs without the expectation of a profitable level of demand. Changes in market conditions can induce firms to change the level and focus of their research and technology investments.

Demand for technology changes as a country develops. In low income countries, market demand causes agricultural research to focus on key industrial crops, export crops, and basic staples. As per capita income rises, consumer demand increases for high value products such as livestock, fruits, and vegetables, and more firms are willing to undertake research on these commodities. In the more affluent countries, new demands affecting agricultural research emphasize food safety, environmental protection, and ethical considerations such as animal welfare.

Research is also influenced by the supply and relative cost of agricultural inputs and resources. If certain resources grow scarce and expensive, innovators respond by seeking substitutes. Hayami and Ruttan (1985) demonstrated the importance of relative resource scarcity in determining the direction of technical change in agricultural development. In densely populated regions of Asia, agricultural growth was constrained by the inelastic supply of land. Countries offset this constraint by developing technology that substituted for land, such as new seeds, fertilizers, and irrigation schemes that raised yields. In land-rich and labor-scarce countries like the United States, Canada, and Australia, however, farm mechanization was the first wave of technological change in agriculture. Mechanical technology was meant not to increase yields but to offer a substitute for relatively expensive labor. The more recent emphasis on yield-increasing technology in these countries was induced by the declining price of fertilizer relative to land. As wages and incomes rise in Asia, the demand for labor substitutes such as chemical weeding and mechanical cultivation and harvesting should also increase—and more private firms should be interested in supplying them.

In some situations markets may not respond adequately to consumer preferences. The market may, for example, undervalue the nonsensory attributes of food products (such as safety, nutrition, and environmental friendliness) as well as those attributes that meet ethical or religious standards. A consumer may be unable to identify these attributes simply by looking at, feeling, smelling, or tasting products. If consumers cannot vote for these product characteristics by changing their consumption patterns, then consumer preferences may be unable to drive new product and technology development in a direction that meets these demands (Fuglie et al. 1996).

Technological Opportunities for Applied Research and Development

Technological opportunity is the likelihood that an investment in research or technology transfer will result in a successful innovation. But technological opportunity is constrained by the prevailing level of basic scientific knowledge and the cost of research inputs. For example, before the rediscovery of Mendel's theory of genetics in 1900, very little progress had been made in improving crop yields. More recently, basic scientific advances in molecular biology and genetic engineering have greatly expanded the technological possibilities for developing new plant varieties and animal breeds.

Basic scientific research at public research institutions and universities is a primary source of new technological opportunities for the private sector. Expanding the number of technologies that have some potential for development increases private incentives to invest in applied research. Public research has in fact offered private researchers many technological opportunities. In the United States, basic research in molecular biology and genetics by public universities and national research institutes helped create the biotechnology industry (Caswell, Fuglie, and Klotz 1994). An example from a developing country is the joint Kasetsart University-Rockefeller Foundation corn breeding program in Thailand. In the 1960s this program identified genetic material that confers resistance to downy mildew, an important corn disease in Southeast Asia. This scientific advance encouraged the private seed industry in Southeast Asia to provide farmers with improved hybrid corn varieties carrying the trait. As a result, private-sector investment in corn breeding expanded significantly (Pray 1987).

Public universities play an indirect role in encouraging private research by increasing the supply of scientific resources available to the private sector. By providing academic consultants, training for scientific and technical personnel employed in private research programs, and specialized facilities for product testing and calibration, public universities offer services that may be prohibitively expensive for a private company to support on its own. Lowering the cost of conducting research increases the expected profitability of a given level of research investment.

Another source of technological opportunity is the exchange of information and knowledge among researchers. One purpose of patents is to ensure that the technology embodied in an innovation is made public for other firms to use in making further innovations. Firms also learn about technological advances made by other researchers through reverse engineering of the developing firm's product, by hiring the scientists and engineers who developed the product, or by purchasing a controlling interest in the developing firm itself.

Transferring technology that has already been commercialized elsewhere depends not only on the environmental specificity of the technology (that is, its adaptability to local climatic conditions and its resistance to local pests and diseases), but also on its adaptability to local markets. Because corn varieties have a high degree of environmental specificity, they are difficult to transfer directly to other regions and often require adaptive research before they can be transferred. In contrast, poultry breeds raised in controlled climate conditions are much easier to transfer across geographic regions. But if the varieties or breeds do not meet local market conditions—for instance, if the broilers have too much white meat in countries where dark meat is preferred—they may have to be modified to meet local consumer preferences.

Foreign technology can provide new ideas and opportunities for local scientists. Modifying foreign technology often provides a very high payoff because the modifications require only a limited amount of local research. For example, public and private plant breeding programs may be able to use disease-resistant or yield-enhancing genetic traits developed by an international agricultural research center (IARC) quickly and easily to produce a profitable new plant variety suited to local environmental and market conditions. Field testing a foreign pesticide to determine its effectiveness against local pests and diseases can be another high-payoff investment. Local firms can then commercialize effective chemicals and earn back the cost of the adaptive research relatively quickly.

Appropriating Gains from New Technology

Appropriability, a key incentive for private investment in research and technology transfer, involves the ability of firms to capture the economic benefits of an invention. Firms will not undertake major investments in developing new machines or crop varieties, for example, if these products can be easily reproduced by rival firms that have not incurred the initial research costs. Firms need some guarantee that they will recoup their investments in new products—and earn a profit.

The degree to which firms can appropriate the benefits of research depends on three main factors: the existence of IPRs, the structure of the industry, and the nature of the technology (Schumpeter 1950). Intellectural property rights, which in agriculture include patents, trademarks, copyrights, and plant breeders' rights, limit the rights of others to make, use, or sell an invention or product without permission from the inventor. Individuals or companies wishing to use a patented invention are obliged to seek a license from the patent owner. These IPRs thus increase the appropriability of the economic benefits of an invention. Most countries recognize some form of patent laws, but not all governments enforce these protections, and some countries have specifically excluded agricultural inventions from patent protection.

Schumpeter (1950) hypothesized that industries with a concentrated market structure are likely to have higher rates of technical change, because it is generally easier for a company to appropriate the gains from research when it has sufficient market power to influence prices. Such market power is usually enhanced when a company gains a large share of a market and where there are significant barriers to entry of potential rival firms. However, too little competition can reduce incentives for private research. A firm facing little or no competition may see little need to devote resources to research and innovation and may be content to charge monopoly prices for old technology. Scherer (1980) summarizes the findings of subsequent research on Schumpeter's early theory:

> A bit of monopoly power in the form of structural concentration is conducive to invention and innovation, particularly when the advances in the relevant knowledge base occur slowly. But very high concentration has a favorable effect only in rare cases, and more often it is apt to retard progress by restricting the number of independent sources of initiative and by dampening firms' incentive to gain market position through accelerated research and development (p.438).

Scherer's observation appears to hold true for agriculture. Research investments by the U.S. seed and agricultural chemical industries expanded significantly during a period of consolidation (Fuglie et al. 1996). However, a high level of concentration and lack of competition in fertilizer manufacturing has been cited as one reason for the low level of innovation in this industry (Ruttan 1982).

Some innovations are by nature more difficult to copy or reverse engineer because of their technical characteristics. Hybrid seed, for example, cannot be reproduced without the parent inbred lines, so developers of hybrids can protect their lines by controlling access to the parent lines. The breeders thus ensure a continued market for their seed, as farmers must buy hybrid seed every year to maintain maximum yields. And, since the developer of a high-yielding hybrid is often the sole source of the improved variety, the company can charge a higher price for its new seed and capture a large share of the economic gains from the initial investment.

Policies and Private Incentives for Research and Technology Transfer

Public policies and regulations can hinder or support the international transfer of technology by influencing market size, technology opportunity, and the ability of firms to appropriate benefits. Table 13.1 describes how different policies may affect these determinants. First, there are macroeconomic policies and the general economic environment. Macroeconomic stability, good transportation and communication infrastructure, functioning capital and insurance markets, and a reasonable level of general education and training, especially farmer schooling, are factors that positively affect private investments in research. These factors reduce the costs of transacting business in an economy, including agricultural research and the delivery of improved inputs to farmers (Evenson and Westphal 1995).

Table 13.1. Policies and incentives for private agricultural research and technology transfer

Private research determinants	Policies affecting determinants
General state of the economy	– Macroeconomic stability – Public infrastructure – General education and training – Development of capital and insurance markets
Size of input markets	– Market share of state-owned enterprises (SOEs) – Restrictions of foreign participation in input markets – Trade restrictions on inputs – Price interventions in input or product markets
Technological opportunity and the cost of research inputs	– Public investment in agricultural research and education – Registration and testing requirements for new seed and agricultural chemicals – Biosafety requirements for biotechnology field trials – Public subsidies for private research, including tax holidays, tax credits, research grants, and technology parks
Appropriability	– Intellectual property laws (patents, plant breeders rights, trademarks, trade secret protection) and enforcement – Technology licensing requirements and regulations affecting technology imports – Competitiveness and antitrust policies

The size of input markets is affected by several specific policies. In some countries, state-owned enterprises (SOEs) maintain a monopoly on the production and/or distribution of certain agricultural inputs. Limiting the access of the private sector to input markets acts as a disincentive to private research. Similarly, some countries may pursue protectionist policies to help national industries and limit the degree to which foreign companies can participate in local markets. These restrictions include outright bans, regulations that require majority control by a local partner in a foreign subsidiary, restrictions on the foreign remuneration of profits, or other regulations on foreign direct investment. These restrictions can reduce the incentives for private research in a country by multinational firms. Moreover, empirical research shows that protectionist policies retard the technological development of national industries by blocking access to critical elements of foreign technology (Evenson and Westphal 1995).

Government intervention in agricultural input or output markets may also take more direct forms. Subsidies that increase farm prices increase farm demand for inputs, and likewise explicit or implicit taxes on agriculture reduce farm demand for inputs. Prices of inputs themselves may be subsidized or taxed, similarly affecting input demand, and therefore the size of the market for improved inputs.

Appropriability is affected by policies toward intellectual property rights, trade secret protection, and market competitiveness (e.g., antitrust policy). Several countries have until recently excluded agricultural inventions from patent protection. And even in countries with legal protection for patents and trademarks, enforcement may be cumbersome or lax. However, under the Uruguay Round of the GATT, member countries of the World Trade Organization are required to provide IPRs for agricultural (and other) inventions, including new plant varieties. Otherwise they may face retaliatory trade measures.

National laws also influence a firm's ability to protect technology through trade secrets. Some countries require foreign companies to import and license their technology to local firms in order to participate in local markets. For example, agricultural chemical companies may be required to manufacture active ingredients locally, or seed companies may be required to import advanced breeding material. While technology importation and licensing requirements are often designed to increase technology transfer, they can also cause some companies to stay out of a market altogether. Since maintaining control over proprietary technology is critical for appropriating gains from research, multinational companies may be reluctant to participate in markets if doing so puts their patented technology and trade secrets at risk (see box 13.1).

The main policies affecting technology opportunity and the cost of research inputs are public-sector investments in agricultural research and higher education. A strong public agricultural research and university system can significantly reduce the cost of private research by expanding the supply of highly skilled scientific and technical personnel available for private agricultural research. Public research can also provide key, enabling technologies that increase the likelihood private research endeavors will be successful.

Regulatory policy also influences technology opportunities and the cost of conducting research in a country. Countries differ in their requirements for efficacy and safety testing for registering new pesticides which affects the time and cost of introducing new products into a market. Countries also differ in their regulations governing the introduction of new seed varieties. Some countries only allow varieties that have been demonstrated to be superior to existing varieties to be released and sold to farmers.

BOX 13.1. INTELLECTUAL PROPERTY RIGHTS AND INTERNATIONAL TECHNOLOGY TRANSFER

The strength of IPRs and a government's willingness to enforce them influence private firms interested in licensing technology in a country, transferring it through their subsidiaries, or conducting research there. In a survey of U.S. multinational firms, Mansfield (1994) asked companies whether strong IPRs (or the lack thereof) influenced the companies' willingness to transfer technology to a country. The companies were also asked to rate the strength of IPR protections in 16 countries.

A full 80 percent of firms said that IPRs had a strong effect on their decision to invest in research and development (R&D) facilities. Only 20 percent of companies said that the strength or weakness of IPRs had a strong effect on their decision to establish sales and distribution outlets, however. Of the 16 countries named in the survey, Brazil, India, Nigeria, and Thailand were seen as having weak laws, while Hong Kong, Japan, Singapore, and Spain were perceived as having relatively strong laws. Only 8 percent of the companies surveyed indicated that they thought Japanese IPRs were too weak to permit licensing of their newest and most effective technology. But 38 percent thought Thailand's IPRs were too weak, and 44 percent thought Indian IPR laws would not protect their newest and best technologies.

Other countries allow companies to market any new variety they develop, relying on market competition to provide an incentive to seed companies to introduce high-quality varieties. With respect to biotechnology, some countries have moved quickly to establish regulatory protocols for conducting field trials with genetically modified plants in order to encourage applications to agriculture. Other countries have not yet developed protocols or used very strict ones to discourage biotechnology.

Finally, governments may provide subsidies to private research in the form of research grants, research tax credits, or tax holidays. Governments may also provide indirect subsidies by making research facilities at universities or technology parks available to private firms at less than their full cost.

QUANTITATIVE INDICATORS OF INTERNATIONAL AGRICULTURAL TECHNOLOGY TRANSFER

The increasingly rapid pace of agricultural technology transfer can be illustrated by examining the international diffusion of genetically modified crops (see box 13.2). Initially developed in the United States, transgenic crops were first grown commercially in China, two years before the first commercial

BOX 13.2. THE PACE OF INTERNATIONAL TECHNOLOGY TRANSFERS: HYBRID CORN VERSUS TRANSGENIC CROPS

A comparison of the spread of two important technologies that originated in the United States—genetically engineered plants and hybrid corn—highlights the increasing speed of agricultural technology transfer. Hybrid corn, developed through a combination of public and private efforts in the 1920s and 1930s, spread across the Cornbelt in the 1930s and 1940s. With the assistance of the Marshall Plan, it spread to Europe in the late 1940s. With the help of foreign aid, the Rockefeller Foundation, and private companies, it reached Asia and Latin America in the 1950s, although successful hybrids were not developed for Southeast Asia until the 1970s. Hybrid corn did not reach most of Africa until the late 1980s.

In contrast, transgenic crops developed in the United States spread almost immediately to the rest of the world. China was growing transgenic virus-resistant tobacco (developed by a university research program using technology from a U.S. firm) commercially in 1992, before transgenic crops were being used commercially in the United States. By 1996 transgenic tobacco covered more than two million acres, or 30 percent of the national tobacco area of China. China also started growing virus-resistant transgenic tomatoes in 1994. Argentina began growing herbicide-tolerant soybeans in 1996. Monsanto, one of the largest private investors in agricultural biotechnology, expects that by 2000 it will have approvals to produce genetically engineered crops commercially in Brazil, China, India, Indonesia, Mexico, South Africa, Thailand, Vietnam, and Zimbabwe (table 13.3). Surveys by the authors indicate these expectations appear to be realistic (Pray and Fuglie 1999). In China, for example, thousands of acres of Monsanto's genetically engineered cotton are being produced under controlled conditions for use as seed.

Hybrid corn was successfully commercialized throughout the world only after decades of efforts by relatively small public-sector programs. Transgenic crops have spread globally in less than a decade and have needed little help from the public sector. Private firms developed all the transgenic varieties that were being grown commercially in the United States in 1996, as well as all the varieties for which applications have been made for commercial release. Big multinational companies such as Monsanto, AgrEvo, Novartis, and Pioneer are largely responsible for the rapid dissemination of transgenic crops. But China's success suggests that the public sector can still play an important role in technology transfer through strong public research programs that move aggressively to exploit foreign technology.

plantings in the United States. Between 1996 and 1998, the area planted to transgenic crops increased from about 5 million acres to more than 70 million acres globally (table 13.2), and is expected to increase severalfold in the next few years. Through the activities of multinational life science firms such as Monsanto, myriad agricultural biotechnology products are being developed for numerous national markets (table 13.3).

Table 13.2. Field experiments and commercialization of transgenic crops by region, 1995–98

	U.S. and Canada	Western Europe	Eastern Europe and Russia	Industrial Asia and Oceania	Developing Asia	Latin America	Africa
Number of field experiments							
1995	834	240	22	26	31	85	9
Applications for commercial use							
Approved	30	6	0	0	2	6	0
Pending	3	12	0	7	n.a.	5	0
Area in commercial transgenic crops (1,000 acres)							
1996	>3,200	0	0	0	>2,000	>375	0
1997	>18,000	0	0	0	>2,000	>3,500	0
1998	57,500	<0.5	0	0.2	>2,000	10,900	<0.2

Sources: Field experiment and application/approval data from James and Krattiger (1996). Area estimates from James (1997).

While the evidence on the diffusion of transgenic crops is suggestive of international technology transfer, more systematic measurement is difficult because technology transfer is embedded in so many activities. Hayami and Ruttan (1985) identified three general types of international technology transfers: material, design, and capacity transfers (table 13.4). In the agricultural sector, *material transfer* is embodied in improved agricultural inputs that are sold or distributed to farmers and takes place through international trade in seeds, breeding stock, chemicals, and machinery. *Design transfer* involves the importing and exporting of plans or blueprints for machines, chemicals, or plants that are used to manufacture or reproduce improved inputs locally. Design transfer also occurs through exchanges of advanced germplasm (parent lines of plants and animals) used to breed superior crop varieties or animal breeds. *Capacity transfer* occurs when a foreign government or private enterprise builds local research capacity to develop or invent new technology. The international diffusion of transgenic crops described above involves elements of material, design, and capacity transfer.

Table 13.3. Expected regulatory approvals of Monsanto's transgenic crops (years indicate earliest expected date of approval)

Argentina
1996
Roundup Ready soybeans
1999
YieldGard insect-protected corn
2000
Bollgard insect-protected cotton
Roundup Ready corn
2001
Roundup Ready cotton

Australia
1996
Ingard insect-protected cotton
Roundup Ready soybeans
1998
Roundup Ready cotton

Brazil
2000
Bollgard insect-protected cotton
Roundup Ready corn
YieldGard insect-protected corn
2001
Roundup Ready cotton

Canada
1995
Roundup Ready canola
1996
Bollgard insect-protected cotton
NewLeaf insect-protected potatoes
Roundup Ready soybeans
1997
NewLeaf Plus insect- and-virus-protected potatoes
NewLeaf Y insect- and-virus-protected potatoes
Roundup Ready cotton
YieldGard insect-protected corn

China
1998
Bollgard insect-protected cotton
1999+
YieldGard insect-protected corn

European Community
1996
Roundup Ready soybeans
1997
NewLeaf insect-protected potatoes
1998
MaisGard insect-protected corn
1999+
Roundup Ready corn
Roundup Ready oilseed rape
Roundup Ready sugar beets

India
1999+
Bollgard insect-protected cotton
YieldGard insect-protected corn

Indonesia
1999+
Bollgard insect-protected cotton
Roundup Ready cotton
YieldGard insect-protected corn

Japan
1996
NewLeaf insect-protected potatoes
Roundup Ready oilseed rape
Roundup Ready soybeans
1997
Bollgard insect-protected cotton
NewLeaf insect-protected potatoes (new varieties)
Roundup Ready cotton
YieldGard insect-protected corn
1998
Roundup Ready corn

Mexico
1996
NewLeaf insect-protected potatoes
1997
Bollgard insect-protected cotton
Roundup Ready cotton
1998
Roundup Ready cotton
YieldGard insect-protected corn
1999
NewLeaf Plus insect-protected potatoes
NewLeaf Y insect-protected potatoes
Roundup Ready corn

South Africa
1998
Bollgard insect-protected cotton
1999+
Roundup Ready cotton
YieldGard insect-protected corn

Thailand
1999+
Bollgard insect-protected cotton
Roundup Ready cotton
YieldGard insect-protected corn

United States
1994
NewLeaf insect-protected potatoes
1995
Bollgard insect-protected cotton
Improved ripening tomatoes
Roundup Ready soybeans
1996
Roundup Ready canola
Roundup Ready cotton
NewLeaf insect-protected potatoes (new variety)
YieldGard insect-protected corn
1998
NewLeaf Plus insect- and virus-protected potatoes
NewLeaf Plus insect- and virus-protected potatoes
Roundup Ready corn

Vietnam
1999+
Bollgard insect-protected cotton
Roundup Ready cotton

Zimbabwe
1998
Bollgard insect-protected cotton
1999+
Roundup Ready cotton

Source: Monsanto Company Web site, July 1997 [www.monsanto.com/ag/_asp/monsanto.asp].

Table 13.4. Mechanisms for international technology transfer

	Public sector	Private sector
Material transfer	Universities, national research institutes, and CGIAR centers exchange varieties and other technologies	Firms import and export crop seeds, livestock breeds, fertilizers, pesticides, and machinery
Design transfer	Universities, national research institutes, and CGIAR centers exchange scientists, germplasm, and scientific publications	Firms license or copy patented technology and transfer technology through foreign direct investment
Capacity transfer	Development assistance projects build public research institutions and train foreign scientists; universities, CGIAR centers, and national research institutes conduct collaborative research	Firms conduct research in other countries; firms reverse-engineer and modify foreign proprietary technology; firms hire expatriate scientists or purchase foreign research firms

Both public and private organizations participate in all three kinds of transfers. Material transfers flow primarily through the private sector in the form of agricultural input trade. Design transfers are transmitted by the public sector via journal articles, information revealed in patents, bilateral exchanges of germplasm, or multilateral organizations such as the international agricultural research centers of the Consultative Group for International Agricultural Research (CGIAR).[2] The private sector also transfers designs through licensing agreements, consulting firms (as part of "packages" of designs and advice), and foreign direct investment in factories that produce agricultural inputs.

Public-sector capacity transfers to developing countries often take the form of projects aimed at creating national and international research institutions. The funding organizations—such as the World Bank, the United States Agency for International Development (USAID), and philanthropic foundations—often channel their support through research institutes, universities, and consulting firms in industrial countries. These groups then provide the technical assistance needed to create research capacity in developing countries. Doctoral and postdoctoral programs, research networks involving national research systems and IARCs, and collaborative research projects serve as important means of transferring research capacity through the public sector.

In the private sector the most important means of transferring capacity is through firms that set up their own research facilities and experiment stations abroad. Many multinational seed and chemical companies operate international networks of research stations to develop and test technology for local farmers. Private firms occasionally support collaborative research programs with foreign public institutions. Increasingly, firms from developing countries are also establishing their own research capacity, often through joint ventures with multinational companies from industrial countries. Firms in developing countries sometimes hire expatriate scientists or even buy controlling shares in foreign companies in order to increase their access to technology and research capacity.

Below we develop some indicators of material, design, and capacity transfer for the United States and for transfers between industrial and developing nations. For the United States, we examine trends in imports and exports of agricultural technology in two ways. First, we examine material transfer through trade in agricultural inputs such as seeds, chemicals, and farm machinery. Second, we measure design transfers of agricultural technology through the number of patents awarded to U.S. inventors in foreign countries (technology exports) and the number of U.S. patents awarded to foreigners (technology imports). For agricultural technology flows between industrial and developing countries, we focus on private-sector investments in agricultural research in developing countries and public-sector transfers of crop germplasm through international agricultural research centers.

U.S. Trade in Agricultural Inputs

Historically, the United States has been a net exporter of agricultural inputs, but the country's reliance on foreign sources of agricultural innovation has grown over time. Most trade in agricultural inputs occurs between countries with similar agricultural and environmental conditions, so that little or no adaptation of the technology is required. The United States conducts most of its trade in agricultural inputs with other industrial countries, especially Canada and the European Community. Exports of agricultural inputs to developing countries are growing, especially to Mexico.

The United States had consistently registered positive trade balances in crop seed, fertilizers, and pesticides, although in recent years there has been a trade deficit in farm machinery (table 13.5). Between 1983 and 1996, total exports of seed, agricultural chemicals, and farm machinery averaged $7.1 billion annually (in 1997 dollars), compared with $5.0 billion in imports. Over this period, exports of agricultural inputs grew by an average of 2.2 percent annually in real terms, while imports grew by 3.8 percent annually. The growth in imports of agricultural inputs, especially in seeds and agricultural chemicals, reflects increased capacity in agricultural research in other countries relative to the United States.

Foreign Agricultural Patents

A firm planning to export a new technology to a foreign market, or to manufacture it in that market, will often patent the technology there. Recent work by Evenson and Johnson (1999) and in chapter 11 of this volume, examining patenting trends, provides further evidence that the United States is both a ma-

Table 13.5. U.S. trade in agricultural inputs, 1983–96

	Exports	Imports	Net trade	Export/import	Real annual growth rate (%)	
	(millions of 1997 dollars, annual avg.)			ratio	Exports	Imports
Seed	577	197	381	2.94	5.41	7.27
Fertilizer	2,899	1,818	1,081	1.59	0.00	1.87
Agricultural chemicals	1,589	544	1,045	2.92	−1.14	12.07
Herbicide	737	198	539	3.72	−2.03	4.12
Insecticide	390	77	313	5.08	−3.58	2.17
Fungicide	203	68	135	2.98	−3.99	8.94
Other pesticide	259	230	29	1.13	6.09	20.92
Farm machinery	2,009	2,491	−482	0.81	7.95	3.76
Total	7,075	5,049	2,025	1.40	2.23	3.81

Annual trade values deflated by producer price index for intermediate materials, supplies and components.

Source: FATUS (Foreign Agricultural Trade Statistics of the United States), Calendar Year Supplement, annual issues.

jor exporter and a major importer of agricultural technology (table 13.6). The share of U.S. patents for agricultural technology awarded to foreigners increased steadily between 1969 and 1993 from 13 percent to 39 percent. The number of patents granted in Canada to U.S. inventors provides a picture of design transfers of technology abroad. American inventors' share of agricultural patents in Canada gradually declined between 1978 and 1993, falling from 58 percent to 43 percent. A more detailed analysis of the patents awarded in 13 agricultural technology fields by 11 countries between 1977 and 1984 found that the United States was a net exporter of technology in every cate-

Table 13.6. U.S. exports and imports of agricultural technology as indicated by patents, 1969–93

	1969	1978	1987	1993
Number of U.S. agricultural patents awarded (all)	1,763	1,735	1,483	2,730
Number of U.S. agricultural patents awarded to foreigners	231	405	481	1,065
Number of Canadian agricultural patents awarded (all)	n.a.	1,855	1,683	2,236
Number of Canadian agricultural patents awarded to U.S. inventors	n.a.	1,067	780	967
Share of U.S. patents awarded to foreign inventors (%)	13	23	32	39
Share of Canadian patents awarded to U.S. inventors (%)	n.a.	58	46	43

Source: Evenson and Johnson (1999).

gory (Evenson, Putnam, and Pray 1985). The primary sources of imported technology were Germany (agricultural chemistry and fertilizer, animal husbandry, and earthworking and planting equipment), France (animal husbandry), and Japan (earthworking and planting equipment).

These data suggest that while the United States is still the global leader in the development of new agricultural technology, its share of the market for agricultural inventions is declining, both in the United States and abroad. Trends in patent data along with the trends in agricultural input trade indicate that the United States relies increasingly on foreign sources of innovation to improve farm productivity.

Private Agricultural Research in Developing Countries

Agricultural research by private firms, especially multinational companies, provides an important indicator of capacity transfer for agricultural technology. Although evidence is fragmentary, it appears that over the past decade private agricultural research has expanded considerably in many developing countries (table 13.7). New surveys of private agricultural research in Asia (Pray and Fuglie 1998) are especially useful in tracking developments over time, as these can be compared to surveys conducted in the mid-1980s (Pray and Echeverria 1991).

The figures in table 13.7 show that private agricultural research grew rapidly in real terms in several developing countries over the last decade. Between 1985 and 1995, private agricultural research grew from essentially zero to $11–16 million/year in China and doubled in real terms in India and Pakistan between 1985 and 1995. Most countries in Southeast Asia also saw rapid growth in private research over this period. In Latin America, the trends were mixed, with significant growth in Colombia and Ecuador but a decline in Chile. Nevertheless, private research is still quite small in developing countries relative to the size of their agricultural sectors. Private research expenditures were less than 0.15 percent of agricultural GDP in Asian countries compared with more than 2 percent in the United States and the United Kingdom (UK). The surprisingly low levels of private research in the relatively advanced economies of Chile and Australia reflect international technology spillovers: These temperate countries can borrow agricultural technology directly from North America and Europe with little or no need to adapt it to local conditions.

Evidence available from seven developing countries in Asia shows that private research grew in real terms in all industries (table 13.8). Research expenditures were largest and increased most rapidly in the chemical industry, tripling in real terms between 1985 and 1995. Half of the growth in chemical research occurred in China where economic reforms opened up the agricultural chemical market to foreign competition. Growth in poultry and livestock research, 10.8 percent per year in real terms, was due in large part to an in-

Table 13.7. Private agricultural research expenditures in selected countries, 1985–95

	1985	1995	Average annual growth rate	Share of ag. GDP in 1995
	(millions of 1995 U.S. dollars)		(%)	(%)
Asia				
India	25.7	55.5	7.7	0.06
China	0.0	11–16.0	Infinite	0.01
Thailand	10.6	17.4	5.0	0.10
Malaysia	14.1	16.6	1.6	0.15
Philippines	6.2	10.5	5.3	0.06
Indonesia	2.8	6.1	7.8	0.02
Pakistan	2.4	5.7	8.1	0.04
Latin America				
Colombia	7.8	11.1[b]	5.9	0.16[b]
Ecuador	3.3[a]	4.6[b]	6.6	0.26[b]
Chile	0.4	0.3[c]	–4.1	0.01[c]
Industrialized countries				
United States	3,050.0	3,824.0	2.3	2.75
United Kingdom	689.0[a]	822.0[d]	3.5	3.75
Australia	n.a.	54.0[e]	n.a.	0.54[e]

[a] 1986
[b] 1991
[c] 1992
[d] 1993
[e] 1994

Sources: Asian countries for 1985 from Pray and Echeverria (1991); Asian countries for 1995 from Pray and Fuglie (1999); Ecuador from Falconi (1992), and Colombia from Falconi (1993), and for comparability only include the share of nongovernment research done by private companies, given in Echeverria (1998, p. 4); Chile from Venezian (1992); Australia from Industry Commission (1995); United Kingdom from Thirtle, Piesse, and Smith (1994); United States from Economic Research Service, unpublished data, agricultural GDP from World Bank (1997).

crease in consumer demand for meat products as per capita incomes grew in the region. Private research in seed, farm machinery and plantations/food processing increased by 6–7 percent per year.

Multinational companies, either through subsidiaries or through joint ventures with local firms, accounted for about half of all of the private-sector research shown in table 13.8, dominating agricultural chemical research in all countries and playing important roles in other input industries. Several multinational seed and agricultural chemical companies operate global networks of field experiment stations linked to their main research laboratories. These experiment stations enable firms to test and adapt new technologies in a variety of agricultural regions. Data on research by multinationals for a larger group of developing countries has been collected for only one period, 1985–90 (table 13.9). These companies conducted a total of about $52 million/year, with relatively little research in Africa, relatively more in Asia, and the most in Latin America. Agricultural research concentrated on seeds, pesticides, and

Table 13.8. Private agricultural research by industry in Asia, 1985–95[a]

	1985	1995	Average annual % growth
	(millions of 1995 U.S. dollars)		
Agricultural chemicals	14.5	47.0	11.8
Crop seed	8.2	16.4	6.9
Livestock and poultry	5.4	15.9	10.8
Farm machinery	3.9	7.5	6.5
Plantations and processing	21.2	40.8	6.5
All input industries	53.2	127.6	8.7

[a]Figures include China, India, Indonesia, Malaysia, Philippines, and Thailand.

Source: Pray and Fuglie (1999).

animal health products, and in some countries, plantations. In a few industries, multinationals provided technology to developing countries but did little research there because the technology is relatively easy to transfer. For example, the supply of new breeds, vaccines, and feed additives for the poultry industry was dominated by multinational firms, but these companies conduct only a small amount of applied research in developing countries. Since 1990, research by multinational companies in agricultural biotechnology has undoubtedly increased in a number of developing countries; this research is not reflected in table 13.9

Public Agricultural Technology Transfer

Public-sector flows of agricultural technology between industrial and developing countries through international agricultural research increased markedly after 1960 but has recently begun to ebb. The most notable development was the establishment of the CGIAR system of international agricultural research centers. This evolved into an effective system for breeding and transferring new crop varieties and germplasm to national agricultural research programs in developing countries. Advanced germplasm provided by IARCs has also benefited industrial countries.

The importance of IARCs for crop varietal development is indicated by the growing share of new varieties, released around the world, that use germplasm from these centers. These include varieties that were bred at IARCs and released through national programs (direct material transfers) and varieties developed by national programs using some parent material and genetic characteristics from an international center (an example of a design transfer).

In wheat, there was a sharp increase in the percentage of new varieties released in developing countries that were either developed by the International Maize and Wheat Improvement Center (CIMMYT) or had some parentage from CIMMYT. During 1966–70, some 30 percent of all new wheat varieties

Table 13.9. Agricultural research by multinational companies in developing countries

	Africa	Asia and Pacific	Latin America and Caribbean	*Total*
	(thousands of 1995 dollars/year, 1985–90 annual average)			
Plant breeding[a]				
Number of stations	5	14	36	55
Expenditures	1,411	3,951	10,160	15,522
Poultry breeding[b]				
Number of programs	0	2	0	2
Expenditures	0	1,288	0	1,288
Veterinary pharmaceuticals[c]				
Number of programs	0	1	0	1
Expenditures	0	192	0	192
Agricultural chemicals[d]				
Number of programs	4	10	12	26
Expenditures	1,315	3,288	3,946	8,549
Farm machinery[e]				
Number of programs	0	0	2	2
Expenditures	0	0	5,480	5,480
Livestock feed processing[f]				
Number of programs	0	2	1	3
Expenditures	0	104	52	156
Food processing				
Number of programs	2	9	6	17
Expenditures	400	1,800	1,200	3,400
Plantations[g]				
Number of programs	5	10	4	19
Expenditures	4,569	9,138	3,655	17,362
Total	7,695	19,761	24,493	51,949

[a] Based on the following companies: Pioneer, Cargill, DEKALB, Continental, Asgrow, Agrigenetics, and Ciba- Geigy/Funk (Novartis). Expenditures include operational and personnel research costs.
[b] Based on the following companies: Arbor Acres, Cobb, and Babcock.
[c] Based on the following companies: American Cyanamid and Monsanto.
[d] Based on the following companies: ICI, Bayer, Hoechst, Ciba-Geigy, Monsanto, DuPont, American Cyanamid, Chevron, and Shell. Includes only research on screening and field testing new chemical compounds.
[e] Based on the following companies: Deere, Massey-Ferguson, and International Harvester.
[f] Based on the following companies: Cargill and Ralston Purina.
[g] Based on the following companies: BAT Industries, Philip Morris, Del Monte, Unilever/Brook Bond, Dole, and United Fruit.

in developing countries were direct material transfers of CIMMYT-bred varieties, about 17 percent had some CIMMYT parentage (design transfers), and more than half had no CIMMYT ancestry. By 1989–90, almost 60 percent of the releases were direct transfers of CIMMYT-bred varieties, another 30 percent of releases had at least some parent material from CIMMYT, and varieties with no CIMMYT parentage accounted for less than 5 percent of all new releases (Byerlee and Traxler 1995).

Rice germplasm developed by the International Rice Research Institute (IRRI) shows a similar pattern, except that design transfers (parent breeding material) are now more important than material transfers (finished varieties)

(Evenson and Gollin 1997). In corn as well, the role of international programs is now to provide inbred lines and germplasm rather than finished varieties for local breeding programs.

The IARCs have also become important sources of crop improvement for the United States. A recent study of wheat and rice varietal development in the United States found that CIMMYT and IRRI germplasm made important contributions to breeding programs and helped significantly raise U.S. crop yields since the 1960s (Pardey et al. 1996). This finding supports the evidence provided by trends in input trade and patenting that foreign sources of agricultural innovation make a significant contribution to U.S. agricultural productivity growth. Nevertheless, funding for international agricultural research has fallen in real terms since 1992 (Alston, Pardey, and Roseboom 1998).

FUTURE TRENDS IN TECHNOLOGY TRANSFER

Most determinants of private investment in agricultural research and technology transfer point to a stepped-up pace for international flows. However, declining public support by some countries for national and international agricultural research, policy disincentives to private research, and regulatory barriers to testing and marketing new technology may constrain international technology transfer in some regions.

Trends in Demand

Demand for new agricultural technology should continue to increase. Growth in per capita income in several developing countries, especially is East and Southeast Asia, has increased demand for high-value food products, particularly meats, fruits, and vegetables. Demand for staples has risen more slowly as consumers devote a larger share of their food budgets to high-value products. These changes in food demand, together with rapid growth in the industrial and service sectors, have driven up prices for agricultural land and labor, increasing the derived demand for substitutes for these inputs, including high-yielding varieties of food crops and feed grains, chemical herbicides, and farm machinery.

Furthermore, many developing countries are moving away from policies that taxed agriculture and subsidized industry to policies that remove biases against agriculture and, in some cases, subsidize farmers. Structural adjustment policies that realign exchange rate often result in increased farm prices for agricultural commodities. Asia provides several good examples of such policies. In Thailand, export taxes on rice and rubber have been significantly reduced, making these crops more profitable for local farmers and encouraging the use of improved technology. South Korea has followed Japan's example by subsidizing farmers through agricultural price supports and trade protections. Even some of the poorest countries in Asia (such as the Philippines) are now subsidizing agriculture (David 1996). China is one of the few countries in Asia that still keeps food prices down, to the detriment of the agricultural sector. While the recent downturn in several Asian economies will curtail demand growth, in some cases this has been more than offset by exchange rate devaluations that significantly increased the competitiveness of the agricultural sector.

Increased Opportunities for Technological Transfer

Two important sources of technological opportunity have expanded significantly in recent years: discoveries in the basic sciences and public-sector investments in agricultural science. Basic scientific advances in molecular genetics and biotechnology have provided a growing number of technological opportunities in agriculture. The private sector has responded by increasing its investments in biotechnology research and technology transfer. Applications of biotechnology have already reached many farmers in both industrial and developing countries, and the amount of research currently being conducted suggests that myriad further advances in biotechnology are in the pipeline.

So far, the number of approved field trials of genetically engineered crops has been dominated by North America and Western Europe (see table 13.3). Although the lack of regulations in some developing countries means that not all field trials are recorded in the official statistics, it is clear that most research in agricultural biotechnology is being conducted in industrial countries. European and U.S. firms have benefited from this situation. As developing countries adopt regulatory protocols for agricultural biotechnology, many of these firms have invested in applied research in these countries so that technologies can be adapted to local environmental and market conditions.

Public-sector investment in agricultural science is one of the principal sources of new technological opportunities for applied R&D. Between 1971 and 1991 public research grew much more rapidly in developing countries than in industrial countries (see table 3.3 in chapter 3). Between 1971 and 1981, public expenditures for agricultural research in all regions except sub-Saharan Africa grew faster than in industrial countries. Between 1981 and 1991, public research funding increased in Asia, North Africa, and the Middle East. In Latin American countries, however, public research declined or stagnated.

Since 1991, most developing countries in Asia have continued to increase their investment in agricultural research. Public investment in agricultural science and technology in China is now increasing again after stagnating in the 1980s. However, because agricultural production has been growing rapidly, research expenditure as a percent of agricultural GDP is still declining (Pray, Rozelle, and Huang 1998). In India, real expenditure on agricultural research has grown since 1991, and research as a percentage of agricultural GDP has increased (Singh, Pal, and Jha 1996).

Funding for the international agricultural research centers, however, has declined in real terms since the early 1990s. For the four original centers (which include IRRI and CIMMYT, two of the most productive), the decline in real funding started in the late 1980s (Alston, Pardey, and Roseboom 1998).

Protections for Intellectual Property

Protection for intellectual property is increasing around the world, but very slowly. The Trade Related Intellectual Property standards (TRIPs) component of the GATT mandates the adoption and enforcement of IPRs, including those for new plant varieties. Many countries, including most Asian countries, already have patent laws with provisions for standard industrial patents. Many of these laws are being strengthened either by extending coverage or eliminating exclusions. Korea and Taiwan extended protection for chemicals in the late 1980s. China and Thailand both revised their patent laws in 1992 to include, for the first time, patents for agricultural chemicals and farm machinery. Some countries also have provisions for petty patents, which protect inventions that represent relatively small changes from previous technology. These patents provide less protection (usually 5–8 years) than standard industrial patents (20 years). Petty patents are available in China, Japan, the Philippines, and Thailand. Mexico passed a new patent law in 1991 that extended patent coverage to chemicals and pharmaceuticals. And many African countries kept the patent systems of their former colonial rulers and thus offer stronger IPR protection than Asian and Latin American countries.

Plant breeders' rights (PBRs) provide patent-like protection for new varieties of plants. In Asia, only Japan currently has an effective PBR law, although several other Asia countries are in the process of developing PBRs, including China, India, Indonesia, the Philippines, and Thailand. In Latin America, Argentina, Chile, and Colombia have already established PBRs, and most other countries in the region are committed to passing and enforcing them.

Structural Change in International Input Industries

During the 1990s research-intensive agricultural input industries have undergone considerable structural transformation and consolidation. Much of the consolidation centers around multinationals in the chemical, pharmaceutical, biotechnology and seed industries that seek to exploit complementarities among biological and chemical sciences.

This consolidation may take one of several forms. Firms may separate or sell off certain manufacturing and marketing components in order to concentrate on high-technology products in pharmaceuticals, veterinary medicine, pesticides, and biotechnology. The British chemical company ICI was one of the first to take this step. It formed a new company, Zeneca, to concentrate on drugs, pesticides, seeds, and agricultural biotechnology. The streamlined original company retained only its traditional bulk chemicals. In 1997, the U.S.-

based Monsanto corporation announced that it would sell off its bulk chemicals business and concentrate on high-technology life sciences.

A second trend is the merger of large chemical and pharmaceutical firms. The Swiss firms Ciba-Geigy and Sandoz recently merged to become Novartis. The German firms Hoechst and Schering formed a joint venture, AgrEvo, for their agricultural and environmental products. The U.S. chemical giant Merck and the French firm Rhone-Poulenc have formed a new joint venture for their animal products (veterinary medicines and poultry genetics) called Merial Animal Health.

Large chemical-based firms are also buying small biotechnology firms and seed companies or, in some cases, negotiating strategic alliances with them. AgrEvo purchased Plant Genetic Systems, the largest European plant biotechnology firm, in 1996. In 1997, DuPont purchased 20 percent of Pioneer Hi-Bred, the world's largest seed company. Monsanto has been the most active of all. It acquired shares in three important biotechnology firms, buying 100 percent of Agracetus and Calgene and gaining a minority holding in Ecogen. Monsanto also purchased the corn and soybean businesses of Asgrow, the second largest soybean seed producer; Delta and Pineland, the leading U.S. cotton seed firm; and Holdens Foundation Seeds, the largest foundation seed firm in the United States; and owns minority shares in DEKALB, the second largest U.S. hybrid corn firm, and Agroceres, the largest Brazilian seed company.

A small but growing trend in technology transfer is the acquisition of technology-based companies by newly emerging multinational corporations based in developing countries. A pioneer in this area is the Thai firm Charoen Pokphand, which has a long history of establishing joint ventures in Asian countries with foreign multinationals such as DEKALB (for seed) and Arbor Acres (for poultry genetics). Charoen Pokphand purchased the U.S. broiler breeding company Avian Farms to give it another source of poultry genetics. More recently, the Mexican firm Empresas La Moderna purchased the U.S. biotechnology company DNA Plant Technology and three vegetable seed companies: Seminis, Peto Seeds, and Asgrow Seeds. It then sold off Asgrow's corn and soybean business to Monsanto, keeping the vegetable business.

These structural changes in the international input industries are likely to facilitate international technology transfer. The mergers and acquisitions link small, science-based seed and biotechnology firms with multinationals that have large worldwide distribution networks. These large firms also have considerable capital to promote new products and make investments in countries that smaller firms might consider too risky.

Shifts in Policies and Regulations

Some of the most significant developments affecting international technology transfer have occurred through changes in national policies toward agriculture. Both India and China have shifted toward greater market orientation in their economic and agricultural policies. China, India, and Indonesia have reduced the role of state-owned enterprises in providing inputs to farmers (Pray and Fuglie 1999). The reform and liberalization process is likely to continue. Trade and investment barriers to technology transfer are coming under both internal pressure from farmers seeking low-priced, high-quality inputs and external pressure from the World Trade Organization. In India, private investment in seed research increased significantly once barriers to foreign investment in the Indian seed industry and restrictions on importing plant germplasm were removed. China has allowed some foreign participation in its agricultural chemical and farm machinery industries, but so far has kept foreign direct investment and imports entirely out of its seed industry (Pray, Rozelle and Huang 1996). In the 1980s, Thailand allowed foreign seed companies to purchase land for experiment stations and reduced tariffs on imported research equipment (Pray and Fuglie 1999). And farmers in the Philippines have been influential in reducing trade barriers to important agricultural inputs (David 1996).

Recently a number of countries have started using tax incentives and other kinds of subsidies to stimulate private research. Tax incentives typically allow firms to write off a certain percentage of their research costs (sometimes more than 100 percent). Since 1985, Australian firms have been able to claim 150 percent of the value of their research investments as costs (the government recently announced that it will reduce the credit to 125 percent). Singapore and Malaysia have 200 percent tax deductions for research. Taiwan has an incremental R&D tax deduction: Firms can claim 20 percent of their research expenditures in excess of the maximum amount spent on research during the previous five years (Hou and Gee 1993). In the 1980s, Thailand offered a special tax holiday to encourage seed companies to set up local research facilities and established a National Center for Genetic Engineering and Biotechnology (BIOTEC) to promote the commercialization of new biotechnology products (Pray and Fuglie 1999).

Significant policy barriers to international technology transfer remain in many countries, however. Regulations on seed and chemical registration can be cumbersome and costly. And the diffusion of transgenic varieties is limited by the lack of effective regulatory protocols and intellectual property rights. The Philippines, for example, prohibits field testing of transgenic varieties, and China does not recognize foreign patents on biotechnology products. And some countries, such as China, continue to restrict the participation of foreign multinationals in the local seed industry.

CONCLUSIONS AND IMPLICATIONS

The evidence presented in this chapter shows that rate of international technology transfer in agriculture is increasing, and that much of this is taking place through the private sector. Multinational agricultural input companies, mostly based in industrial countries, are conducting substantial amounts of agricultural research in developing countries in order to adapt new technology for these markets. Incentives for private research have been enhanced by the expanding global demand for improved agricultural technology, falling barriers to foreign participation in local agricultural markets, increased ability of companies to capture a share of the benefits of research, and new technological opportunities made possible by advances in biotechnology and other sciences.

While private agricultural research in developing countries is growing, investments are nevertheless small in relation to the size of the agricultural sector. In the least developed countries, private research may be narrowly focused on one or two commodities, such as plantation crops, or may be entirely nonexistent. Even in middle-income countries, private research tends to be concentrated on certain commodities and technologies where appropriability is strongest, such as hybrid seeds and poultry, agricultural chemicals, and farm machinery. In all but the most advanced industrial countries, most applied agricultural research still requires substantial support from the public sector. And the public sector continues to be the principal provider of basic agricultural research in all countries.

For governments, however, international spillovers can reduce incentives to fund agricultural research. Countries may choose to "free ride" on research conducted in other countries, relying on technology developed elsewhere to improve farm productivity. Further, countries may be less willing to fund agricultural research if a substantial share of the benefits appear to accrue to farmers in other countries. In addition, there is some fear that international technology spillovers may adversely affect a country's international competitiveness in trade.

Despite these concerns, there are persuasive reasons why the United States should continue to support international agricultural research and technology transfer, especially transfers to developing nations. First, there is little evidence that economic growth in other countries actually harms the U.S. economy. Foreign economic growth can reduce prices paid for U.S. imports and expand demand for U.S. exports, thus improving U.S. terms of trade and benefiting the U.S. economy (Krugman 1994). Second, there is substantial evidence that agricultural growth in developing nations actually increases U.S. agricultural exports to these countries (Vocke 1990; Pinstrup-Andersen, Lundberg, and Garrett 1995). This occurs because agricultural development contributes to broad-based income growth, which greatly expands the demand for high-value food products. Third, U.S. producers benefit directly from international technology transfer. Between 1970 and 1993, the wheat and rice germplasm obtained from IARCs for U.S. crop breeding programs added between $3.4 billion and $14.7 billion to the value of U.S. agricultural production, compared with only $134 million in U.S. support of these IARC programs over this period (Pardey et al. 1996). Moreover, as the indicators of international agricultural technology transfer showed, U.S. reliance on foreign sources of agricultural innovation is growing. And finally, promoting economic development in poor nations contributes to humanitarian objectives and to political and economic stability in these regions.

Several important policy implications emerge from these findings. Efforts by the United States to reduce global barriers to trade and foreign direct investment and to strengthen international protections for intellectual property enhance private incentives for agricultural research and international technology transfer. Support by the United States for greater harmonization of food and agricultural regulations and insistence on science-based regulatory standards for food products are also likely to enhance these incentives.

Another implication is that strengthening international collaboration in agricultural research is likely to be beneficial, if not necessary, to maintaining past rates of progress in agricultural science and technology. International collaboration among governments in support of high-spillover, basic scientific research is particularly critical.

Increasing the level of collaborative research between U.S. and foreign scientists allows the United States to benefit more from the many excellent agricultural research institutions around the world that are sources of new ideas, new research methods, and new technology. To obtain access at a time when more and more research in public institutions is becoming proprietary, U.S. scientists need to be involved in the research project. Another advantage of collaborative international research is that the costs of conducting certain types of research can be reduced. Each partner can emphasize those research activities in which they have a comparative advantage. For example, collaboration can be built on the U.S. strength in basic biological research and the cost advantage of developing countries in labor-intensive research activities, such as plant breeding or poultry breeding. Collaboration could also combine the elite germplasm developed in the industrial countries with the underutilized exotic germplasm in developing countries to produce technology that benefits both parties.[3] The U.S. support of IARCs has already proven to be extremely useful to U.S. farmers and illustrates the value of international collaboration in agricultural research.

A good example of international collaboration in basic agricultural research is the Rockefeller Foundation's Rice Biotechnology Network. Individual scientists at U.S. and Asian universities have actively exploited these opportunities. For example, Cornell University scientists have developed a relationship with the leading rice breeding program in China, the Hunan Hybrid Rice Research Center. Chinese and U.S. scientists have already reported a breakthrough in combining the use of molecular markers with conventional breeding to significantly increase the yield potential of rice and other major grains (Xiao et al. 1996). The U.S. government has also solicited international cooperation in carrying out the National Plant Genome Initiative, an undertaking to improve understanding of the structure and function of genes in plants of economic importance (National Science and Technology Council 1998).

Nevertheless, the free-rider problem presents a significant institutional challenge to public agricultural research. In the late 1980s and 1990s, for example, U.S. support for international agricultural research steadily declined (Pardey et al. 1996). New institutional approaches may be required to strengthen international collaboration in agricultural research. Some alternatives are

- *The CGIAR model* (in which national governments and other partners jointly fund research that contribute to common goals). The CGIAR model can be extended to the basic agricultural sciences (Ruttan 1986). To strengthen international commitments to this research, a funding formula can be developed according to the likely distribution of economic benefits. The actual research can be conducted at international research institutes such as the IARCs or awarded competitively to agricultural scientists at universities and national research institutes.
- *The research consortia model* (see chapter 9). Multinational companies can jointly fund generic research at universities or national research institutes. Because much basic research is not a profitable investment for these firms, sharing the costs of long-term research is an attractive option.
- *The technology licensing or joint-venture model* (see chapter 8). Public universities and national agricultural research institutes could charge higher royalties or require larger matching funds for joint research projects with multinational or foreign firms than with firms that only serve domestic markets. This would require public institutions to more carefully assess the potential of their technologies in foreign markets.

Considerably more work needs to be done to examine ways of financing and organizing public agricultural research in the presence of large international technology spillovers. It is important to know, for instance, how benefits from basic and applied research are distributed among producers and consumers in the countries where they are conducted and in other regions. Alston, Norton, and Pardey (1995) suggest some economic models for setting priorities for research on commodities and other broad categories of technology. The increased rate of international technology transfer also raises new questions on the allocative efficiency of public research. Byerlee and Traxler (1996) argue that the allocation of public research in developing countries could be improved by more explicit consideration of international technology spillovers and economies of scale in research activities. Answers to such questions will likely have important implications not only for agricultural research policy but for science policy in general.

NOTES

1. Thirtle and Echeverria (1994) note that a research organization can be classified as public or private according to (1) ownership and control, (2) sources of funds, (3) economic motives, and (4) whether technology produced enters the public domain or is proprietary. In this chapter, we focus on organizations that are private by all four criteria. However, we also include companies in which governments may be the major shareholder, such as commercially oriented, state-owned enterprises (SOEs).

2. The CGIAR is a network of 16 international agricultural research centers sponsored by the FAO, UNDP, and World Bank and supported by a consortium of industrial and developing countries, development banks, and philanthropic foundations. These centers include the International Rice Research Institute (IRRI) located in the Philippines and the International Maize and Wheat Improvement Center (CIMMYT) in Mexico.

3. However, provisions of the Convention on Biodiversity, which recognizes national sovereignty of genetic resources, complicate international exchanges of plant germplasm. New institutional arrangements may be required that explicitly take into account benefit sharing.

REFERENCES

Alston, J., G. Norton, and P. Pardey. 1995. *Science under Scarcity: Principles and Practices for Agricultural Research Evaluation and Priority Setting.* Ithaca, NY: Cornell University Press.

Alston, J., P. Pardey, and J. Roseboom. 1998. "Financing Agricultural Research: International Investment Patterns and Policy Perspective." *World Development* 26(6): 1057–72.

Byerlee, D., and G. Traxler. 1995. "National and International Wheat Improvement Research in the Post-Green Revolution Period: Evolution and Impacts." *American Journal of Agricultural Economics* 77 (May): 268–78.

______. 1996. "The Role of Technology Spillovers and Economies of Size in the Efficient Design of Agricultural Research Systems." In *Global Agricultural Science Policy for the Twenty-First Century.* Melbourne, Australia: The Conference Secretariat of Global Agricultural Science Policy for the Twenty-first Century, pp. 391–432.

Caswell, M., K. Fuglie, and C. Klotz. 1994. *Agricultural Biotechnology: An Economic Perspective.* Agricultural Economics Report 687. Washington, DC: Economic Research Service, U.S. Department of Agriculture.

David, C.C. 1996. "Country Paper—Philippines." Paper presented at the Asia-Pacific Economic Cooperation Conference on Food and Agricultural Policy Challenges in the Asia-Pacific, Manila, Philippines, October 1–3.

Echeverria, E. 1998. "Agricultural Research Policy Issues in Latin America: An Overview," *World Development* 26(6): 1103–12.

Evenson, R.E., and D. Gollin. 1997. "Genetic Resources, International Organizations, and Improvement in Rice Varieties." *Economic Development and Cultural Change* 45(3): 471–500.

Evenson, R.E., and D.K.N. Johnson. 1999. "R&D Spillovers to Agriculture: Measurement and Application." *Contemporary Economic Policy* 17(4):432–456.

Evenson, R.E., and L.E. Westphal. 1995. "Technological Change and Technology Strategy." In J. Behrman and T.N. Srinivasan eds. *Handbook of Development Economics, Volume III.* Amsterdam: Elsevier Science, B.V.

Evenson, R.E., J. Putnam, and C.E. Pray. 1985. *The Potential for Transfer of U.S. Agricultural Technology.* Report prepared for the Office of Technology Assessment, U.S. Congress.

Falconi, C. 1992. "Public and Private Sector Interactions in Agricultural Research in Less-Developed Countries: The Case of Ecuador," Discussion Paper 92-13. The Hague, Netherlands: International Service for National Agricultural Research.

______. 1993. "Public and Private Sector Interactions in Agricultural Research in Less-Developed Countries: The Case of Colombia." Discussion Paper 93-14. The Hague, Netherlands: International Service for National Agricultural Research.

Foreign Agricultural Trade of the United States (FATUS) Annual Calendar Year Supplement. Various years. Washington, DC: U.S. Department of Agriculture.

Fuglie, K., N. Ballenger, K. Day, C. Klotz, M. Ollinger, J. Reilly, U. Vasavada, and J. Yee. 1996. *Agricultural Research and Development: Public and Private Investments under Alternative Markets and Institutions.* Agricultural Economics Report 735. Washington, DC: Economic Research Service, U.S. Department of Agriculture.

Hayami, Y., and V.W. Ruttan. 1985. *Agricultural Development: An International Perspective.* Baltimore: Johns Hopkins University Press.

Hou, C.-M., and S. Gee. 1993. "National Systems Supporting Technical Advance in Industry: The Case of Taiwan." In R.R. Nelson, ed. *National Innovation Systems: A Comparative Analysis.* New York: Oxford University Press.

Industry Commission. 1995. *Research and Development.* Canberra: Australian Government Publishing Service.

James, C. 1997. "Global Status of Transgenic Crops in 1997." *ISAAA Briefs.* No. 5. Ithaca, NY: International Service for the Acquisition of Agri-biotech Applications (ISAAA).

James, C., and A.F. Krattiger. 1996. "Global Review of the Field Testing and Commercialization of Transgenic Plants, 1986–1995: The First Decade of Crop Biotechnology." *ISAAA Briefs.* No.1. Ithaca, NY: International Service for the Acquisition of Agri-biotech Applications (ISAAA).

Krugman, P. 1994. "Does Third World Growth Hurt First World Prosperity?" *Harvard Business Review* 72 (July/August): 113–21.

Mansfield, E. 1994. "Intellectual Property Protection, Foreign Direct Investment, and Technology Transfer." Discussion Paper 19, International Finance Corporation. Washington, DC: World Bank.

National Science and Technology Council. 1998. *National Plant Genome Initiative.* Washington, DC: Office of Science and Technology Policy, Executive Office of the President.

Pakes, A., and M. Schankerman. 1984. "The Rate of Obsolescence of Patents, Research Gestation Lags, and the Private Rate of Return to Research Resources." In Zvi Griliches, ed. *R&D, Patents, and Productivity.* Chicago: University of Chicago Press, pp. 209–32.

Pardey, P., J. Alston, J. Christian, and Shenggen Fan. 1996. *Hidden Harvest: U.S. Benefits from International Research Aid.* Food Policy Report. Washington, DC: International Food Policy Research Institute.

Pinstrup-Andersen, P., M. Lundberg, and J. L. Garrett. 1995. *Foreign Assistance to Agriculture: A Win-Win Proposition.* Food Policy Report. Washington, DC: International Food Policy Research Institute.

Pray, C.E. 1987. "Private Sector Agricultural Research in Asia." In V.W. Ruttan and C.E. Pray, eds. *Policy for Agricultural Research.* Boulder, CO: Westview Press.

Pray, C.E., and R. Echeverria. 1991. "Private-Sector Agricultural Research in Less-Developed Countries." In P. Pardey, J. Roseboom, and J.R. Anderson, eds. *Agricultural Research Policy: International Quantitative Perspectives.* Cambridge, UK: Cambridge University Press, pp. 343–64.

Pray, C.E., and K. Fuglie. 1999. "The Private Sector and International Agricultural Technology Transfer to Developing Countries." Staff Paper. Washington, DC: Economic Research Service, U.S. Department of Agriculture.

Pray, C.E., S. Rozelle, and J. Huang. 1996. "Incomplete Commercialization of the Chinese Seed Industry: The Need for Further Reform." Paper presented at the Conference on Global Agricultural Science Policy for the Twenty-first Century, Melbourne, Australia, August 29.

______. 1998. "Can China Feed Itself: The Role of Agricultural Research," Unpublished. Department of Agriculture Food and Resource Economics, Rutgers University.

Ruttan, V.W. 1982. *Agricultural Research Policy.* Minneapolis: University of Minnesota Press.

______. 1986. "Toward a Global Agricultural Research System: A Personal View." *Research Policy* 14: 307–27.

Scherer, F. M. 1980. *Industrial Market Structure and Economic Performance.* Boston: Houghton Mifflin Company.

Schumpeter, J.A. 1950. *Capitalism, Socialism, and Democracy.* 3rd edition. New York: Harper and Row.

Singh, A., S. Pal, and D. Jha. 1996. "Public Investment in Agricultural Research and Education in India." New Delhi: National Center for Agricultural Policy. Mimeo.

Thirtle, C., and R. Echeverria. 1994. "Privatization and the Roles of Public and Private Institutions in Agricultural Research in Sub-Saharan Africa," *Food Policy* 19(1): 31–44.

Thirtle, C., J. Piesse, and V. Smith. 1994. "Agriculture and Food R&D and Technology in the United Kingdom." Discussion Paper, University of Reading, United Kingdom.

Venezian, L.E. 1992. "Agricultural Research in a Growing Economy: The Case of Chile 1970–90." In *Getting Beyond the National Institute Model for Agricultural Research in Latin America. Latin America and the Caribbean.* Washington, DC: World Bank.

Vocke, G., ed. 1990. *Trade and Development: Impact of Foreign Aid on U.S. Agriculture.* AGES 9004. Washington, DC: Economic Research Service, U.S. Department of Agriculture.

Xiao, J., S. Grandillo, S.N. Ahn, S.R. McCouch, S.D. Tanksley, J. Li, and L. Yuan. 1996. "Genes From Wild Rice Improve Yield." *Nature* 384 (21 November): 223–24.

14

ASSESSING U.S. BENEFITS OF TRAINING FOREIGN AGRICULTURAL SCIENTISTS

NICOLE BALLENGER AND
CASSANDRA KLOTZ-INGRAM[1]

STUDENTS FROM other nations choose to study at the advanced level in the United States because of the excellence of U.S. graduate programs, but the benefits to the country of training these students are an ongoing source of debate. Relatively little literature tackles head-on the costs and benefits of training foreign graduate students, and no rigorous cost-benefit analyses have been done. Some state legislatures have raised concerns about the numbers of foreign university students in state schools, and some universities, such as the University of Tennessee, have set targets for foreign student enrollment (University of Tennessee 1982). These concerns do not disguise the fact that many U.S. graduate programs have come to depend heavily on foreign student demand, however. By 1995, one-third of all U.S. doctorate degrees were awarded to non-U.S. citizens (Food and Agricultural Education Information System, FAEIS).

This paper looks at ways to assess the benefits to the United States of training agricultural scientists from other countries. In many ways, the issues related to training foreign agricultural scientists are similar to those surrounding the training of foreign citizens in any other scientific or technical field. But agriculture has some unique characteristics that warrant an independent analysis—for example, its relative significance to the economies of developing countries. Given this fact, our objectives here are

- To provide a conceptual framework for assessing the short- and long-term benefits to the United States of training foreign agricultural scientists;
- To derive from the economics literature conceptual approaches and empirical estimates of benefits and costs; and
- To present data on foreign graduate students in agricultural science in the United States and offer some comparisons with foreign graduate students in other scientific fields.

The United States benefits not only from having foreign graduate students in U.S. agricultural programs, but also from the postgraduate work foreign graduates undertake both in this country and abroad.[2] The first section of the paper presents data on the numbers of these students and the fields of study they choose. The next three sections deal with the benefits that accrue to the United State while the students are in school, the benefits from postgraduate employment in the United States, and the benefits from postgraduate employment outside the United States. For the purposes of this paper, those students who remain in the United States are called *stayers* and those who return to their native countries *returnees*.

Separating the benefits in this way illuminates the different aspects of the debate. State taxpayers and university administrators tend to focus on the costs and benefits of having foreign students in graduate programs. Employers in the United States, including universities, are interested in foreign students as potential additions to the pool of U.S. job candidates. And foreign governments and international organizations are concerned with the potential loss of human capital if foreign graduate students decide to stay in the United States. These divisions are also useful in discussing public policy options. For example, the marginal benefit to the United States of a stayer can be compared to the marginal benefit to the United States of a returnee (see appendix at the end of this chapter).

FOREIGN GRADUATE STUDENTS IN THE UNITED STATES

Today, the percent of foreign students in U.S. graduate programs is higher than the percent in U.S. undergraduate programs (Davis 1996). Chishti (1984a) predicted this shift, arguing that demand for U.S. education would move to the graduate level as developing countries built satisfactory undergraduate facilities. Agarwal and Winkler (1985) analyzed demand for higher education (graduate and undergraduate) by students from the 15 low- and middle-income Eastern Hemisphere nations that had the highest rates of enrollment growth at U.S. colleges and universities.[3] They found that the number of students from these countries studying in the United States had increased and would continue to increase, largely because of improved secondary and undergraduate educational opportunities at home. Like Chishti, they found that the highest rates of growth would be at the graduate level. Pardey, Roseboom, and Beintema (1997) support this trend, noting that while almost all African countries now provide some undergraduate training, few provide adequate graduate-level training.[4]

We reviewed data on numbers of degree recipients by degree level, field of study, and student demographic characteristics, using information from the Survey of Earned Graduates (NSF and NRC) and USDA's Food and Agricultural Education Information System (FAEIS). (FAEIS is a USDA-supported database, housed at Texas A&M University, that integrates statistical information from a variety of sources such as colleges of agriculture and schools of forestry and veterinary medicine, professional associations and societies, the U.S. Department of Education, the NRC, and NSF.) According to these sources, the number of doctoral degrees awarded by U.S. universities rose from about 32,000 in 1986 to about 42,000 in 1995, a 30-percent increase. The number of Ph.D. degrees awarded annually in science and engineering (S&E) fields increased 36 percent, and the number in food and agricultural sciences (F&A) by 28 percent (table 14.1).

Table 14.1. Number of doctorate recipients in food and agricultural sciences, science and engineering, and all fields in the United States, 1986–95

	1986	1987	1988	1989	1990	1991	1992	1993	1994	1995
Total doctorates in all fields	31,902	32,370	33,501	34,326	36,067	37,522	38,856	39,771	41,017	41,610
Total foreign doctorates	6,709	7,190	7,817	8,274	9,791	11,169	11,932	12,189	13,154	13,113
Total doctorates in the food and agricultural sciences	2,606	2,673	2,767	2,928	3,100	3,121	3,174	3,230	3,435	3,334
Total foreign doctorates	646	699	812	818	1,025	1,158	1,242	1,269	1,450	1,412
Total doctorates in science and engineering	19,437	19,894	20,933	21,731	22,867	24,019	24,673	25,441	26,202	26,515
Total foreign doctorates	5,154	5,557	6,066	6,515	7,768	8,927	9,474	9,754	10,543	10,493

Source: FAEIS; NSF and NRC, Survey of Earned Graduates.

Over the same period, the increase in the number of foreign citizens who earned doctorates in the United States each year outpaced the growth in the total number of Ph.D. recipients. The number of foreign students earning doctorates annually in all fields increased by 95 percent, and the number of foreign students earning doctorates in S&E by 104 percent (table 14.1). The agricultural sciences saw proportionally larger increases in foreign students compared with other scientific fields: Between 1986 and 1995 the number of foreign students awarded Ph.D. degrees annually in the F&A fields increased 119 percent. In total, 10,531 foreign citizens received U.S. Ph.D. degrees in the food and agricultural sciences between 1986 and 1995.

Consequently, foreign citizens receive a growing share of all U.S. doctoral degrees. In 1986 foreign citizens composed slightly more than 20 percent of all Ph.D. recipients, 27 percent of Ph.D. recipients in S&E, and 25 percent of Ph.D. recipients in F&A. In 1995 one-third of all Ph.D. degrees went to foreign citizens. Foreign citizens earned 40 percent of the doctoral degrees granted in S&E and over 40 percent of those granted in F&A fields (calculated from table 14.1).

In terms of actual numbers, the F&A fields produced 1,412 of the 13,113 doctorates received by foreign citizens in 1995. Within this category, most of the doctoral degrees were awarded in the animal sciences (27.1 percent), the plant, horticultural, and soil sciences (21.8 percent), and related biological and physical sciences (21.3 percent). Food science and human nutrition accounted for 10.5 percent of doctoral degrees. The remainder of the doctoral degrees in the agricultural sciences were in agricultural business, education, management, and social sciences (8.1 percent), natural resources and forest sciences (7.8 percent), and agricultural engineering and mechanization (3.4 percent). (For an earlier summary of data on foreign students in the agricultural sciences, see Swanson 1986).

MEASURING BENEFITS WHILE STUDENTS ARE IN GRADUATE SCHOOL

How do we assess the most immediate costs and benefits of awarding over 40 percent of Ph.D. degrees in agricultural science to foreign citizens? Chishti (1984b) makes the most complete calculations we could find of the measurable costs and benefits to the United States of training foreign students.[5] In this section we extrapolate from Chishti's analysis to estimate the annual costs and benefits to the United States of training the 1,412 foreign citizens who received F&A doctorates in 1995. Table 14.2 shows Chishti's results and our extrapolations.

Chishti calculates the costs to the United States of educating foreign students based on the following cost components:

- Education and general expenditures (E&G), excluding research and public services;
- The user cost of capital, including the interest foregone and depreciation incurred on university buildings, land, and equipment;
- Indirect costs due to tax exemptions;
- Maintenance costs for those foreign students receiving allowances from U.S. sources;
- Travel costs financed from U.S. sources; and
- Other administrative costs related to the special needs of foreign students.

We used Chishti's E&G costs for training doctoral-level students at public universities because, according to the NRC (1995), almost all agricultural scientists are trained at public land-grant institutions. We then adjusted these E&G costs and all other cost components by the GDP deflator (1995/1980 =

Table 14.2. U.S. costs and benefits of training foreign students: extrapolations to foreign doctorate recipients in agricultural sciences

	All foreign students enrolled in 1980/81[a]	Only foreign students receiving doctorate degrees in agricultural sciences in 1995
	(millions in 1980 dollars)	(millions in 1995 dollars)
Total costs	1,973.4	27.23[b]
E&G expenditures[c]	1,509.2	23.54
Capital usage costs	155.0	1.24
Tax exemption costs	177.0	1.42
Living expenses	104.7	.83
Travel costs	25.3	.18
Other administrative costs	2.2	.018
Total benefits	$3,173.7	$12.6 ($18.1)
Tuition revenues	739.2	5.0 (10.5)[d]
Additional aggregate demand	630.5	7.6[e]
Stayers' value of embodied capital	1,804.0	NA
Benefts minus costs	$1,200.3	NA
Excluding value of stayers	–603.7	–14.6 (–8.6)
Ratio of cost over benefits	.6	NA (most likely less than one)
Excluding value of stayers	1.4	2.2 (1.5)

NA = not available.

[a]From Chishti (1984b) based on 311,880 foreign students.

[b]Costs estimates are extrapolated from Chishti based on 1,412 foreign students receiving agricultural science doctorates in 1995, a GPD deflator of 1.77, and using Chishti's average E&G costs for doctoral students.

[c]Education and general expenditures.

[d]Tuition revenues are estimated two ways: (1) using Chishti's estimates and inflating by the GDP deflator and (2) using estimated actual tuition in 1994/95 for nonresident students at NASULGC universities. The second estimate is shown in parentheses.

[e]Additional aggregate demand is estimated using estimated cost-of-living expenses for 1994/95 and multiplying those by the estimated number of students whose primary source of support was either personal or foreign funds.

1.77) and multiplied the inflated per-student costs by 1,412 (the number of foreign students who received F&A doctorates in 1995). Using this method, we estimate that training this group of F&A Ph.D. recipients cost the United States approximately $27 million a year (in 1995 dollars).

This crude cost estimate rests on important assumptions. For example, in Chishti's analysis the living expenses of 7.44 percent of foreign students were subsidized by the U.S. government, private sponsors, or university fellowships and scholarships (as opposed to research or teaching assistantships, which require work in return). We carry this figure over as an assumption into our calculation of "maintenance costs." The Institute of International Education (Davis 1996) reported that in 1995–96, 3.3 percent of foreign graduate students received their primary support from the U.S. government or private

U.S. sponsors, and 33.9 percent were supported primarily by U.S. universities. Some 49 percent of foreign graduate students were supported primarily by personal funds and 11 percent by foreign sources.[6]

Turning to the benefit calculations, Chishti derives the economic benefits of having foreign students in the United States from tuition payments from non-U.S. sources and additions to aggregate demand by foreign students supported primarily by personal or foreign sources. Using his figures for average out-of-state tuition and fees for four-year public institutions, adjusting by the GDP deflator, and multiplying by the number of foreign Ph.D. recipients in 1995 supported by non-U.S. sources, we estimate tuition benefits of approximately $5 million (table 14.2). However, actual nonresident tuition at land-grant universities in 1995—around $8,000—appears to have risen considerably faster than the general inflation rate. Using the actual tuition figure, we estimate annual tuition benefits of approximately $10.5 million.[7]

The Institute of International Education (Davis 1996) calculates a student's cost-of-living expenses at around $9,000 for the nine-month academic year. If this figure holds true, foreign citizens who received doctoral degrees in agricultural science in 1995 spent about $12.7 million on goods and services in the United States during the academic year. Using Chishti's approach, we adjusted this figure by the percentage (59.8) of foreign students supported primarily by personal or foreign sources (Davis 1996). Adding tuition and aggregate demand benefits together ($5 million plus $7.6 million) and then comparing them to total costs ($27 million) yields a ratio of costs to benefits of slightly more than 2. Using the actual tuition rates rather than the estimated tuition rates yields a cost-benefit ratio of about 1.5. Our cost-benefit ratio is somewhat higher than Chishti's, however, because estimated E&G costs for graduate students are much higher than they are for undergraduates.

Chishti notes that his calculations leave out two very important benefits of training graduate students: the likelihood that the viability of many graduate programs may actually depend on foreign students, and the fact that graduate students perform a significant amount of unpaid research (that is, the value of their research contribution is greater than their salaries as research assistants and may in fact approximate the student's salary plus the tuition subsidy). The same argument might be made for their contributions as teaching assistants. These omissions are crucial to our crudely adjusted estimates since our focus is graduate students. In addition, Chishti's approach fails to capture intangible benefits, such as cultural diversity, which can result in important international contacts at a later date.

ASSESSING THE BENEFITS OF STAYERS

A sizable portion of foreign citizens who receive graduate degrees in the United States elect to stay. Concern about this "brain drain" from developing countries has generated a number of studies (Bratsberg 1995; Borjas 1987; Finn, Pennington, and Anderson 1995). Using data from the Immigration and Naturalization Service, Bratsberg (1995) expanded Borjas's (1987) work on the reasons foreign students elected to stay in or leave the United States. Bratsberg found that the rates at which students returned varied significantly across countries and that differences in economic and political conditions accounted for the variations. Rates at which graduates returned home were positively correlated with both per capita GDP and rates of return to education. Financial aid to students from home countries had a positive but statistically insignificant impact on return rates. Ties with immigrant communities in the United States and poor civil rights were found to be deterrents to returning home.

Finn, Pennington, and Anderson (1995) narrowed the study of stayers and returnees to foreign nationals receiving doctorates in S&E from U.S. universities in 1984. Some 42 percent of these 3,653 Ph.D. recipients were still working in the United States in 1992 as temporary residents. The stay rate was highest for aeronautical engineers (64 percent) and lowest for agricultural and biological engineers (23 percent). Graduates from China, India, and Iran had stay rates well above average; graduates from Brazil, Japan, and Korea had stay rates well below average. The study went on to look at differences between those who worked for a few years in the United States and then left and those who remained longer. Those who left sooner earned lower salaries than those who remained longer and appeared to be working in fields that typically command low salaries—for example, earth or social sciences.

The Survey of Earned Doctorates (NSF and NRC) details the postgraduate plans of foreign Ph.D. recipients in S&E, including agricultural scientists.[8] Those with temporary visas returned home at much higher rates than those with permanent visas, and agricultural scientists had the highest return rates. In 1995 about 42 percent of S&E graduates with temporary visas reported definite plans to go abroad. In contrast, 65 percent of foreign Ph.D. recipients in the agricultural sciences who held temporary visas had definite plans to go abroad. The number of agricultural scientists with permanent visas who planned to return home was also significant—more than 25 percent (table 14.3).[9]

As these percentages show, stay rates for the agricultural sciences have been significantly lower than for other scientific fields, and return rates have been significantly higher. The reasons for this distinction have not been studied. One factor may be the relatively large amounts of foreign support graduate students in the agricultural sciences receive. In 1995 graduate students in agricultural sciences received 4.4 percent of their financial support from foreign sources, as opposed to 0.1–2.3 percent for students in other fields (table 14.4). Another deciding factor may be employment opportunities in the international agricultural research system and national agricultural research centers.

Table 14.3. Postgraduation plans of foreign doctorates in science and engineering and agricultural sciences, 1995

	Total number	Total with definite plans	Post-doctorate study	Academic employment	Industry employment	Other	Abroad	Location unknown
	(number)		(percent)					
Temporary visas								
Science and engineering	6,992	3,698	32.4	8.3	14.3	2.4	41.8	0.8
Agricultural sciences	429	230	23.5	3.9	4.8	1.3	65.2	1.3
Permanent visas								
Science and engineering	3,501	1,705	51.8	10.8	24.6	5.2	6.8	0.9
Agricultural sciences	106	47	53.2	6.4	8.5	6.4	25.5	0

Source: NSF, *Survey of Earned Doctorates*; NRC.

How can the benefits of stayers be assessed? Chishti's analysis includes the benefits to the United States of the value of capital embodied in this group. Embodied capital consists of child-rearing costs (which are not measured in his analysis), educational costs, and the opportunity cost of time while in school—in other words, the savings to the United States of not having had to bear these costs for stayers from birth. The value of capital embodied in stayers in 1980–81 was estimated at more than twice the tuition revenue from all foreign students. This calculation was based on a stay rate of only 5 percent of all foreign students—those who received both undergraduate and graduate degrees. When the value of capital embodied in stayers is factored into Chishti's calculation, he finds the benefits to the United States of training foreign students significantly outweigh the costs (table 14.2).[10]

A benefit not considered by Chishti is the contribution by stayers to the nation's knowledge base in scientific and technical fields. In further analyses of training foreign students in F&A fields, this potential benefit should be taken into account. A considerable body of literature finds substantial returns historically to public investment in agricultural research in the United States (Fuglie et al. 1996). But data also suggest that agricultural science fields are having difficulty competing with other S&E fields for the next generation of scientists (NRC 1995). If stayers are enhancing the quantity and quality of human capital in U.S. agricultural science, they may have an important role in sustaining rates of return on the nation's agricultural research investments.

Table 14.4. Full-time graduate students in science, engineering, agriculture, and health fields at land-grant institutions, by field of study and major source of financial support, 1995

	Total, all sciences	Physical sciences	Earth, atmospheric, and ocean sciences	Mathematical sciences	Computer sciences	Agricultural sciences	Biological sciences	Psychology	Social sciences
Total students	78,907	11,666	4,979	5,535	5,006	7,818	19,037	6,481	18,385
Percent of total with support from									
Federal gov., total	23.6	38.5	36.5	8.1	24.3	22.1	33.8	13.5	8.6
USDA	3.1	0.2	1.1	0.2	0.2	13.1	4.8	0.2	2.1
All other federal	20.5	38.3	35.4	7.9	24.1	8.9	29.1	13.4	6.5
Institutional support	48.0	50.5	37.1	73.6	39.9	41.5	43.4	53.9	49.3
All other U.S.	6.5	5.7	6.2	2.4	5.7	11.4	9.2	4.5	4.5
Foreign	1.8	0.8	2.3	1.5	1.0	4.4	1.7	0.1	2.1
Self-support	20.1	4.5	17.9	14.4	29.2	20.6	11.9	28.0	35.6

Source: National Science Foundation/SRS, Survey of Earned Graduates and Postdoctorates in Science and Engineering.

ASSESSING AND MEASURING BENEFITS OF EMPLOYMENT ABROAD

As we have seen, foreign students trained in agricultural science are more likely to leave the United States than foreign students in other fields. Examining the type of work returning agricultural scientists undertake and the impact of their efforts is important in assessing the benefits that accrue to the United States from training these students. Few studies have been done on this subject, either in the United States (by the educating institutions) or at the international level, and data collection to underpin such analyses is inadequate. This section assesses the available literature and suggests approaches to assessing the benefits to the United States of returnees.

Following the Students

Sterns, Oehmke, and Schwartz (1993) conducted a survey of West African researchers and administrators who graduated from the Michigan State University College of Agricultural and Natural Resources after 1965.[11] The researchers collected data on monetary and nonmonetary gains from education, with nonmonetary gains including health care and nutrition, access to education, quality of housing, and leisure time. The nominal increase in respondents' salaries from pretraining levels averaged 341 percent. This figure actually represents a decline in real terms of 21 percent, but this decline is less severe than that suffered by their countries' GDPs, which fell an average of 43 percent. Respondents also reported additional opportunities in the form of consulting, personal enterprises, increased job responsibilities, and international networking opportunities such as invitations to present papers at conferences. A majority of respondents also reported improvements in their ability to provide education, health care, and other amenities for their families.

The researchers used regression analysis to calculate a private benefits stream and compared the private benefits to personal and public costs (including tuition, fees, living costs, and state appropriations). The result was an estimated private rate of return of 22 percent from increased income and 31 percent from enhanced income and nonmonetary benefits combined. Although these private returns cannot be counted as direct benefits to the United States, the success of these graduates may translate into indirect benefits in forms we discuss in the next section.

Henderson and Shibano (1990) took a different approach, tracking foreign students in order to assess their perceptions of the relevance of U.S. training to their job responsibilities. The researchers followed foreign students from 14 developing countries who graduated from the Department of Agricultural Education at Ohio State University between 1980 and 1988. (In 1988, 45 percent of graduate students in the agricultural sciences at Ohio State were foreign citizens, compared with 22 percent for the university as a whole.) Out of

29 responding graduates, 17 were in university positions and 6 were serving as extension officers in agriculture ministries. The majority were working in positions for which they had been educated, and most found that the bulk of their graduate courses had been "very useful" or "useful" in relation to their job responsibilities. The courses cited as being most useful were those in research methods, instrumentation and data collection, analysis techniques, and program planning and evaluation. Cashman and Plihal (1987) obtained similar findings from a survey of international agricultural students that had graduated from the University of Minnesota.

An alternative to following the careers of returnees involves assessing the characteristics of the global (or a regional) pool of agricultural scientists, including their training backgrounds. Again, however, systematic data collection is lacking. Judd, Boyce, and Evenson (1987) provide global estimates of agricultural research and extension manpower for 1959, 1970, and 1980, but these data have not been updated. They also do not distinguish among researchers by degree level or other qualifications. Pardey, Roseboom, and Anderson (1991) reported on the human capital component of research conducted within national agricultural research systems in the first half of the 1980s. They found that expatriate researchers represented a significant portion of the human capital of these systems in developing countries, particularly in sub-Saharan Africa, the Pacific, the Caribbean, and West Asia—12 percent on average, and about 29 percent for sub-Saharan Africa. They also found that fewer than one-half of these nationals had graduate degrees (doctorates or M.Sc. degrees); the shares ranged from 17 percent in West Asia to about 55 percent in South America and southeast Asia. The study did not report on the source of graduate degrees held by the national researchers.

In a recent update of investments in Africa's public agricultural research system, Pardey, Roseboom, and Beintema (1997) describe characteristics of African agricultural research personnel. They find that the numbers of employed researchers in African agricultural research systems (measured in full-time equivalents) increased substantially between 1961 and 1991, growing at an annual rate of 5.1 percent.[12] They also found that the composition of the scientific workforce has changed markedly. The share of expatriates among the research staff decreased from 90 percent in the early 1960s to 29 percent in the 1980s. Today, it stands at 11 percent, suggesting domestic and foreign training opportunities have led to a stronger role for African citizens in the African agricultural research system. The level of training of researchers has also improved. The percentage with graduate degrees has risen from 45 percent to nearly 65 percent in the last decade. Twenty-two percent of the researchers hold Ph.D. degrees, although this group is concentrated at a few institutions. Like its predecessor, this follow-up study does not detail the institutions where the African scientists received their graduate degrees.

Information on the degrees researchers have and the granting institutions should be relatively easy to compile, at least for the international agricultural research centers (IARCs). A recent survey of human resources at the IARCs in 1994 reported 887 staff members with Ph.D. (or equivalent) degrees, 213 with masters-level degrees, and 124 with "other" degrees (Merrill-Sands 1995). (The analysis of the data focused on the role of women and included, among other things, breakdowns by nationality and scientific discipline.) The Internet Home Page of the International Council for Research on Agro-Forestry (ICRAF) contains detailed biographical information on its staff. At ICRAF, 37 percent of the researchers with Ph.D. degrees received their doctorates in the United States. Of these researchers, slightly more than 50 percent are not U.S. citizens.

Assessing the Benefits to the United States

In general, only indirect benefits accrue to the United States from returnees. These indirect benefits, a number of which are hypothesized in table 14.5, are difficult to calculate, but it is important to begin to understand them. Institutions in the United States have trained many returnees over the years, so that the cumulative effects of this group may be increasingly tangible.

One direct impact, however, is the importance of returnees as overseas contacts for U.S.-based scientists. Surveys of both institutions that sponsor international research collaborations, such as the U.S. Agency for International Development (USAID) and the U.S. Department of Agriculture (USDA), and U.S. agricultural scientists could offer insights into the professional relationships between returnees and U.S.-based agricultural scientists. Returnees may also be an important source of the next generation of Ph.D. candidates, since professors and senior scientists often recommend their own alma maters to potential students.

A second important potential effect of U.S. training is the increased output and productivity of research organizations that employ returnees. It seems likely to us that research productivity is influenced greatly by the quality of the most important scientific input, the scientist.[13] The United States supports agricultural research and development in developing countries under the assistance-to-agriculture component of Official Development Assistance. A particular focus of agricultural research investments in other countries has been IARCs. Between 1972 and 1996 the United States was the largest single contributor to these institutions, accounting for 28 percent of contributions from industrial countries and 20 percent of all contributions (CGIAR 1997).[14] For this reason, the United States should have a keen interest in the productivity of these and other agricultural research entities that benefit from U.S. funding.

A number of economists have used knowledge production functions to quantify the relationship between research inputs and outputs, where outputs are measured by patents (Pakes and Griliches 1980) or research publications

Table 14.5. Potential benefits to the United States of training foreign graduate students who return to their home countries

Potential U.S. benefit	Measurement approach	Illustrative literature
Most direct		
—Enrollment —Tuition revenue —Spending in U.S. —Unpaid or underpaid research	Cost-benefit analyses relate tuition proceeds and other benefits to costs of supporting foreign grads	—Chishti (1984b)
—Culturally diverse learning environment —The opportunity to develop personal contacts among overseas professionals —Scientific collaboration and and exchange	Surveys have been used to assess relevance of U.S. training to foreign grads. Survey methods could also be used to assess benefits to U.S. students and faculty of cultural diversity in graduate programs and of postgraduate connections to other nations' researchers	—Education research literature
—Enhanced productivity of international research —Knowledge spillovers that contribute to U.S. science	Knowledge production functions controlling for training of scientists and quality of research output	—Adams and Griliches (1996a,b) —Pakes and Griliches (1980) —Pardey (1989)
—Enhanced trade linkages through technology transfer	Adoption models Technology transfer models	—Foster and Rosenzweig (1996) —Echeverria (1991)
—More open agricultural policy regime		
—Technological spillovers that benefit U.S. agriculture	Models of returns to U.S. investments in international research	—Pardey, Alston, Christian, and Fan (1996)
	Models of research and technological spillovers	—Schimmelpfennig and Thirtle (1999)
Least direct		

and citations (Pardey 1989; Adams and Griliches 1996a). When U.S. funders want to know how scientific knowledge generated in research centers in other countries contributes to scientific knowledge in the United States, research publications and citations might be adjusted by the percentage found in U.S. or internationally recognized journals. [However, Echeverria (1991) suggests that for developing countries, the number of new crop varieties released would be a better measure of crop research output than publications and citations.]

These measures have sometimes been adjusted for the quality of research outputs (Pardey 1989), but studies have paid little or no attention to the quality of research inputs as measured by the level and quality of training of the research staff.[15] One study that does come close to considering quality of scientific training is Adams and Griliches (1996b). The researchers compare productivity across types of universities (public versus private and "top ten" versus others). They find evidence that costs per publication are nearly equal for top-ten and lower-ranked universities but that costs per citation are 30 percent lower at the top ten. Although other factors may also be responsible, it seems likely that the primary reason for this difference is that top universities attract top scientists whose work is widely cited by peers.

The least direct benefits to the United States of returnees are those that accrue to the agricultural economy as a result of the findings of overseas scientists and related professionals such as extension personnel and government economists and advisers. Among such potential benefits are the adoption by U.S. farmers of improved crop varieties developed at research centers in other countries; enhanced U.S. exports stemming from the adoption and importation of modern agricultural inputs such as seeds, chemicals, and machinery; and possibly the restructuring of trade policy regimes that are more open to U.S. agricultural exports. These areas are all amenable to empirical investigation and are likely sources of the spillovers of agricultural knowledge discussed by Schimmelpfennig and Thirtle in chapter 12.

Pardey et al. (1996) found, for example, that wheat breeding successes at the International Maize and Wheat Improvement Center (CIMMYT) brought economic benefits to U.S. agriculture in the form of higher wheat yields. The increased yields generated economic benefits 190 times the total U.S. monetary contribution to CIMMYT's wheat program. The researchers also found that economic benefits to U.S. agriculture from rice research at the International Rice Research Institute (IRRI) were 17 times the total U.S. contribution to IRRI's rice program. Even if such successes are not produced directly by U.S.-trained scientists, U.S. contributions to a strong pool of international scientists are important to ensuring that these types of gains continue.

The adoption of improved production technologies in developing countries—and the role of U.S.-trained professionals in that process—can be analyzed in terms of the impact on U.S. agricultural exports. Foster and Rosenzweig (1996) found that villages in India with agricultural extension programs were more likely to adopt imported high-yielding seed varieties that require a higher level of fertilizer than traditional seeds. Echeverria (1991) found that U.S. exports of seeds—as a measure of the direct transfer of biological technology—were important in explaining maize yield growth in 45 countries (especially temperate countries) between 1960 and 1985. His analysis suggests that seed trade complemented private research in temperate countries, although trade substituted for maize research in tropical countries. Echeverria also included a quality-of-research variable that measures quality by staff training in maize breeding. He found that the quality of training had a significant positive effect on maize productivity in both temperate and tropical countries. Ultimately, the question of whether global agricultural productivity growth through technology transfer provides a net benefit to U.S. agriculture is an empirical one and depends on changes in the prices and volumes of U.S. agricultural exports in relation to the prices and volumes of agricultural imports.

A significant share (about 8 percent) of foreign Ph.D. recipients in agricultural science are agricultural economists. Although we do not have comprehensive data on the employment of these returnees, it is likely that a significant number join the staff of agricultural ministries or become university faculty members with advisory roles. In those capacities they are likely to have the opportunity to influence the making or reform of agricultural trade policy. Economists trained in the United States have had a role in influencing the "openness" of countries such as Chile, for example.

STRATEGIES FOR FUTURE RESEARCH

The increase in the number of foreign students receiving doctorates in S&E from universities in the United States over the last decade reflects a comparative advantage in graduate education. At the same time that foreign students have become relatively more important to university agricultural science departments, however, Ph.D. recipients in agricultural fields are returning home at a faster rate than graduates of most S&E programs. These trends suggest that it is particularly important to understand more about the ways stayers contribute to U.S. human capital in the agricultural sciences and how research conducted by returnees benefits the United States. If the pool of U.S.-based agricultural researchers becomes increasingly dependent on foreign citizens, competition for human capital between agricultural research entities in the United States and in other countries may increase. Given this situation, one of the most important areas for future research on the costs and benefits of training foreign agricultural scientists will be the factors that drive Ph.D. recipients either to stay in the United States or to return home.

The costs of training foreign graduate students are more transparent and more easily calculated than the benefits. When the economic costs of training are compared to the economic benefits while foreign students are in graduate school, the costs appear to outweigh the benefits. However, important benefits that are not readily measured—such as the role of these students in sustaining demand for U.S. agricultural science programs and their unpaid contributions to research and teaching programs—must be taken into account. Formulating accurate measures of the costs and benefits of training foreign agricultural scientists is another important task for future research.[16]

Additionally, although the most important benefits may accrue to the United States following graduation, little attention has been paid to documenting or measuring these contributions. Over the last decade alone, more than 10,500 foreign students have received agricultural science doctorates, substantially enhancing the level of training of the global pool of agricultural scientists (FAEIS). These developments need to be formally assessed in a number of dimensions—for instance, in terms of how they enhance the body of knowledge and contribute to technological spillovers that augment the science base and benefit agriculture in the United States. Such assessments will require that the necessary data be systematically collected, including information on where foreign graduate students go, the type of work they perform, and the contributions of their work to science, technology transfer, and agricultural policy.

Finally, educating scientists at the graduate level requires a public-private partnership because of the level of public support involved. Yet the literature on public research investments (where investments are typically measured by funds expended or budgeted for research and development) has paid little attention to the graduate education of those who conduct research. The interactions and complementarities between human capital investments and other research inputs need to be analyzed. In an era of little or no growth of public funds for agricultural research, the productivity of individual agricultural scientists and scientific teams is especially important. Policy makers may want to look more closely at incentives for high-quality graduate training in agricultural sciences, a commodity in which the U.S. land-grant university system has a strong comparative advantage.

APPENDIX: COMPUTING U.S. BENEFITS FROM TRAINING FOREIGN AGRICULTURAL SCIENTISTS

Three types of net benefits accrue to the United States from training foreign graduate students: benefits from having foreign students in the country, benefits from the postgraduate work of those who stay in the United States, and benefits from those who return to their home countries. Assuming a constant interest rate, r, the present value of the net benefits to the United States of training a group of foreign agricultural scientists, n, starting at time $t=0$, can be written as

$$PV_n = \sum_{t=0}^{T} \frac{B_{Gt}}{(1+r)^t} + \sum_{t=T+1}^{L} \frac{[\alpha B_{St} + (1-\alpha) B_{Rt}]}{(1+r)^t} - \sum_{t=0}^{L} \frac{N_t}{(1+r)^t}$$

where B_G is the net benefit to the U.S. while the foreign students remain in graduate school, T (see Chishti 1984b); αB_S is the net benefit from the percentage of foreign doctorates who stay in the United States; $(1\text{-}\alpha)B_R$ is the net benefit gained from the proportion of foreign doctorates who leave the United

States during their scientific productive life, L; and N represents the opportunity costs associated with the net benefits that would have been realized if the student had not come to the United States to earn a doctoral degree. The values of PV_n, B_G, B_S, or B_R can be either positive or negative, depending on whether costs outweigh benefits.

To estimate the benefits to the United States from foreign agricultural scientists trained here, we estimate the factors that influence the decision to remain after graduation. The stay rate, α, for foreign agricultural science doctorates is a function of many factors, including

$$\alpha = f(I, \rho, S, R, \nu, \gamma),\ 0 \leq \alpha \leq 1$$

where I is a ratio of the expected income in the United States and the expected income in the home country in a particular field of study; ρ is the field of study; S is the source of support for graduate study; ν is the type of visa (permanent or temporary); R is the rate of return to agricultural research in the home country; γ is a catchall for the economic and political conditions that may affect the decision to stay or leave, such as the GDP of the home country, since agriculture's share of the GDP may be an important determinant. (Other factors that could influence a foreign student's decision to stay or go include immigration and visa policies; requirements attached to scholarships or financial aid; and employment opportunities.)

A logistic probability function can be used to estimate the proportion of foreign Ph.D. recipients who choose to remain in the United States (see Bratsberg 1995; Finn, Pennington, and Anderson 1995). For policy analysis, an estimated stay rate can then be incorporated into equation 1 in order to predict the net benefits of training foreign agricultural scientists.

This framework for conceptualizing the benefits to the United States from training foreign agricultural scientists is fairly general and can easily be adapted to other uses. These could include measuring the net benefits to the country of training students in all fields or of training U.S. citizens who choose to conduct research abroad.

NOTES

1. They share equally in the authorship of this chapter.

2. The benefits this paper discusses accrue only to the United States, not to individual students, the students' home countries, or the global community.

3. Ethiopia, Greece, India, Indonesia, Iran, Israel, Kenya, Korea, Kuwait, Lebanon, Nigeria, Philippines, Saudi Arabia, Taiwan, and Turkey.

4. Higher education opportunities in many parts of Africa are still relatively few. In 1990 African universities had facilities to train only 139 students per

100,000 population; 100,000 Africans were studying outside Africa, including somewhat less than one-fourth of these in the United States; and over 7,000 Africans were enrolled in U.S. graduate programs (Sterns, Oehmke, and Schwartz 1993).

5. A second study by Mehrabi (undated) used the same methodology and derived almost identical results.

6. We do not know how much of the university support was in the form of fellowships and grants and how much in teaching or research assistantships, although it is likely that fellowships and grants were only a small portion. We also do not know how U.S. financial aid to foreign Ph.D. students in agricultural science compares with U.S. financial aid to all foreign graduate students.

7. Some might argue that demand for U.S. graduate education by foreign students raises tuition costs for or even displaces some domestic students. Our approach assumes a perfectly elastic supply of graduate education. Also, foreign students in the agricultural sciences represent too small a share of university students to have a significant impact on tuition.

8. The Survey of Earned Graduates (NSF and NRC) finds a smaller number of agricultural scientists than the FAEIS. The NSF and NRC survey shows degree fields, while the FAEIS reports all degrees awarded by agricultural colleges and programs. A number of agricultural colleges are the academic homes of basic biological and physical science departments, and many students in these specializations will not identify their degrees as being in agricultural science.

9. Data from the Survey of Earned Graduates (NSF and NRC) indicates that the region with the fastest rate of increase in Ph.D. recipients in science over the last 10 years has been Asia, especially China. Over this period, the percent of Asian students that received doctorates in science increased from 38.9 to 62.1 percent of all foreign science doctorate recipients in the United States. Since Chinese students have among the highest stay rates (the rates have varied markedly with the political situation), we can expect these rates to increase over time. The percent of European students earning science doctorates in the United States also increased from 12.1 to 13.6 percent. The percent of African, Latin American, Middle Eastern, and other North American students receiving science doctorates decreased. However, not only the percent but also the number of African students receiving doctoral degrees in the United States dropped between 1986 and 1995.

10. We have not extrapolated from Chishti's calculations for all foreign students to derive estimates of the capital embodied in nonreturning agricultural scientists, because too many adjustments would have to be made with insufficient information. However, because the stay rate is dramatically higher for Ph.D. recipients than for all degree recipients (including for agricultural scientists), it seems likely that the benefits are substantial.

11. The most common fields of study of the respondents were agricultural economics, agricultural education and extension, and food science.

12. The study also finds, however, that spending per scientist was lower in 1991 than in 1961, leading the authors to suggest that "research throughout Africa is severely curtailed because of inadequate operational resources" (415). In a sense, their point is the flip side of ours: We emphasize the need for well-trained

scientists to complement spending in research and development; they emphasize the need for research and development resources to complement scientific staff.

13. We liked a quote in Sterns, Oehmke, and Schwartz (1993) that is attributed to Vern Ruttan: "[My] concerns about the development of national agricultural research systems [include] the apparent presumption in some national systems that it is possible to do research in agricultural science without scientists."

14. United States' contributions have declined significantly during the 1990s, and since 1994 Japan's contributions have been larger (CGIAR 1997).

15. The literature concerned with returns to education does contain examples of attempts to account for differences in the quality of schooling, but the focus is on precollege education. Card and Krueger (1992), for example, find that quality of education—measured by student-teacher ratios, the length of terms, and relative teacher pay—is positively correlated with the rate of return to education for men born between 1920 and 1949.

16. A paper published by Johnston, Baker, and Creedy (1997) provides a list of potential long-term benefits to Australia from training foreign students. This list is similar to the one presented in our paper.

REFERENCES

Adams, James D., and Zvi Griliches. 1996a. "Measuring Science: An Exploration." *Proceedings of the National Academy of Sciences* 93:12664–70.

———. 1996b. *Research Productivity in a System of Universities.* Working Paper 5833. Cambridge, MA: National Bureau of Economic Research.

Agarwal, Vinod B., and Donald R. Winkler. 1985. "Foreign Demand for United States Higher Education: A Study of Developing Countries in the Eastern Hemisphere." *Economic Development and Cultural Change* 33(3): 623–44.

Borjas, George J. 1987. "Self-selection and the Earnings of Immigrants." *American Economic Review* 77:531–53.

Bratsberg, Bernt. 1995. "The Incidence of Non-Return Among Foreign Students in the United States." *Economics of Education Review* 14(4): 373–84.

Card, David, and Alan B. Krueger. 1992. "Does School Quality Matter? Returns to Education and the Characteristics of Public Schools in the United States." *Journal of Political Economy* 100(1): 1–40.

Cashman, Kristin, and Jane Plihal. 1987. "International Agriculture Students' Perceptions of the Relevance of Their U.S. Education." *Journal of Agronomic Education* 16:61–65.

CGIAR (Consultative Group on International Agricultural Research). 1997. *1996 Annual Report.* Washington, D.C.: CGIAR Secretariat.

Chishti, Salim. 1984a. "International Demand for American Higher Education." *Research in Higher Education* 20(3): 329–44.

———. 1984b. "Economic Costs and Benefits of Educating Foreign Students in the United States." *Research in Higher Education* 21(4):397–415.

Davis, Todd M., ed. 1996. *Open Doors 1995/96: Report on International Educational Exchange.* New York: Institute of International Education.

Echeverria, Ruben G. 1991. "Impact of Research and Seed Trade on Maize Productivity." In Philip Pardey, Johannes Roseboom, and Jock Anderson, eds. *Agricultural Research Policy: International Quantitative Perspectives*. Cambridge, UK: Cambridge University Press.

Finn, Michael G., Leigh Ann Pennington, and Kathryn Hart Anderson. 1995. *Foreign Nationals Who Receive Science or Engineering PhDs from U.S. Universities: Stay Rates and Characteristics of Stayers.* Oak Ridge, TN: Oak Ridge Institute for Science and Education, U.S. Department of Energy.

Food and Agricultural Education Information System (FAEIS). Historical data collected from the FAEIS Website, Texas A&M University, College Station, TX.

Foster, Andrew D., and Mark R. Rosenzweig. 1996. "Technical Change and Human-Capital Returns and Investments: Evidence from the Green Revolution." *American Economic Review* 86(4): 931–53.

Fuglie, Keith, Nicole Ballenger, Kelly Day, Cassandra Klotz, Michael Ollinger, John Reilly, Utpal Vasavada, and Jet Yee. 1996. *Agricultural Research and Development: Public and Private Investments under Alternative Markets and Institutions.* Agricultural Economic Report No. 735. Washington, D.C.: Economic Research Service, U.S. Department of Agriculture.

Henderson, Janet L., and Gayle Shibano. 1990. "Follow-up of International Graduate Students in Agricultural Education at The Ohio State University." *Journal of Agricultural Education* Spring:71–75.

Johnston, David, Meredith Baker, and John Creedy. 1997. "The Cost and Benefit of International Students." *Economic Papers* 16(3): 44–54.

Judd, M. Ann, James K. Boyce, and Robert E. Evenson. 1987. "Investment in Agricultural Research and Extension." In Vernon W. Ruttan and Carl E. Pray, eds. *Policy for Agricultural Research.* Boulder, CO: Westview Press.

Mehrabi, Shah. Undated. "Educating Foreign Students in the U.S.A.: A Cost-Benefit Analysis." Paper prepared for the Annual Meeting of the Association for the Advancement of Science, Policy and Research in the Third World.

Merrill-Sands, Deborah. 1995. *CGIAR Human Resources Survey: 1991, 1994: Key Observations on International Staffing with a Focus on Gender.* CGIAR Gender Program Working Paper No. 9. Washington, D.C.: CGIAR Secretariat.

National Research Council (NRC). 1995. *Colleges of Agriculture at the Land Grant Universities: A Profile.* Washington, DC: National Academy Press.

National Science Foundation (NSF)/Science Resources Studies and National Research Council (NRC). Various years. *Survey of Earned Doctorates.*

National Science Foundation (NSF)/Science Resources Studies. Various years. *Survey of Earned Graduates and Postdoctorates in Science and Engineering.*

Pakes, Ariel, and Zvi Griliches. 1980. "Patents and R&D at the Firm Level: A First Report." *Economic Letters* 5:377–81.

Pardey, Philip G. 1989. "The Agricultural Knowledge Production Function: An Empirical Look." *Review of Economics and Statistics* 71(3): 453–61.

Pardey, Philip G., Johannes Roseboom, and Jock R. Anderson. 1991. "Topical Perspectives on National Agricultural Research." In Philip G. Pardey, Johannes Roseboom, and Jock Anderson, eds. *Agricultural Research Policy: International Quantitative Perspectives.* Cambridge, UK: Cambridge University Press.

Pardey, Philip G., Johannes Roseboom, and Nienke M. Beintema. 1997. "Investments in African Agricultural Research." *World Development* 25(3): 409–23.

Pardey, Philip G., Julian M. Alston, Jason E. Christian, and Shenggen Fan. 1996. *Hidden Harvest: U.S. Benefits from International Research Aid.* Food Policy Report. Washington, D.C.: International Food Policy Research Institute.

Schimmelpfennig, D., and C. Thirtle. 1999. "The Internationalization of Agricultural Technology: Patents, R&D Spillovers and Their Effects on Productivity in the European Union and United States." *Contemporary Economic Policy* 17(4): 457–468.

Sterns, James, James Oehmke, and Lisa Schwartz. 1993. *Returns to Education: The Impacts of M.S.U. Training on West African Scientists.* Agricultural Economics Report Number 567. East Lansing: Department of Agricultural Economics, Michigan State University.

Swanson, Burton E. 1986. "The Contribution of U.S. Universities to Training Foreign Agricultural Scientists." In Lawrence Busch and William B. Lacy, eds. *The Agricultural Scientific Enterprise: A System in Transition.* Boulder, CO: Westview Press.

University of Tennessee. 1982. *Guidelines on Foreign Student Enrollment.* Approved by the Graduate Council, Undergaduate Council, and Faculty Senate.

World Bank. 1997. World Development Report. New York, NY: Oxford University Press.

Conclusion

15

PUBLIC-PRIVATE COLLABORATION IN AGRICULTURAL RESEARCH: THE FUTURE

JOHN M. REILLY AND
DAVID E. SCHIMMELPFENNIG

THE PREVIOUS chapters consider specific issues in the funding and performance of agricultural research in an effort to better understand how public- and private-sector research institutions can better collaborate and more effectively complement one another. Ongoing changes in agricultural research institutions need, however, to be considered in the context of broader changes in agriculture and the economy; globalization, changing diets and increasing disposable income, reform of farm policy, changes in the structure of agriculture, and pressure to lessen impacts on the environment are forces that will carry along with them the agricultural research enterprise.

This chapter will lay out the broader forces facing agriculture, consider themes that cut across the previous chapters, and draw some implications for what these changes mean for public agricultural research institutions. Before doing so, a brief review of the basic economics of research is in order. In the jargon of economics new knowledge that results from research and innovation is a true public good—once available, knowledge can be applied over and over again unlike the loaf of bread or pound of bacon that must be produced anew for each consumer. Therefore, economists generally recognize that the private market economy will tend to underinvest in research and overprice new products seeking to recoup research investment costs. Hence comes the role for public-sector involvement either as a funder of agricultural research directly or as an enforcer of intellectual property rights. And while the latter approach—intellectual property rights protection—is an essential mechanism for encouraging innovation, it has its drawbacks and limits. The legally granted monopoly rights that come with a patent and other forms of intellectual property rights (IPR) protection afford the innovator the opportunity for substantial profits as an incentive to pay for the research, but in earning those profits, higher priced products do not spread as far and as fast as they might without monopoly pricing. Futhermore, the profit motivation for research may limit investigation into both truly long-term, risky projects and more basic types of scientific investigation. The shortcoming of intellectual property rights protection for many types of research, particularly more basic R&D, remains a strong justification for federal R&D programs. Private companies are unable to appropriate all the benefits of basic research and are therefore unlikely to fully fund it.

Despite the case for public involvement, under the influence of declining real resources for public projects in general, competition for scientific research funding has intensified worldwide. Basic scientific public research is under increasing pressure in the United States to pass the same Agency, Departmental, and Congressional scrutiny, as do expenditures on infrastructure, education, and defense projects. The evidence on public agricultural research is that the return on this public investment is high—at least 35 percent, and much higher for more basic types of research.

A high rate of return means that more should be invested. Against the pressure to reduce federal spending, however, agricultural research has been at best able to hold steady, with essentially flat spending in real terms since the late 1970s. The private agricultural sector in the United States has stepped up to fill some of the gap between desirable levels of funding and the reality of public budgets, with significant increases in funding over the last decade for largely applied agricultural research. This is occurring largely through multinational companies such as Monsanto and DuPont/Pioneer Hi-Bred, responding to improved IPR protection and emerging technologies such as biotechnology. Reflecting this trend toward the private provision of agricultural R&D, the coordination of public and private research in the United States has become an active area of economic research (Fuglie et al. 1996). What makes this coordination difficult is that neither public nor private agricultural research communities operate in isolation in the global economy. And at this point, it is unclear whether public research funding is either adequate or as focused as it might be on the enabling basic research that is the basis for applied development of new products. In addition, the orderly flow in which basic research feeds applied research which in turn leads to diffusion of innovations may increasingly be accompanied by less orderly mechanisms in which applied research leads basic research.

More recently, globalization of the world economy and increased international trade, with the cultural and scientific interactions that it brings, has led to foreign public and private R&D both becoming sources of intellectual capital that can influence domestic agricultural productivity. One of the reasons that some of the largest and most profitable companies are multinationals is their ability to exploit international knowledge flows. Intensifying this effect is the fact that bidirectional flows of intellectual capital are taking place between most U.S. universities and foreign countries.

Like a dye marker for international knowledge flows, biotechnology is moving through national and international, as well as public and private, channels, fundamentally changing the structure of the agricultural enterprise worldwide. Other basic innovations outside agricultural research, from improved computer processing and information management technologies to advances in global positioning systems, have also been key sources of new tools for agricultural research. The Internet has become the backbone of modern agricultural research; it has an increasing role in agricultural extension, and the networking, both formal and informal, between researchers, government agencies, universities, and private labs will grow as more and more information is stored and retrieved digitally. These developments help create opportunities for dealing with competitive pressures resulting from increased global competition. At the same time, technological advances are improving agriculture's ability to meet the nutritional needs of a growing world population, while aiding in the wise use of natural resources and enabling the development of new products to meet the changing tastes of a more diverse and wealthier domestic population.

Structural change in agriculture is manifested in many forms including both the size and types of agricultural enterprises. These include farms and agribusinesses that produce agricultural commodities, firms that produce inputs for farming, and firms that purchase, process, and distribute farm commodities. The relationships among these entities have changed, and greater industrial concentration and vertical integration are changing both the roles for and the nature of federal government commodity, conservation, and research policies.

Agricultural research policy affects some of these trends and forces but largely must respond to these forces for change. The remainder of this chapter considers again the fundamental forces for change that are coming from sources that interact with agriculture. The trends in agricultural research discussed in previous chapters as well as broader economic trends point toward the future while answering the following question with respect to the relationship between public and private research systems:

Agricultural research—what might the future hold?

Six broad economic trends likely to shape the agricultural system in the future are:

1. The size of farms and agribusiness firms is increasing, both in sales and number of employees, particularly among firms supplying technology to farmers (seeds, pesticides, and machinery inputs) that embody the knowledge developed from agricultural research.

Questions about public versus private funding of agricultural research and the split between basic and applied research (chapters 5, 6, and 7) are further complicated by buyouts and mergers that have concentrated control of the agricultural input supply industry in the hands of a few multinational firms. Consumers of monopoly products often face reduced product quality at higher prices. Here the effect may be different, with intensification of the input industry allowing greater appropriation of research benefits and leading to quicker development of new technologies. In the agricultural inputs industry, a company that stops innovating to extract monopoly rents may become obsolete. For example, several large pharmaceutical companies have sold, or are selling, their agricultural divisions.

2. *Structural change continues in U.S. agriculture from small farms in diversified rural communities and viable limited resource farming to a smaller number of large, vertically integrated agribusinesses placing greater emphasis on contract production of farm commodities (Hallam 1993, Fig. 15.1).*

As this trend continues the clients that agricultural research serves will change. The level of technological sophistication required to serve rural producers will increasingly diverge from the needs of the agribusiness community. The land-grant universities and state agricultural experiment stations are expected to respond to these changing pressures by re-

Figure 15.1. Farm size increased from 1930 to 1992 as farm numbers declined from their 1935 peak

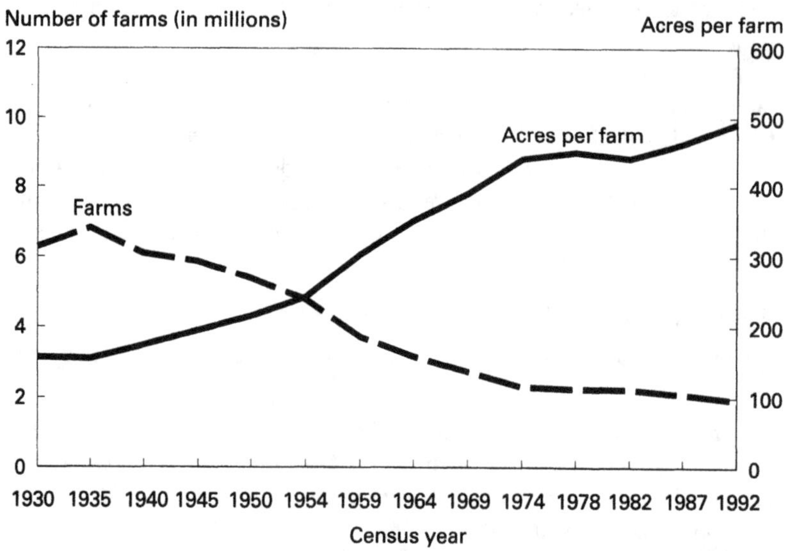

Source: Rural Economy Division, Economic Research Service, USDA.

organizing the structure of their research programs to operate in concert with USDA and private efforts.

3. The 1996 version of the Farm Bill puts greater emphasis in farm production decisions on market forces.

The FAIR Act phases out commodity support programs that reduced incentives for farmers to respond to market forces. Annual acreage idling programs are being eliminated, giving farmers the freedom to plant any crop on contract acres, with limitations on fruit and vegetables. In return farmers bear income fluctuation risks that they were not exposed to when payments were fixed. One side effect of this legislation may be increased reliance on timely, accurate agricultural research and weather information.

4. Companies face an increasing need to internalize both environmental costs and the costs of consumer health and food safety requirements that result from new technology into the research and technology development stages of the new product life cycle.

Increased federal, state, and local regulation coupled with rising public pressure are behind this force for change, which is being experienced in other industries as well. It has even led to the complete transformation of some American cities, particularly in the South and Midwest, over the last few decades, from industrial centers to cultural-, sports-, and entertainment-based economies.

5. Globalization of agricultural markets is creating opportunities for input suppliers, as well as commodity purchasing, processing, and distribution companies, to take advantage of international knowledge spillovers.

We argue that this is a function of the international nature of agricultural expertise and knowledge. Agricultural inputs take on the character of pure public goods because of possibilities for scrutiny, resulting in the imitation of technologies embodied in inputs (chapter 11, and also chapter 12). A by-product of trade liberalization is increased international participation in the U.S. agricultural education system and increased representation of Americans within the National Agricultural Research Systems (NARS) in foreign countries. Chapter 14 discusses U.S. benefits of training foreign agricultural scientists, largely at the land-grant agricultural institutions.

6. The demand for niche, specialty, and altered characteristic products has risen among consumer groups both domestically and internationally.

Rising per capita incomes and standards of living in many parts of the world are strengthening demand for agricultural commodities. The agricultural system is responding by making a wider selection of fruits and vegetables available, as well as locally produced, organically grown commodities at farmers' markets. Labeling of genetically modified and nonmodified foods is increasing. Multinational U.S. food processing companies have experienced sharp increases in demand, with exports of snack foods, beverages, and desserts doubling between 1990–91 and 1993–94. Four countries received over half of U.S. processed food and beverage exports: Japan, Canada, Mexico, and South Korea; but many of the fastest growing export destinations are less-developed countries (Henderson, Handy, and Neff 1996).

These broader economic trends should be considered in conjunction with some of the changes taking place in agricultural R&D as identified in previous chapters:

1. Public research expenditures are likely to continue to stagnate, while research activity by the private sector will grow as long as market opportunities exist.

Private industry is rapidly increasing its research investment in areas that were long the purview of the public sector, such as plant breeding and livestock improvement. The private sector in the United States now invests more in the breeding of new crops than the public sector (chapter 2). In the European Community, the United Kingdom and the Netherlands, have responded to fiscal belt-tightening by privatizing those areas of public agricultural research showing the largest rates of return. These countries have more centralized governments than the United States and therefore have fewer opportunities for regional agricultural support to exert influence on budget decisions (chapter 3). Quasi-privatization is also taking place as public research is increasingly carried out by researchers on fixed-term contracts. Unfortunately, renewed public commitment to basic research and "less profitable" research areas in these countries often had to succumb to general budget pressures (chapter 12). Therefore, not surprisingly, the private sector has been unable to sustain agricultural research at historical levels on its own.

2. Private support for applied or near-market research (with commercial benefits that can be appropriated by funders) is increasing over basic research that has high social returns (but with benefits that flow outside the borders of political units or beyond the specific crop it was developed for) (chapter 3).

Private nonappropriability of the benefits of basic research is a primary motivation for federal funding of basic research. At the state level, where agricultural research at the land-grant universities and state agricultural experiment stations is funded and carried out, there are even more severe problems appropriating benefits of research than at the national level. Care needs to be taken that adequate funding for basic agricultural research, and basic scientific research in general, does not suffer unnecessarily from downsizing and party politics. When basic research pays off, the benefits tend to be longer lived, but basic research has a longer gestation period. The lag between funding and productivity improvement is shorter for applied near-market research, but may depend on a stock of basic research accomplishment.

3. *The internal organization of research funding is also changing.*

   Along with fiscal belt-tightening around the world has come the demand for greater accountability in the use of available research funds. In the United States Department of Agriculture (USDA), many intramural research projects are being replaced by competitively awarded extramural projects through regional organizations like the state experiment stations.

4. *Intellectual property rights (IPR) protection has been strengthened.*

   The legal system has been strengthened, with both the Plant Variety Protection Act, PVPA, in the United States and the International Union for the Protection of New Varieties of Plants, UPOV, gaining wide acceptance alongside plant utility patenting. New technologies (particularly biotechnology) are increasing the rate at which new varieties are developed and are increasing applications for protection. The U.S. Environmental Protection Agency (U.S. EPA) now grants and requires registreation of transgenic crop varieties that have pesticide properties. Examples are *Bt* varieties that are resistant to certain insects.

Responding to these forces for change, agricultural enterprises are preparing for the future in four predictable ways:

1. *With private companies taking a larger role in agricultural research, the public sector is increasingly able to focus its attention on areas of research that are neglected by the private sector.*

   Public agricultural research will need to focus resources on basic research where there is the greatest potential for spillovers and hence the least incentive for private funding. Public research institutions and funding also have an important role where research is needed to provide unbiased evaluation of product claims or to serve as the basis for private or public choices. Understanding how agriculture and the environment interact or how health and diet are related, even if quite applied, are types of research that do not generate an easily marketed product, and hence there is little private-sector incentive to invest in such research. While claims for health-related or environmentally friendly products can generate sales, there is a need to have research funded by a source that does not have a financial stake in the outcome. The public sector may also have a role in the development of technologies that internalize environmental costs. But here a preferred course is to seek to remedy the environmental externality in the first place, thereby providing a market incentive to develop new technologies that have improved environmental performance. Publicly funded research heavily weighted toward improved environmental performance can generate environmentally friendly technologies that do not compete economically with conventional technologies.

2. *Belt-tightening has led to a reexamination of the agricultural research funding process.*

Whether because of increased private funding for applied agricultural research or in response to a worldwide call for privatization, funding for agricultural research is evolving to place greater emphasis on competitive processes versus block grants or formula-funding (chapter 3). Nontraditional private sources of funding are also being considered for applied research as industry or commodity groups in Australia, the Netherlands, and the United Kingdom routinely fund commodity-specific agricultural research out of commodity checkoffs and possibly matching funds. In the United States, these programs are usually operated at the state level and tend to be voluntary, so they have had less impact on the research agenda (chapter 4). Mandatory national programs have emphasized commodity advertising and promotion more than research funding. Consumer groups, environmental and conservation lobbies, and even agribusiness lobbies are attempting to influence agricultural research programs, and have been at least partially successful in the Netherlands, the United Kingdom and the United States in relation to sustainable agriculture, wetland maintenance, public health, and food safety (chapter 3).

3. *The existence of large international spillovers of agricultural knowledge weakens the incentives for individual countries to pursue research and puts greater emphasis on mechanisms for international collaborative funding and development of multinational firms that can capture the spillovers.*

The Consultative Group for International Agricultural Research (CGIAR), for example, works closely with individual country NARS to coordinate research efforts (chapter 13). Agricultural research benefits spillover internationally (chapter 11). Basic research may suffer from the largest international spillovers. Such spillovers make it difficult to estimate the benefits of agricultural research correctly. Measured rates of return have historically been at least 35 percent (chapter 2). But with many of these returns flowing outside of the country and unmeasured in these estimates, the true social benefits may be higher. Smaller countries that are primarily beneficiaries of spillovers, may have inflated measured rates of return to research when spillins are not included in research costs. Broad geographic presence through international organizations may allow firms and funders of research to partially offset the problems. The CGIAR has other problems of its own, following a series of unproductive reorganizations, but international agricultural knowledge spillovers will continue to command responses from agricultural enterprises. It is likely that multinational firms will continue to compete for dominant market positions through cycles of mergers, acquisitions, and spinoffs as they refocus and redefine their products and services in the years ahead. While it is common to think of DuPont and Monsanto as *the* agricultural multinationals, a Mexican firm, Empresas La Moderna (ELM), has purchased the U.S. vegetable seed company Peto Seeds, the vegetable seed portion of Asgrow, and the European vegetable seed company, Sluis. Before that, the Thai firm Charoen Pokphand purchased Avian Farms to complement its joint venture with Arbor Acres for poultry research.

4. Changes in IPR legislation and the availability of other mechanisms for technology transfer has created incentives for public and private institutions to engage in cooperative research.

Private companies are taking advantage of these opportunities through "research consortia" in which private firms and public agencies jointly engage in basic and applied research, jointly own products from their collaborative work, and sometimes jointly share the profits (chapter 8). One of the most prominent of these research consortia, is the Genetic Enhancement of Maize project, in which Pioneer Hi-Bred and DEKALB Genetics work with the USDA to enhance the maize gene pool under agreements to market the resulting varieties (chapter 9). Technology transfer legislation from 1986 has also allowed for cooperative research and development agreements (CRADAs) that offer firms the opportunity to capture IPRs. In some cases the CRADA itself may substitute for formal property rights. Since the inception of the agricultural program, over 700 agreements have been put in place. Although the share of total public or private research in cooperative agreements is small, the USDA is expanding the availability of these agreements through an Office of Technology Transfer (chapter 10). As the U.S. EPA continues to regulate transgenic varieties in the future, the effect could be a tendency to concentrate corporate involvement in agricultural research because larger companies may have the expertise and resources to better deal with a new set of regulations.

REFERENCES

Fuglie, K., N. Ballenger, K. Day, C. Klotz, M. Ollinger, J. Reilly, U. Vasavada and J. Yee. 1996. *Agricultural Research and Development: Public and Private Investments under Alternative Markets and Institutions.* Agricultural Economics Report 735. Washington, DC: Economic Research Service, U.S. Department of Agriculture.

Hallam, A., ed. 1993. *Size, Structure, and the Changing Face of American Agriculture.* Boulder, CO: Westview Press.

Henderson, D.R., C.R. Handy, and S.A. Neff, eds. 1996. *Globalization of the Processed Foods Market.* Agricultural Economics Report 742. Washington, DC: Economic Research Service, U.S. Department of Agriculture.

CONTRIBUTORS

Julian M. Alston is a professor in the Department of Agricultural Economics of the University of California at Davis. He teaches graduate and undergraduate classes in microeconomic theory and the analysis of agricultural markets and policies. Prior to beginning in his current position in 1988, Dr. Alston was Chief Economist in the Department of Agriculture in Victoria, Australia, where he had been employed in various capacities since 1975. His experience in public policy analysis and advice and in the administration of a large scientific organization shaped Dr. Alston's research interests in the economic analysis of agricultural markets and public policies concerning agricultural incomes, prices, trade, and agricultural research and promotion. He is a coauthor of *Making Science Pay: The Economics of Agricultural R&D Policy* and *Science under Scarcity: Principles and Practice for Agricultural Research Evaluation and Priority Setting*.

Alston was raised on the family farm in northern Victoria, Australia. He has a Bachelor's degree in Agricultural Science from the University of Melbourne, completed in 1974; a Master's degree in Agricultural Economics from La Trobe University, completed in 1978; and a Ph.D. in Economics from North Carolina State University, completed in 1984.

Nicole Ballenger is a deputy director for research of the Market and Trade Economics Division of USDA's Economic Research Service. She recently directed the National Research Council's study of the future of the land-grant colleges of agriculture. She has a Ph.D. in agricultural economics from the University of California at Davis. Her areas of research have included agricultural policy and trade, Mexican agriculture, and linkages between trade and the environment. During 1990–91 she was senior staff economist for agriculture and trade for the President's Council of Economic Advisers.

Kelly Day-Rubenstein is an agricultural economist at USDA's Economic Research Service. As a member of the Research Development and Technology Policy team, her work focuses on public-sector research, natural resource research, public-private research agreements and technology transfer programs. She coauthored the USDA report *Agricultural Research and Develop-*

ment: Public and Private Instruments under Alternative Markets and Institutions. She has coauthored several articles on cooperative research efforts including "Medical Research and Genetics Resources Management: The Case of Taxol" in *Contemporary Policy Issues* and "Cooperative Research Bolsters War on Cancer" in *Forum for Applied Research and Public Policy*. Her work on natural resources research has appeared in *Evolutionary Trends in Plants* and *Bioscience*. She is also a member of the working group responsible for revising USDA's Current Research Inventory System (CRIS) which classifies and collects agricultural research data. Day-Rubenstein received an M.S. in agricultural and applied economics from the University of Minnesota and a B.A. from St. Thomas College in St. Paul, Minnesota.

Robert E. Evenson is the director of the Economic Growth Center and a professor in the Department of Economics at Yale University. Evenson is a fellow of the American Association for the Advancement of Science and the American Agricultural Economics Association. He received a Ph.D. in economics from the University of Chicago in 1968 and an M.S. in agricultural economics from the University of Minnesota in 1964.

Jose Falck-Zepeda is a post-doctoral research fellow in the Department of Agricultural Economics and Rural Sociology at Auburn University. He received his M.S. and Ph.D. in agricultural economics from Auburn University and a B.S. from Texas A&M.

Kenneth J. Frey, Distinguished Professor Emeritus, received his collegiate education at Michigan State and Iowa State universities. From 1953 to 1993, when he retired from Iowa State University, he was a plant breeder and graduate student educator. He and his colleagues published more than 300 scientific papers and edited five books. Ken served as major advisor for 75 Ph.D. and 42 M.S. degree students and 25 visiting scientists during his career. He was a Fulbright scholar and visiting professor in Australia, Germany, Yugoslavia, Egypt, Japan, and Norway. He served as president of the Crop Science Society of America, the American Society of Agronomy, and the International Crop Science Congress.

George B. Frisvold joined the faculty in the Department of Agricultural and Resource Economics at the University of Arizona in 1997. He has been a visiting scholar at the National Institute of Rural Development in Hyderabad, India, a lecturer at The Johns Hopkins University and Chief of the Resource and Environmental Policy Branch of USDA's Economic Research Service. His research interests include domestic and international environmental policy as well as the causes and consequences of technological change in agriculture. In 1995–96, Dr. Frisvold served as a senior economist on the President's

Council of Economic Advisers with responsibility for agricultural, natural resource, and international trade issues.

Keith O. Fuglie has been an agricultural and resource economist with the International Potato Center in Bogor, Indonesia, since 1998. His research interests include the economics of science policy, technical change in agriculture, and economic development. During 1997–98, he served as senior economist for agriculture and natural resources on the President's Council of Economic Advisors in Washington, DC. He worked with the Economic Research Service from 1992 to 1997, where he directed a research program on agricultural research policy. During 1989–91 he was a Rockefeller Foundation research fellow in social sciences in Tunisia. He received a Ph.D. and M.S. in agricultural and applied economics from the University of Minnesota and a B.A. from Concordia College, Moorhead, Minnesota.

Richard Jones is Dean for Research and Director of the Florida Agricultural Experiment Station at the University of Florida. He was previously Dean of the College of Agriculture at the University of Minnesota, where he held various positions since 1977. He has a B.S. and M.S. in entomology from Mississippi State University, and a Ph.D. in entomology from the University of California, Riverside. He was an NIH Scholar (1965-1968), a Fulbright scholar (1980), and is presently a member of the Gamma Sigma Delta Honor Society.

Cassandra Klotz-Ingram is an agricultural economist with Economic Research Service of USDA. Her research on agricultural R&D and technology issues includes examining the nature and extent of private-sector agricultural R&D, the effect of intellectual property rights on R&D, and the socioeconomic effects associated with farmer adoption of new agricultural technologies, such as biotechnology. She is coauthor of several USDA publications including *Agricultural Research and Development: Public and Private Investments Under Alternative Markets and Institutions*; *Private-Sector Agricultural Research Expenditures in the United States, 1960–92*; and *Agricultural Biotechnology: An Economic Perspective*. Klotz-Ingram also coauthored an article in *Choices* entitled "Intellectual Property Rights Encourage Investment in Plant Breeding" and one in the *Review of Agricultural Economics* entitled "The Role of Information in Technology Adoption: The Case of bST in the California Dairy Industry." She received an M.S. from the University of California at Davis.

Mary K. Knudson received her B.S. in agronomy from the University of Minnesota in 1981, her M.S. in plant breeding and plant genetics from the University of Wisconsin in 1983, and her Ph.D. in agricultural and applied

Economics from the University of Minnesota in 1988. From 1988 to 1990 she worked as a research economist in the Resources and Technology Division of the Economic Research Service, USDA, where she focused on issues of intellectual property rights, agricultural research and development, regulation of biotechnology, and technological change. From 1990 to 1995 she was affiliated with the School of Public Policy and the School of Public Health at the University of Michigan, where she taught the economics of science and technology and participated in the Project on Global Change. She was also a research economist at Rutgers University before her current position as a research associate at Purdue University, continuing her work on intellectual property rights and agricultural biodiversity, and participating in a study of the Indiana equine industry.

Richard L. Lower has an undergraduate degree from the University of Illinois, a Master's degree from Purdue University, and a Ph.D. in plant genetics and breeding from the University of Illinois. After graduate school, he was employed in private industry for five years as a vegetable breeder and geneticist and then joined the faculty at North Carolina State University as a professor in the Department of Horticultural Sciences. He remained at NCSU for 11 years and then left to become Professor and Chairman of the Department of Horticulture at the University of Wisconsin-Madison in 1978. From 1984 to 1994, he was Associate Dean and Director in the College of Agricultural and Life Sciences. In December 1994, he was appointed Executive Director of the North Central Regional Association of State Agricultural Experiment Station Directors. He has authored numerous publications on genetics and plant breeding, primarily vegetable crops. Other research and administrative interests have been in plant germplasm and integrated pest management. He served on the National Plant Germplasm Committee, National Plant Genetic Resources Board, and presently is the chair of the National Genetic Resources Advisory Council. He is a member of numerous professional societies and is a Fellow and past president of the American Society for Horticultural Science.

Clare A. Narrod is an American Association for the Advancement of Science Fellow at USDA's Food Safety Inspection Service. Her research has focused on livestock policy and the economics of food safety. In the past she has worked as a consultant for USDA's Economic Research Service and the Extension Service, the Inter-American Institute for Cooperation on Agriculture, and the World Bank. She has written numerous papers on public and private agricultural research, manure management, and sanitary and phytosanitary trade barriers. She has a Ph.D. in Energy Management and Policy from the University of Pennsylvania.

Catherine Neumeyer is an adjunct professor at St. Thomas in Minnesota. She has a Ph.D. in Agricultural Economics from the University of Minnesota.

Her research focus is on agricultural research policy. In the past she has worked for Ag. Canada, the University of Minnesota, and Rutgers University. In addition, she has been a consultant at the World Bank. In 1994, she aided in the establishment of an agricultural economics program in Albania. She has written numerous papers dealing with public and private research in the seed industry.

Philip G. Pardey is a research fellow at the International Food Policy Research Institute, where he leads a global and regional program of research on agricultural research, extension, and education policy. He also holds an appointment as an associate professor in the Department of Applied Economics at the University of Minnesota. Before joining IFPRI in 1994, Pardey was a senior research officer at the International Service for National Agricultural Research in The Hague, Netherlands. His research interests include the economics of technical change in agriculture, research policy, and economic development. He has published numerous articles in these areas, is coeditor of the volume *Agricultural Research Policy: International Quantitative Perspectives*, and coauthor of *Making Science Pay: The Economics of Agricultural R&D Policy* and *Science under Scarcity: Principles and Practice for Agricultural Research Evaluation and Priority Setting*. Pardey is an Australian national. He obtained a Bachelor's degree in economics and a Master's degree in agricultural science from the University of Adelaide and a Ph.D. in agricultural economics from the University of Minnesota.

Carl E. Pray is a professor of Agricultural Economics and Marketing at Rutgers University, where he specializes in science and technology policy and economic development. He has been a visiting fellow at the Economic Growth Center, Yale University, and a research associate in the Department of Agricultural and Applied Economics, University of Minnesota. He is coeditor of *Research and Productivity in Asian Agriculture*, and *Policy for Agricultural Research*, and has more than 40 scholarly articles in refereed journals. He received a Ph.D. in economic history from the University of Pennsylvania in 1978 and a B.A. from Carleton College, Northfield, Minnesota.

John M. Reilly is Associate Director for Research in the Joint Program on the Science and Policy of Global Change, Massachusetts Institute of Technology. Prior to joining MIT, he spent 12 years with Economic Research Service of the USDA, where he served in various capacities including acting director, deputy director, branch chief, and section leader. His research at USDA focused on agricultural technology, R&D, and economic impacts of climate change. Other past positions include visiting senior scientist at the Massachusetts Institute of Technology and research positions at Battelle's Pacific Northwest National Laboratory and at the Institute for Energy Analysis, Oak Ridge Associated Universities. He received his Ph.D. in economics from the

University of Pennsylvania in 1983 and holds a B.S. in economics and political science from the University of Wisconsin. He has published numerous articles, books, and reports on the economics of natural resources, technology, energy, and climate change and has served in various capacities on federal government and international panels.

David E. Schimmelpfennig is a Program Leader in agricultural research and productivity at USDA's Economic Research Service. He also led an ERS team investigating climate change impacts on agriculture. He has published journal articles on public-sector plant breeding in the United Kingdom, returns to South African R&D and supply response in South Africa, and R&D spending and productivity in the United States and the European Community; as well as numerous popular and professional articles on climate change. He has chaired sessions on agricultural research spillovers and endogenous growth at meetings of the American Agricultural Economic Association, the American Economic Association, and the Western Economic Association and is a member of the NC-208. He teaches a class at the USDA Graduate School, is a Team Leader on the Strategic Planning Task Force for USDA's Research Facilities, and serves on a National Science Foundation Tiger Team. His Ph.D. and M.A. in economics are from Michigan State University, and he has an honors economics B.S. from Purdue University. He matriculated from public high school in Johannesburg, South Africa, in 1977.

Vincent H. Smith is an Associate Professor of Agricultural Economics at Montana State University in Bozeman. He specializes in agricultural policy analysis and trade policy. He holds a B.A. and an M.A. from the University of Manchester, England, and received his Ph.D. from North Carolina State University in 1987.

Colin G. Thirtle is a Londoner who worked on the stock exchange and in sales management before studying economics at the London School of Economics, Southern Illinois, and Columbia, where he completed a Ph.D. on induced innovation in U.S. agriculture. Following employment with Columbia University, Planning Assistance, Inc., and the University of San Francisco, he returned to the United Kingdom, to work at the University of Manchester and later the University of Reading, as a professor in Agricultural Economics. His publications on productivity and R&D cover the United States, the United Kingdom, the European Community, Cameroon, India, Nepal, Bangladesh, Zimbabwe, South Africa, Kenya, Sub-Saharan Africa, Slovenia, and Hungary. He is a coauthor of *The Role of Demand and Supply in the Generation and Diffusion of Technical Change*. There is no truth to the rumor, started by Charlie Cobb of the USDA, that he has to keep moving because he is a horse thief.

Greg Traxler is an associate professor in the Department of Agricultural Economics and Rural Sociology at Auburn University, and an affiliate scientist in the Economics Program at the International Maize and Wheat Improvement Center (CIMMYT). He has a Ph.D. from Iowa State University and an M.S. from the University of Minnesota. Traxler's research has focused on the economics of technical change and the impacts of agricultural research. He has published articles on the economics of crop management research, the effect of plant breeding research on production risk and on measuring sustainability. Current research interests include the coordination of research roles across state, national, and international institutions; economies of scale in plant breeding; and identifying international flows of germplasm. At Auburn he teaches econometrics, production economics, and farm management.

Stephen J. Vogel joined the Economic Research Service in 1992 after completing his Ph.D. in agricultural economics at the University of California at Berkeley. His primary research interests center on structural change, growth, and economic development at the regional and national levels. His past work in applied policy modeling has dealt with food assistance reform, commodity export subsidies, and implementing the Endangered Species Act.

INDEX

Iowa State University Press
Ames, IA 50014
Web site: www.isupress.edu

ISBN 0-8138-2789-2

9 780813 827896

90000

www.ingramcontent.com/pod-product-compliance
Lightning Source LLC
LaVergne TN
LVHW011908200926
848484LV00002B/561